Lecture Notes in Control and Information Sciences

Edited by M. Thoma and A. Wyner

For information about Vols. 1-96 please contact your bookseller or Springer-Verlag

Lecture Notes in Control and Information Sciences

Edited by M. Thoma and A. Wyner

165

G. Jacob,
F. Lamnabhi-Lagarrigue (Eds.)

Algebraic Computing in Control

Proceedings of the First European Conference
Paris, March 13–15, 1991

Springer-Verlag Berlin Heidelberg GmbH

Editors
Professor Gérard Jacob
LIFL Bât. M3 Informatique
Université de Lille I
59655 Villeneuve d'Ascq Cedex
France

Dr. Françoise Lamnabhi-Lagarrigue
Laboratoire des Signaux et Systèmes, CNRS
Ecole Supérieure d'Electricité
91192 Gif-sur-Ivette Cedex
France

ISBN 978-3-540-54408-1   ISBN 978-3-540-47603-0 (eBook)
DOI 10.1007/978-3-540-47603-0

Preface.

This conference is devoted to algebraic and symbolic computing methods in systems and control, and represent a milestone in the evolution in the field, bringing together for the first time in Europe, an international group of scientists and engineers, working in many related facets of the subject.

The importance of symbolic computation, first in mathematics and computer science, and later in engineering and systems' and control's, has been recognized for many years. However, the ability to successfully utilize it's power is only now beginning to be realized, and so this conference comes at an opportune time for researchers to step back and take a look at the progress that has been made to date and to discuss the options for further work.

If one briefly examines the evolution of the role of numerical methods in linear systems' and control's theory, one recognizes some important steps, such as the early realization that linear algebra is a powerful conceptual tool in the development of the theory, and the subsequent realization that the theory posed some unusual problems in numerical linear algebra, requiring special purpose algorithms to be developed. Many of these algorithms have been so successful that they are used in 'real time' environments as elements of control system feedback loops. Finally came the realization that the engineer needs the corresponding software to be carefully packaged and presented so that the design process can make full and efficient use of the calculations that can be performed. Several outstanding examples of these software packages exist today, such as MATLAB, and their development has included extensive and important use of visualization techniques.

This conference is aimed at developing the same progression of ideas with numerical methods replaced by symbolic computing methods. This is of course a very much wider field of study; since in the first place the conference is not limited to linear time invariant control systems, by includes papers from other areas, such as discrete event systems, game theory and most importantly nonlinear systems theory. In the second place, symbolic computation's role is often that of reducing a problem to one where numerical methods can be applied, so that the interface between numeric and symbolic computation problem must also be tackled.

The first theme of the conference is therefore aimed at the study of the mathematics tools that can be applied to systems' and control's theory which will enable solutions of systemss and control's problem to be easily manipulated and conceptualized, in the same *way in which linear algebra was used so effectively in the linear theory. Among the many tools that are being actively considered are algebra and combinatorics. Although the importance of algebra in general to the study of systems theory has been recognized for a long time, for example Lie algebras in nonlinear systems theory, other algebraic techniques, such as differential algebra, are now receiving more prominence. An outstanding problem is made harder·because these tools are usually be application dependent, in contrast to the almost universal application of linear algebra in linear systems' and control's theory.*

Many professional symbolic manipulation software packages, such as Macsyma, Mathematica, Reduce and Maple, are now available, not to mention the even greater range of 'student versions'. These programs have mostly been written within the Mathematics and Computer Science communities. However it is now clear that the traditional applications of symbolic computation, such as calculus and manipulation of polynomial expressions, are not as useful in solving the problems of the systems' and control's community, as originally might have been expected; the output of the programs is typically characterized by long lists of expressions, which are not easily automatically interrogated, and the computations are not expressions, which are not easily automatically interrogated, and the computations are not usually programmed at the level of the mathematical tool being employed, such as particular algebraic or combinatorial object. The next generation of 'high level' symbolic software package, such as Scratchpad, must be capable of doing the types of calculation that are intrinsically useful to the systems' and control's community and must be developed by that community if they are to be successful. These discussions constitute the second theme of the conference.

Finally, the conference is concerned with the end product, in the form of software packages dedicated to the solution of systems' and control's problems. At this stage in the development of the field, packages with the wide applicability and appeal of MATLAB are certainly not available. However some papers describe the progress that is being made in specific application areas, such as economic systems, biochemical systems, and in some specific analysis techniques, such as linearization and stabilization. Recording and documenting these programs will provide a tangible result of the conference, which can be used by engineers and scientists alike.

Peter E. Crouch
Tempe, Arizona, U.S.A.

TABLE OF CONTENTS

II. ALGEBRAIC METHODS

II.a. Mathematics

II.b. Systems

II.c. Control

AUTHORS INDEX

LIST OF PARTICIPANTS

ABOU-KANDIL H.   LUPRA, ENS Cachan, France
ALEXANDRE P.   Sce Automatique, ULB, Bruxelles, Belgique

BARBOT J.P.   LSS, SUPELEC, Gif-sur-Yvette, France
BENAMARA A.   LAG, ENSIEG, Grenoble, France
BIRK J.   Universität Stuttgart, Allemagne
BLANKENSHIP G.L.   University of Maryland, USA
BRADLEY E.   MIT, Cambridge, USA

CALVET J.L.   LSSSA, Toulouse, France
CAPRASSE H.   Institut de Physique, Liège, Belgique
CHAFOUK H.   ESIGELEC, Rouen, France
CHENIN P.   IMAG, Grenoble, France
CHOU C.T.   Dept. of Engineering, Cambridge University, U.K.

DELLA DORA J.   IMAG, Grenoble, France
DEWAR M.C.   University of Bath, U.K.
DEZA F.   SHELL Recherche, Rouen, France
DIOP S.   LAGEP Automatique, Lyon, France

FAYAZ M.A.   LSS, SUPELEC, Gif-sur-Yvette, France

GAUBERT S.   INRIA, Le Chesnay, France
GIUSTI M.   LIX, Palaiseau, France
GLEYSE B.   INSA, Rouen, France
GRAGERT P.K.H.   University of Twente, Netherlands
GRIMM J.   INRIA, Valbonne, France
GRIZZLE J.W.   University of Michigan, USA
GROSSMAN R.   University of Illinois at Chicago, USA
GUICHETEAU P.   ONERA/DES, Chatillon, France

HARCAUT J.P.   Complexe Aérospatiale de Lespinet, France
HEINIG G.   Universität Leipzig, Allemagne
HESPEL C.   INSA, Rennes, France
HOANG NGOC MINH V.   LIFL, Université de Lille , France

JACOB G.   LIFL, Université de Lille , France

KAWSKI M.   Arizona State University, USA
KRENER A.   University of California, Davis, USA
KROB D.   LIR, Université de Rouen, France
KUCHIN B.L.   Moscow Institute of Oil and Gaz, Soviet Union

LAMNABHI-LAGARRIGUE F.	LSS, SUPELEC, Gif-sur-Yvette, France
LeBORGNE M.	IRISA, Rennes, France
LE VEY G.	SIMULOG, St Quentin, France
LOUERAT M.M.	FAST, Université Paris XI, France
MARINO R.	University of Rome, Italie
MARTIN A.	LSS, SUPELEC, Gif-sur-Yvette, France
MASSE J.	SIMULOG, St Quentin, France
MEHDI D.	LAII, ENSEPS, Strasbourg, France
NORGUET	UFR de Maths., Université Paris VII, Paris, France
OLLIVIER F.	LIX, Palaiseau, France
OTCHVAROV L.A.	Moscow Institute of Oil and Gaz, Soviet Union
OUSSOUS N.E.	LIFL, Université de Lille, France
OUTBIB R.	INRIA, Loraine, France
PANTALOS N.	LSS, SUPELEC, Gif-sur-Yvette, France
PETITOT M.	LIFL, Université de Lille, France
PINCHON D.	IBM Paris, France
POMMARET J.F.	CERMA, Ecole des Ponts, Noisy le Grand
RIDEAU P.	Aérospatiale, Cannes, France
ROTELLA F.	Institut Industriel du Nord, France
SORDALEV	LAG, ENSIEG, Grenoble, France
SZYMKAT M.	Academy of Mining and Metallurgy, Pologne
TALAGRAND O.	Lab. Météorologie National de Paris, France
TOFFIN P.	Dept. Math., Université Caen, France
VALIBOUZE A.	LITP, Université Paris VII, France
WALTER E.	LSS, SUPELEC, Gif-sur-Yvette, France
XIAOYUN LU	UMIST, University Manchester, U.K.
ZERONG YU	LSS, SUPELEC, Gif-sur-Yvette, France
ZEITZ M.	Universität Stuttgart, Allemagne

The symbolic computation of vector field expressions

Robert Grossman* and Richard Larson[†]
University of Illinois at Chicago

Abstract

This is an expository paper explaining how trees can be used to compute effectively the vector field expressions which arise in nonlinear control theory. It also describe the mathematical structure that sets of trees carry.

1 Introduction

This is an expository paper explaining how trees can be used to compute effectively the vector field expressions which arise in nonlinear control theory. It also describe the mathematical structure that sets of trees carry.

This paper is based upon [7] and [15]. It also has been influenced by joint work with P. Crouch described in [5] and [6]. The mathematical structure underlying sets of trees was worked out in [10]. This exposition derives, in part, from [8], which is an expository account of symbolic expressions which arise in the study of differential equations. The figures are taken from [9].

2 The basic idea

In this section, we describe the basic idea of how trees can be used to organize computations involving vector fields following [7] and [8]. Consider a control system

$$\dot{x}(t) = E_1(x(t)) + u_1(t)E_2(x(t)) + u_2(t)E_3(x(t)), \quad x(0) = x^0 \in \mathbf{R}^N, \tag{1}$$

where E_1, E_2 and E_3 are vector fields defined in a neighborhood of $x^0 \in \mathbf{R}^N$ and $t \mapsto u_i(t)$ are controls. Our goal is to describe a class of algorithms for the effective symbolic computation of expressions built from the vector fields that describe the local behavior of the control system. These expressions include iterated Lie brackets and the generating series of the system.

The starting point is to assign trees to vector fields as illustrated in Figure 1, and then to impose a multiplication on the trees which is compatible with the composition of vector fields. Assume that the vector fields E_j have the form:

$$E_j = \sum_{\mu=1}^{N} a_j^\mu D_\mu, \quad j = 1, 2, 3, \quad j = 1, \ldots, M, \tag{2}$$

where a_j^μ are smooth functions on $\mathbf{R}^N$ and $D_\mu = \partial/\partial x_\mu$. Now

$$E_2 \cdot E_1 = \sum b_j(D_j a_i)D_i + \sum b_j a_i D_j D_i$$

*This research is supported in part by NASA grant NAG2-513, NSF Grant DMS-8904740, and the Laboratory for Advanced Computing.

†This research is supported by NSF Grant DMS-8904740.

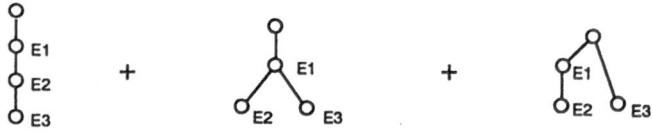

Figure 1: The trees associated with the vector fields E_j.

Figure 2: The trees associated with Equation 3.

and $E_3 \cdot E_2 \cdot E_1$ is equal to

$$\sum a_3^{\mu_k}(D_k a_2^{\mu_j})(D_j a_1^{\mu_i})D_i + \sum a_3^{\mu_k}a_2^{\mu_j}(D_k D_j a_1^{\mu_i})D_i + \sum a_3^{\mu_k}a_2^{\mu_j}(D_j a_1^{\mu_i})D_k D_i$$

$$+ \sum a_3^{\mu_k}a_2^{\mu_j}a_1^{\mu_i}D_k D_j D_i + \sum a_3^{\mu_k}a_2^{\mu_j}(D_k a_1^{\mu_i})D_j D_i + \sum a_3^{\mu_k}a_1^{\mu_i}(D_k a_2^{\mu_j})D_j D_i. \qquad (3)$$

Here the sum is for $i, j, k = 1, \ldots, N$ and hence involves $O(N^3)$ differentiations. It is convenient to keep track of the terms that arise in this way using labeled trees: we indicate in Figure 2 the trees that are associated with the six sums in this expression.

An iterated Lie bracket such as

$$[E_3, [E_2, E_1]] = E_3 E_2 E_1 - E_3 E_1 E_2 - E_2 E_1 E_3 + E_1 E_2 E_3 \qquad (4)$$

gives rise in this fashion to 24 trees corresponding to the $24N^3$ differentiations that a naive computation of this expression requires. On the other hand, 18 of the trees cancel, saving us from computing $18N^3$ terms. We are left with $6N^3$ terms of the form $(junk)D_{\mu_1}$. A careful examination of this correspondence between labeled trees and expressions involving the E_j's shows that the composition of the vector fields E_j's, viewed as first order differential operators, corresponds to a multiplication on trees. This multiplication is illustrated in Figure 3. It turns out that this construction yields an algebra, which we call the *algebra of Cayley trees*.

Let R denote a space of smooth, that is C^∞, observation functions and let A denote the space of differential operators generated by E_1 and E_2. For many applications it is important to have efficient algorithms to compute the actions of these differential operators on observation functions:

$$p \cdot f, \quad p \in A, \quad f \in R.$$

For example, the generating series associated to the system is an infinite sum of terms of this type. Depending upon the representation of the function f, additional cancellations can be expected.

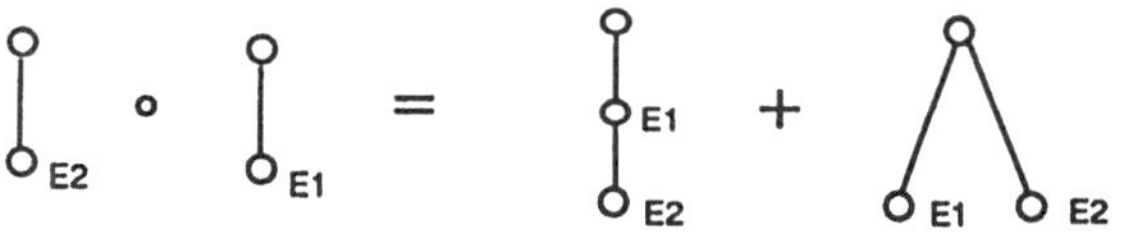

Figure 3: An example of multiplying two trees.

3 Bialgebras

As we will see in the next section, the space of trees has the structure of a bialgebra, which gives the ring of smooth test functions R the structure of a module algebra. By exploiting this algebraic structure, it is possible to derive more efficient algorithms. In this section we review the basic facts about bialgebras and module algebras which will we will need, following [5] and [6].

At the end of this section, we give the definition of *differentially produced* which is fundamental the theorem of Section 6.

Let k denote any field of characteristic 0. By an *algebra* we mean a vector space A over the field k with an associative multiplication and unit. The multiplication can be represented by a linear map $\mu : A \otimes_k A \to A$; the unit can be represented by a linear map $k \to A$ (the map sending $1 \in k$ to $1 \in A$). The facts that the multiplication is associative, and that $1 \in A$ is a unit, can be expressed by the commutativity of certain diagrams. For example, the commutativity of the diagram

$$
\begin{array}{ccc}
A \otimes_k A \otimes_k A & \longrightarrow & A \otimes_k A \\
\downarrow & & \downarrow \\
A \otimes_k A & \longrightarrow & A
\end{array}
$$

where the upper horizontal arrow is the map $\mu \otimes I$, the left vertical arrow is the map $I \otimes \mu$, and the remaining two arrows are the map μ, expresses the associativity of multiplication.

The dual notion to an algebra is a *coalgebra*: a vector space C over the field k together with a coassociative coproduct $\Delta : C \to C \otimes_k C$ and a counit $\epsilon : C \to k$. The fact that Δ is coassociative and that ϵ is a counit is expressed by diagrams which are dual to the diagrams which express the facts that the multiplication of an algebra is associative, and that $1 \in A$ is a unit: they are the same diagrams, with the direction of all arrows reversed. For example, coassociativity is expressed by the commutativity of the diagram

$$
\begin{array}{ccc}
C \otimes_k C \otimes_k C & \longleftarrow & C \otimes_k C \\
\uparrow & & \uparrow \\
C \otimes_k C & \longleftarrow & C
\end{array}
$$

where the upper horizontal arrow is the map $\Delta \otimes I$, the left vertical arrow is the map $I \otimes \Delta$, and the remaining two arrows are the map Δ. Often the element $\Delta(c) \in C \otimes_k C$ is written $\sum_{(c)} c_{(1)} \otimes c_{(2)}$.

A *bialgebra* is a vector space H over k which has both an algebra and a coalgebra structure, such that the coalgebra structure maps are algebra homomorphisms, or equivalently, the algebra structure maps are coalgebra homomorphisms. (This equivalence can be seen by expressing the assertion that the coalgebra structure maps are algebra homomorphisms as a set of commutative diagrams: this set of diagrams is self-dual.)

Some examples of bialgebras are the following:

1. Let G be a group, and let kG be the group algebra of G: the vector space kG has the elements of G as a basis, with multiplication defined by extending the multiplication on

G linearly. The coproduct and counit of kG are defined by

$$\left. \begin{array}{rcl} \Delta(g) &=& g \otimes g \\ \epsilon(g) &=& 1 \end{array} \right\} \quad g \in G.$$

2. Let G be an affine algebraic group, and let $k[G]$ be the algebra of representative functions on G. The algebra structure of $k[G]$ is the usual algebra structure of functions with pointwise multiplications. The coproduct arises from the group multiplication $G \times G \to G$, which induces the map $k[G] \to k[G \times G] \cong k[G] \otimes_k k[G]$. The counit arises from the map $\{e\} \to G$, where $\{e\}$ is the single-element group.

3. Let L be a Lie algebra over k, and let $U(L)$ be the universal enveloping algebra of L. The coproduct and counit of $U(L)$ are defined by

$$\left. \begin{array}{rcl} \Delta(x) &=& 1 \otimes x + x \otimes 1 \\ \epsilon(x) &=& 0 \end{array} \right\} \quad x \in L,$$

and extended to all of $U(L)$ using the fact that Δ and ϵ are algebra homomorphisms.

Usually, in studying bialgebras, an additional condition is imposed which is analogous to the assertion that a semigroup is a group. Such bialgebras are called *Hopf algebras*. The bialgebras which we consider in this paper (such as the universal enveloping algebra of a Lie algebra) satisfy this condition automatically.

A coalgebra is said to be *cocommutative* if it satisfies $\Delta = T \circ \Delta$, where T is the map $T : C \otimes_k C \to C \otimes_k C$ defined by $T(x \otimes y) = y \otimes x$. Note that the bialgebras in Examples 1 and 3 are cocommutative.

A vector space V over k is said to be *graded* if there is a sequence of subspaces $V_0, V_1, \ldots$ such that

$$V \cong \bigoplus_{n=0}^{\infty} V_n.$$

A graded vector space V is said to be *connected* if $V_0 \cong k$.

Let H be a bialgebra. A *H-module algebra* is an algebra R which is an H-module such that the action satisfies

$$h \cdot (fg) = \sum_{(h)} (h_{(1)} \cdot f)(h_{(2)} \cdot g), \qquad \text{for all } h \in H, \quad f, g \in R.$$

An *augmentation* of an algebra R over k is an algebra homomorphism $\epsilon : R \to k$. If $R = k[[x_1, \ldots, x_n]]$ is a power series algebra, the map $f \mapsto f(0)$ is an augmentation. If H is a bialgebra, then the dual space H^* is an algebra with multiplication defined by $pq(h) = (p \otimes q)\Delta(h)$ for $h \in H$ and $p, q \in H^*$. The algebra H^* is a H-module algebra with respect to the action $\rightharpoonup$ of H on H^* defined by $(h \rightharpoonup p)(k) = p(kh)$ for $h, k \in H$ and $p \in H^*$.

We close this section with a remark and two fundamental definitions:

- If $g \in H$ satisfies $\Delta(g) = g \otimes g$ and R is an H-module algebra, then g acts as an endomorphism of R; if $x \in H$ satisfies $\Delta(x) = 1 \otimes x + x \otimes 1$ and R is an H-module algebra, then x acts as a derivation of R.

- If H is a bialgebra, we say that $p \in H^*$ is *differentially produced by the algebra R with the augmentation ϵ* if there is a H-module algebra structure on R and there exists $f \in R$ satisfying

$$p(h) = \epsilon(h \cdot f).$$

- If $H = U(L)$ is a bialgebra which is the universal enveloping algebra of a Lie algebra L, we say that $p \in H^*$ has *finite Lie rank* if $L \rightharpoonup p$ is finite dimensional.

4 The algebra of Cayley trees

In this section, we follow [13] and define a bialgebra structure on spaces of trees. The relation between trees and differential operators goes back at least as far as Cayley [3] and [4]. Of this literature, the work most closely related to the view point taken here is Butcher's use of trees to analyze Runge-Kutta algorithms [1] and [2].

Let k will denote a field of characteristic 0 such as the real numbers or complex numbers. By a *tree* we will mean a finite rooted tree. Let $\mathcal{T}$ be the set of finite rooted trees, and let $k\{\mathcal{T}\}$ be the k-vector space which has $\mathcal{T}$ as a basis.

We first define the coalgebra structure on $k\{\mathcal{T}\}$. If $t \in \mathcal{T}$ is a tree whose root has children $s_1, \ldots, s_r$, the coproduct $\Delta(t)$ is the sum of the 2^r terms $t_1 \otimes t_2$, where the children of the root of t_1 and the children of the root of t_2 range over all 2^r possible partitions of the children of the root of t into two subsets. The augmentation ϵ which sends the trivial tree to 1 and every other tree to 0 is a counit for this coproduct. It is immediate that comultiplication is cocommutative.

We next define the algebra structure on $k\{\mathcal{T}\}$. Suppose that $t_1, t_2 \in \mathcal{T}$ are trees. Let $s_1, \ldots, s_r$ be the children of the root of t_1. If t_2 has $n+1$ nodes (counting the root), there are $(n+1)^r$ ways to attach the r subtrees of t_1 which have $s_1, \ldots, s_r$ as roots to the tree t_2 by making each s_i the child of some node of t_2. The product $t_1 t_2$ is defined to be the sum of these $(n+1)^r$ trees. It can be shown that this product is associative, and that the trivial tree consisting only of the root is a right and left unit for this product. It can also be shown that the maps defining the coalgebra structure are algebra homomorphisms, so that $k\{\mathcal{T}\}$ is a bialgebra. For details, see [10].

The bialgebra $k\{\mathcal{T}\}$ is graded: $k\{\mathcal{T}\}_n$ has as basis all trees with $n+1$ nodes. Because the bialgebra $k\{\mathcal{T}\}$ is graded connected, it is a Hopf algebra. We summarize the above discussion in the following theorem.

Theorem 4.1 *The vector space $k\{\mathcal{T}\}$ with basis the set of finite rooted trees is a cocommutative graded connected Hopf algebra.*

Assume now that each node of the tree (except for the root) is labeled with a symbol from the set $\{E_1, \ldots, E_M\}$. We can define the product and coproduct as above, and, once again, the resulting space is a bialgebra. See [10] for details. Let $k\{\mathcal{LT}\}$ denote this algebra.

Let R denote a subring of the commutative ring of smooth functions on $\mathbf{R}^N$. We now define an action of the algebra of Cayley trees

$$B = k\{\mathcal{LT}\}$$

on the ring R, making R a B-module algebra, which captures the action of trees as higher derivations. This requires that we interpret the formal symbols E_j as derivations of R using Equations 2. The action is defined using the map

$$\psi : k\{\mathcal{LT}\} \longrightarrow \mathrm{End}_k R,$$

as follows:

1. Given a labeled, ordered tree t with $m+1$ nodes, assign the root the number 0 and assign the remaining nodes the numbers $1, \ldots, m$. We identify the node with the number assigned to it. To the node k associate the summation index μ_k. Denote $(\mu_1, \ldots, \mu_m)$ by μ.

2. For the labeled tree t, let k be a node of t, labeled with E_{γ_k} if $k > 0$, and let $l, \ldots, l'$ be the children of k. Define

$$
\begin{aligned}
R(k; \mu) &= D_{\mu_l} \cdots D_{\mu_{l'}} a_{\gamma_k}^{\mu_k}, \quad \text{if } k > 0 \text{ is not the root;} \\
&= D_{\mu_l} \cdots D_{\mu_{l'}}, \quad \text{if } k = 0 \text{ is the root.}
\end{aligned}
$$

Note that if $k > 0$, then $R(k; \mu) \in R$.

3. Define

$$\psi(t) = \sum_{\mu_1, \ldots, \mu_m = 1}^{N} R(m; \mu) \cdots R(1; \mu) c(0; \mu).$$

4. Extend ψ to all of $k\{\mathcal{LT}\}$ by linearity.

It is straightforward to check that this action of B on R makes R into a B-module algebra.

We summarize with the following theorem.

Theorem 4.2 *Let R the algebra of smooth functions on $\mathbf{R}^N$. Let B denote the algebra of Cayley trees $k\{\mathcal{LT}\}$. Then R is a B-module algebra with respect to the action defined by ψ.*

5 Symbolic evaluation of vector field expressions

When expressions involving vector fields, such as Lie brackets, are written out in coordinates, there is typically a lot of cancellation, as we have seen above. Similar cancellation occurs in expressions involving Poisson brackets and when flows are concatenated, as in Campbell-Baker-Hausdorf expansions. In this section, we use the algebra of Cayley trees to exploit this cancellation in order to compute more efficiently formal expressions involving vector fields.

The standard action of the algebra A of differential operators generated by $E_1, \ldots, E_M$ on the algebra of smooth functions R gives R the structure of a A-module algebra. It is easy to relate these two H-module algebra structures on R and this observation is the basis for our algorithms.

Let

$$\phi : A \longrightarrow B$$

denote the map sending the generator E_j of the algebra A to the tree consisting of two nodes: the root and a single child labeled E_j. This map is illustrated in Figure 1. Extend ϕ to be an algebra homomorphism. Let χ denote the map

$$A \to \mathrm{End}_k R$$

defined by using the substitution (2) and simplifying to obtain an endomorphim of R. We have the following diagram:

$$
\begin{array}{ccc}
A & \to & B \\
 & \searrow \downarrow & \\
 & \mathrm{End}_k R &
\end{array}
\tag{5}
$$

Theorem 5.1 *(i) The maps χ, ϕ and ψ are related by $\chi = \psi \circ \phi$. (ii) Fix a function $f \in R$ and a differential operator $p \in A$. Then*

$$p \cdot f = \phi(p) \cdot f.$$

Here the action on the left views R as an A-module algebra, while the action on the right views R as B-module algebra.

The first assertion is proved in [14] and the second assertion follows from the first assertion and the definitions. From this theorem, we get:

No. of terms	Form of terms
$8N^3$	coeff. D_{μ_1}
$12N^3$	coeff. $D_{\mu_2} D_{\mu_1}$
$4N^3$	coeff. $D_{\mu_3} D_{\mu_2} D_{\mu_1}$

Table 1: Naive computation of the differential operator corresponding to p.

Algorithm 5.1 *To rewrite expressions in the first order differential operators E_j in terms of the basis*

$$\frac{\partial}{\partial x_{\mu_1}}, \quad \frac{\partial^2}{\partial x_{\mu_1} \partial x_{\mu_2}}, \ldots, \quad \mu_1, \mu_2, \ldots = 1, \ldots, N,$$

compute the composition of the rightward and downward pointing arrows in the diagram above.

In [11], [12] and [14], we show that the algorithm is much more efficient than naive substitution, which corresponds to computing the diagonal arrow directly. In some common cases, the improvement in efficiency is exponential. We have implemented this and related algorithms in Maple, Mathematica and Snobol.

We illustrate this algorithm by working the example of the last section following [7]: consider a higher order derivation of the form

$$p = E_3 E_2 E_1 - E_3 E_1 E_2 - E_2 E_1 E_3 + E_1 E_2 E_3.$$

Naive simplification requires computing $24N^3$ terms of the form described in Table 1. The image of p in the algebra of Cayley trees contains 24 trees, six for each of the four terms of p. For example, the six labeled trees corresponding to the first term are given in Figure 2. Eighteen of these trees cancel, leaving the six trees in Figure 4. The corresponding differential operator is equal to

$$\sum a_3^{\mu_3}(D_{\mu_3} a_2^{\mu_2})(D_{\mu_2} a_1^{\mu_1}) D_{\mu_1} - \sum a_3^{\mu_3}(D_{\mu_3} a_1^{\mu_2})(D_{\mu_2} a_2^{\mu_1}) D_{\mu_1}$$

$$- \sum a_2^{\mu_3}(D_{\mu_3} a_1^{\mu_2})(D_{\mu_2} a_3^{\mu_1}) D_{\mu_1} + \sum a_1^{\mu_3}(D_{\mu_3} a_2^{\mu_2})(D_{\mu_2} a_3^{\mu_1}) D_{\mu_1}$$

$$+ \sum a_3^{\mu_3} a_2^{\mu_2}(D_{\mu_3} D_{\mu_2} a_1^{\mu_1}) D_{\mu_1} - \sum a_3^{\mu_3} a_1^{\mu_2}(D_{\mu_3} D_{\mu_2} a_2^{\mu_1}) D_{\mu_1},$$

and contains $18N^3$ fewer terms than does the naive computation of p. Notice that all the terms of degree 2 and 3 cancel, as well as some of the first order terms. An example of the cancellation of labeled trees is given in Figure 5.

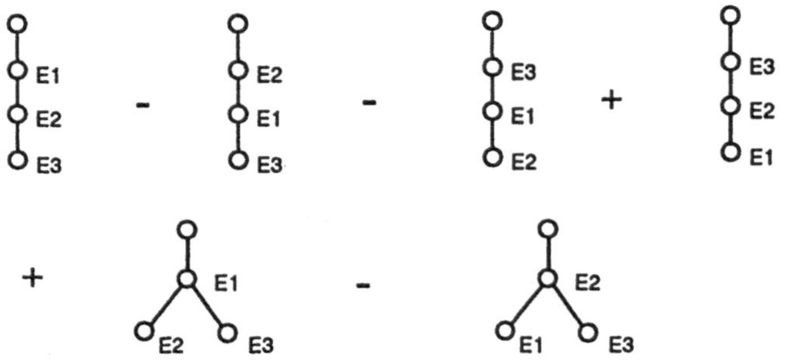

Figure 4: The surviving labeled trees.

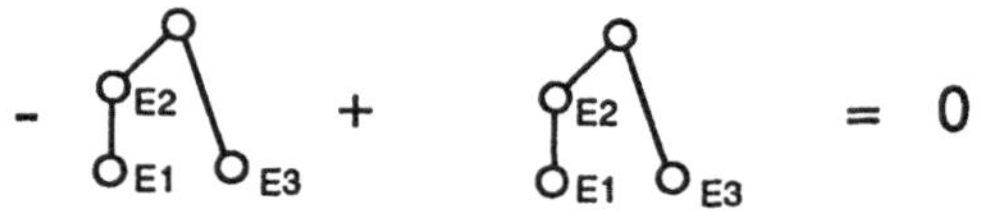

Figure 5: The term $E_3 E_1 E_2$ contributes the first labeled tree and the term $E_1 E_2 E_3$ contributes the second, which cancel.

6 The algebraic realization of input-output maps

In this section we state a theorem which uses bialgebras to describe the realizability of input-output maps of control systems. If L is a Lie algebra of derivations of R which is generated by $\{E_1, \ldots, E_M\}$, then $H = U(L)$ has a basis of monomials in the E_i. If $x(t)$ is a solution to the control system

$$\dot{x}(t) = \sum_{i=1}^{M} u_i(t) E_i(x(t)), \tag{6}$$

and $f \in R$ is an observation function, then the formal generating series

$$p = \sum p_\mu \mu, \quad \text{where } p_\mu = E_{\mu_1} \cdots E_{\mu_k} f(x(0)),$$

can be thought of as a functional on the bialgebra $H = U(L)$. Using the terminology introduced above, p is differentially produced by R. The following theorem gives a condition which is stated in terms of the action of H on H^* for $p \in H^*$ to be the generating series associated with an input-output map. (See [15] for a complete exposition.)

Theorem 6.1 *Let L be a Lie algebra, and let $H = U(L)$. Then p is of finite Lie rank if and only if p is differentially produced by an augmented algebra R with $\ker \epsilon/(\ker \epsilon)^2$ finite dimensional.*

It can be shown that R can be taken to be isomorphic to a power series ring in finitely many variables.

References

[1] J. C. Butcher, "An order bound for Runge-Kutta methods," *SIAM J. Numerical Analysis*, Vol. 12, pp. 304–315, 1975.

[2] J. C. Butcher, *The Numerical Analysis of Ordinary Differential Equations*, John Wiley, 1986.

[3] A. Cayley, "On the theory of analytical forms called trees", *Collected Mathematical Papers of Arthur Cayley*, Cambridge University Press, Vol. 3, pp. 242–6, 1890.

[4] A. Cayley, "On the analytical forms called trees, second part", in *Collected Mathematical Papers of Arthur Cayley*, Cambridge University Press, Vol. 4, pp. 112–5, 1891.

[5] P. Crouch, R. Grossman, and R. G. Larson, "Computations involving differential operators and their actions on functions," *Laboratory for Advanced Computing Technical Report*, Number LAC91-R2, University of Illinois at Chicago, January, 1991.

[6] P. Crouch, R. Grossman, and R. Larson, "Trees, bialgebras, and intrinsic numerical integrators," *Laboratory for Advanced Computing Technical Report*, Number LAC90-R23, University of Illinois at Chicago, May, 1990.

[7] R. Grossman, "The evaluation of expresssions involving higher order derivations," *Journal of Mathematical Systems, Estimation, and Control*, Vol. 1 (1991), pp. 91–106.

[8] R. Grossman, "Using trees to compute approximate solutions of ordinary differential equations exactly," *Computer Algebra and Differential Equations*, M. F. Singer, editor, Academic Press, New York, 1991, in press.

[9] R. Grossman, *Analytic Computation: An Introduction*, Laboratory for Advanced Computing Technical Report, Number LAC90-R17, University of Illinois at Chicago, April, 1990.

[10] R. Grossman and R. G. Larson, "Hopf algebraic structures of families of trees," *J. Algebra*, Vol. 26, pp. 184–210, 1989.

[11] R. Grossman and R. G. Larson, "Labeled trees and the algebra of differential operators," *Algorithms and Graphs*, B. Richter, editor, American Mathematical Society, Providence, pp. 81–87, 1989.

[12] R. Grossman and R. G. Larson, "Labeled trees and the efficient computation of derivations," in *Proceedings of 1989 International Symposium on Symbolic and Algebraic Computation*, ACM, pp. 74–80, 1989.

[13] R. Grossman and R. G. Larson, "Hopf-algebraic structure of combinatorial objects and differential operators," *Israeli J. Math.*, to appear.

[14] R. Grossman and R. G. Larson, "The symbolic computation of derivations using labeled trees," *J. Symbolic Comptutation*, to appear.

[15] R. Grossman and R. G. Larson, "The realization of input-output maps using bialgebras," to appear.

An implementation of tools for graded Lie–algebras in Reduce

P.K.H. Gragert   G.H.M. Roelofs

University of Twente, Department of Applied Mathematics
PO Box 217, 7500 AE Enschede, The Netherlands

1   Introduction

The first thing to state is that this article has no direct connection to control theory. It comes to this conference because of some expertise in using REDUCE, a language for symbolic and algebraic computations.

This article will divided roughly into three parts:

- Underlying mathematical research. We will indicate shortly two research areas, where 'algebraic computations' may grow to such an extent that it is worthwhile to do them as much as possible automatically by using a suitable computational environment.

- Literate programming. Though all modern languages for 'symbolic and algebraic computations' are very powerful already, depending on the application there is some programming necessary. Here we want to tell about a special attitude of a programmer by which the result of his programming effort is likely not only more reliable but even more readable and understandable too.

- Graded Lie algebras. The two formentioned items lead to an environment embedded in REDUCE to do algebraic computations in Lie algebras, extended to $\mathbf{Z}_2$-graded (or super) Lie algebras. Here we will tell about considerations, which lead to a satisfactory implementation of all required operations. This will include some rather technical remarks with respect to REDUCE. Here we will try too to give you an impression of 'working Literate Programming'. The total 'package' will be published and will be available in the near future elsewhere.

2 Coverings and Lie algebras

Our interest in an computational (specialized) environment for Lie algebras stems from two research areas.

Nonlinear partial differential equations may be analyzed using the theory of coverings see [Vinogradov], containing the ideas of Wahlquist and Estabrook about prolongations see [Estabrook]. If a (suitable) covering can be computed it may give rise to (non local) conservation laws, (non local) symmetries, Bäcklund transformations and the like. The conditions to find a covering arise in the form of Lie algebra relations. Such relations are often embedded in infinitedimensional Lie algebras (sometimes of KacMoody type). If one succeeds in identifying such a Lie algebra, known Lie algebra theory may be used to derive properties of the underlying differential equation. To realize such an identification process for given (computed) Lie relations we needed a tool to reduce the amount of algebraic computations. Essentially this is the computation and evaluation of Jacobi identities. In a Lie algebra of dimension N $\binom{N}{3}$ Jacobi identities have to be checked. Imagine the amount of work to check all Jacobi identities (easy but boring and error prone) if N is in the order of 70! And the application mentioned one is forced to really check all Jacobi identities. First results of such computations may be found in [Roelofs].

Another field of application occurs in the analysis of generalized Cartan matrices see [Kac]. A finite dimensional simple Lie algebra may be characterised by its Cartan matrix together with the corresponding Serre relations. Kac Moody Lie algebras appear if the theory of Cartan matrices is generalized. If all elements of a (generalized) Cartan matrix are (negative) integers the theory is nearly complete, see the classification of Kac Moody Lie algebras [Kac]. If this is not the case, less is known. In the construction of a Kac Moody like Lie algebra corresponding to a generalized Cartan matrix with rational numbers on not diagonal elements in the first place an algebra occurs, which often is too big. One looks for a quotient algebra derived using an ideal with special properties, which is described by special relations. In the classical case (and some Kac Moody algebras) these relations are exactly the well known Serre relations. In the general case only a necessary condition is know see [Gabber], where the sought relations may occur. To find these relations we do not know better than really compute them. This again calls for computations in Lie algebras, where besides the analysis of Jacobi identities a grading of elements plays an important role. This research has just started in 1990. In the general case of a 2 by 2 Cartan matrix we found, that at places where the necessary condition allows relations there are sometimes no, one or two relations dependant on the entries of the Cartan matrix. A suitable theory is missing up to now.

3 Literate programming

The authors of this article are using TeX to prepare papers, letters etc. All TeXware is written by using a development system: called WEB. Using the WEB system in an appropriate way is called 'Literate Programming' see e.g. [Knuth]. Using such a system has a number of advantages (and maybe some disadvantages). A WEB system exists for a number of programming languages. Originally it was built for PASCAL programs. Later on it was built for the C programming language and recently a WEB system is available for FORTRAN [Krommes] too. Furthermore it is worthwhile to mention 'Spidery WEB', a system to build WEB-systems for other languages in a comfortable way, which has been done for e.g. AWK (a UNIX tool) and others. Last year we converted the original WEB for PASCAL into a WEB for REDUCE, though we are planning to use Spidery WEB. This conversion was possible as a consequence of 'Literate Programming'. Using Spidery WEB for REDUCE is yet not directly possible, because some adjustments in the used expression reader are necessary too, due to the fact of an escape character in REDUCE (and LISP).

Now let us tell what programming with a WEB system means, for all the details one should read the corresponding 'users guide', which accompanies each WEB.

One writes in one file documentation as well as programming text. Processing this 'source' file by WEAVE, the first half of a WEB system, delivers a file suitable as input for TEX, this is the documentation part. Processing the 'source' file by TANGLE, the second half of a WEB system, delivers a file suitable for the corresponding programming language.

What are the advantages of this approach:

- The power of TEX is available to describe the technical or mathematical background the program is aimed at. This is the place to be 'literate' about your programming!

- The module concept of a WEB allows you to program 'top down'. This means that a WEB system gives you the possibilily to begin a program very abstract and to fill in more and more details later. The TANGLE program will fiddle all parts together throwing away the documentation parts, such that you get an executable (or compilable) file.

- There is some preprocessing included in a WEB system, such that the documentation part may become more readable then the real program would be.

- A complete WEB source file will be split up (and later be printed) as a (big) number of sections, where each section may or may not contain a programming part. The system will automatically create a 'cross-reference' of all the 'variables' used in the programming parts. All 'WEB modules' will occur in an index too.

- An index of the document is generated automatically.

- There exists the concept of a 'CHANGE file'. For TEX it is used for example as follows. To implement a system as TEX on different computers, from mainframe to personal computer, is done as follows: The core of the TEX source will be always the same. If a construction e.g. file access or command line parameters are different on a special machine the machine specific programming can be written in a 'CHANGE file' replacing the original code. In the case of TEX this approach was very successful.

The 'CHANGE file' concept allows experiments with different versions of a program, one may for instance want to run a program where a number of statements is changed.

By the way, the number of of 'real' WEB commands is rather limited (about 25). Nevertheless one has to familiarize to this way of programming. Our experience with the WEB for REDUCE is very positive. We would not like to work without it anymore!

Disadvantages of a WEB system may be:

- You are confronted with 'three languages': WEB, TeX and a 'real' programming language you are writing for. All this three languages may give error messages and you have to understand them all!

In the next section we include some examples from our Lie algebra package indicating hopefully the advantages.

4 Graded Lie algebras

In principle the document of Lie (super)algebras could be included here, because it contains not only all procedures written together with all considerations done, but too a short summary of the mathematical background. As stated earlier, this document will be available elsewhere. Here we will extract some parts, which show 'Literate Programming' and which are interesting in view of 'algebraic computations' in REDUCE:

1. The index of the document:

2. The complete sixth section of the (WEB) document.

This is in a nutshell the mathematical background of the procedures written. Because we are using T$_E$X, all the material will be printed as usual:

———————————————— Begin RWEB section————————————————

6. Implementing free Lie superalgebras.

For $m, n \geq 0$ let $Lib = Lib(x_1, \ldots, x_m, \xi_1, \ldots, \xi_n)$ be the free algebra on generators $x_1, \ldots, x_m, \xi_1, \ldots, \xi_n$. We introduce a $\mathbf{Z}_2$-grading $|.|$ on Lib by defining $|x_i| = 0$ ($i = 1, \ldots, m$), $|\xi_j| = 1$ ($j = 1, \ldots, n$) and $|xy| = |x| + |y|$ for all homogeneous $x, y \in Lib$. We define $L = L(x_1, \ldots, x_m, \xi_1, \ldots, \xi_n)$ to be the quotient algebra Lib/I where I is the ideal, which for all homogeneous $x, y, z \in Lib$ is generated by the elements $xy + (-1)^{|x| \cdot |y|} yx$ and $(-1)^{|x| \cdot |z|} x(yz) + (-1)^{|y| \cdot |x|} y(zx) + (-1)^{|z| \cdot |y|} z(xy)$.

On L we define a bracket $[x, y] \equiv xy$. Then from the definition above it is clear that this bracket satisfies the graded skew-symmetry

$$[x, y] = -(-1)^{|x| \cdot |y|} [y, x]$$

and the graded Jacobi identity

$$(-1)^{|x| \cdot |z|} [x, [y, z]] + (-1)^{|y| \cdot |x|} [y, [z, x]] + (-1)^{|z| \cdot |y|} [z, [x, y]] = 0.$$

Moreover it is bilinear because of the bilinearity of the multiplication in Lib. Therefore L defines a Lie superalgebra, the so called *free Lie superalgebra* on even generators $x_1, \ldots, x_m$ and odd generators $\xi_1, \ldots, \xi_n$. It is obvious that for $m > 1$ or $n > 1$ L is infinite dimensional.

¿From this free Lie superalgebra we can get some specific Lie (super)algebra by imposing additional relations on top of the graded skew-symmetry and the graded Jacobi identity. For instance, we can get a finite dimensional simple Lie algebra by imposing appropriate Serre relations.

As a last point we have to mention gradings of Lie (super)algebras, because these can be very helpful when working on Lie (super)algebras. A Lie (super)algebra can admit more than one grading, for example, the free Lie (super)algebra on n generators admits a $\mathbf{Z}_2$-grading, but also admits the length of "words" as a grading, or a multigrading where the degree of x_i is the n-tuple $(0, \ldots, 1, \ldots, 0)$ (1 on the i-th place).

There is one fact about free Lie superalgebras which is very useful if we want to implement a free Lie superalgebras in REDUCE. To explain this, let $L_1 = L(x_1, \ldots, x_n)$ be the free Lie superalgebra on generators $x_1, \ldots, x_n$ for some $n > 1$. Then it is easy to prove that L_1 is isomorphic to $L_2 = L(x_1, \ldots, x_{n+1})/I(x_{n+1} - [x_1, x_2])$ where $I(x_{n+1} - [x_1, x_2])$ is the ideal in L_2 generated by $x_{n+1} - [x_1, x_2]$. This means that we can avoid expressions containing commutators like $[x_1, x_2]$ just by introducing a new generator x_{n+1} and imposing one additional relation $[x_1, x_2] = x_{n+1}$.

———————————————— End RWEB section————————————————

In the next RWEB-sections first all demands with respect to Lie algebras are surveyed and brought into relation with REDUCE. Here a number of decisions are taken, resulting in (informal) 'specifications', which will be obeyed in the implementation part.

To be honest, this was not a 'one way road', but experience and some experimentation was included in the generation of satisfactory decisions. It means, that there exist several versions, and maybe further optimization is possible. In this light being 'literal' is very fruitful, because one has written down the ideas resulting in specifications.

At this place we only summarize some specifications, to be able to describe some REDUCE specifique implementation considerations:

- To any Lie algebra two (algebraic operator) names are related: The name of the Lie product and the name of a the generators. The defaults are 'lie' and 'x'.

- $x(i)$ with $i > 0$ will denote an even generator (degree 0). $x(i)$ with $i < 0$ will denote an odd generator (degree 1).

- The maximal number of odd and even generators has to be given by the user. It means, that in an infinite dimensional algebra computations are possible only in a 'finite part' of the algebra.

- $x_i \cdot x_j$ will be denoted in REDUCE by $lie(x(i), x(j)) = lie(i, j) = [i, j]$. It means that you may chose, what is most convenient.

- We assume the Lie product to be right associative: $x_i \cdot (x_j \cdot x_k) = x_i \cdot x_j \cdot x_k$ and written in REDUCE as $lie(i, j, k)$ or $[i, j, k]$. By this notation a Jacobi identity of even generators may be written

$$[i, j, k] + [j, k, i] + [k, i, j]$$

One of the main activities is the computation and analysis of Jacobi identities, so these computations should be in some sense optimal and some ideas will be described now to reach this goal. This is as far as we know not available in any 'Symbolic Algebra Package' and has to be programmed by oneself.

In REDUCE the value of algebraic operators are stored as follows: The name of the operator gets a property 'KVALUE', which is an association list: $(...(key.value)...)$. Values of an algebraic operator are retrieved by a linear search on its KVALUE list. If the 'key' is found, the corresponding 'value' is returned, if not the key itself is the value.

Because the number of Lie products in an algebra (or part of an algebra) of dimension N is about $\frac{N^2}{2}$ and in our applications N becomes as large as 100 and even more, computations of Jacobi identities would suffer from 'slow' access of the Lie products (the entries of the product table). By the way, the number of Jacobi identities is slightly greater than $\binom{N}{3}$ in graded Lie algebras. Therefore another, more efficient, storage scheme is needed.

The 'vector' objects of LISP allow direct access of each of its elements and are suitable in this case. One may choose e.g. a (long) one dimensional vector with some index calculations to access a Lie product table element 'lie(i,j)' or a vector of vectors, where the 'i', selects a vector and than the 'j' is used to access 'lie(i,j)'. (Index calculation in this case is more simple). Experiments on two different REDUCE implementations showed, that the second choice needs less time to access an element of the product table and is therefore chosen.

But deciding for such an internal representation of Lie products, a 'great' number of supporting functions have to be written, to really get a bargain:

- A Lie algebra is built upon a linear space. So we need a procedure to deliver coefficients with respect to the basis of a general element.

- A Lie product is multilinear (bilinear). In combination with the foregoing point, we implemented 'multilinearity' of algebraic operators.

Stuff belong to these two items are in itself important and are gathered in a 'TOOLS' package and will be described on the ISSAC 91 conference in Bonn (Germany) July 1991.

- We need a simplification function for the basic Lie products, taking into account the different possibilities, which may occur.

- To clear 'our' Lie products needs a slight adjustment of the 'normal' 'CLEAR' function of REDUCE.

- To give values to Lie products easily, we need some supporting function (setliebracket).

To implement all this properly, one has to know much about the internal working of REDUCE. Because there exists (to our knowledge at least) no 'literate' discription of REDUCE, one has to get this by inspecting the REDUCE source, which is more or less readable (written in REDUCE itself).

To give you an idea again a part of the code is included here:

———————————————— Start RWEB section ————————————————

31. **Assignment to commutators.** With the simplification procedure written above, we are able to retrieve values of commutators. As we have explained in the introduction, we need a set-element-function *set_liebracket* to assign values to commutators.

This seems to be the appropriate moment to explain how one can assign *value* to *kernel*. This is done by calling the procedure *setk(kernel,value)*. If *kernel* is of the form $f(a_1, \ldots)$, where f possesses the property *rtype*, which on its turn possesses a property *setelemfn* (the set-element-function for that rtype), the assignment is done by this set-element-function. In all other cases the procedure *setk1* takes care of it. So to make the construction work in our case, we have to declare any liebracket to be of rtype *liebracket* and assign to *liebracket* the property *setelemfn*.

32. There is, however, one more thing to explain about rtypes: commutators should not be recognized as objects of rtype *liebracket* since this will lead to type mismatch problems throughout REDUCE. To get the rtype of an object REDUCE almost anywhere uses the procedure *getrtype*, which, if provided, uses a rtypefn, to determine the rtype of an object. Rtypefn's have one argument, which are the arguments of the object, if this is not an atom, nil otherwise. So if we do not want commutators to be recognized as objects of rtype *liebracket*, we can simply return nil in all cases;

⟨Lisp initializations⟩ ≡
put('liebracket, 'rtypefn, 'liebracket_rtypefn)$
put(liebracket, 'setelemfn, 'set_liebracket)$

See also sections 41, 73, 75, 88, 107, 124, 138, 139, and 142.

This code is used in section 4.

33.
lisp procedure *liebracket_rtypefn u;*
nil$

──────────────── End RWEB section ────────────────

At the end of the 'RWEB section 32' you can see how RWEB automatically keeps track of where an 'RWEB module' is used. These 'modules' allow for programming 'top down'. At places where the code would become not readable, one creates a module with a discriptive name in place. This module may be defined later. (Here it is already mentioned, thus used, in section 4 and really given in section 32). For the real source code the TANGLE program puts all parts in at the good place.

The last part we will try to bring under your attention is about the 'efficiency' of analyzing Jacobi identities (sections 66 and 67):

──────────────── Start RWEB section ────────────────

66. Analysis of relations in Lie superalgebras.

If we have a relation in a Lie superalgebra we have a few possibilities to solve it:

1. The relation contains a commutator, for which we can solve the relation.

2. The relation contains only generators, we have found a linear dependency which we can solve.

3. The defining relations of the Lie superalgebra contained some parameters, which also occur in the relation to solve. In this case we can proceed as in case 1. and 2., but more carefully. For instance, suppose that we have found the relation $a(1)*lie(1,2)+$

$a(2) * x(1) + x(2) = 0$. Then it is dangerous simply to solve for $lie(1, 2)$ because $a(1)$ eventually may become 0, in which case the relation becomes a linear dependency between $x(1)$ and $x(2)$. Also a relation like $(a(1) - 1) * x(1) + a(2) * x(2) = 0$ can be solved in two ways: we can put $a(1) = 1$ and $a(2) = 0$ or solve the linear dependency in case $a(1) \neq 1$ or $a(2) \neq 0$.

To be able to recognize these parameters we will add to each liebracket the property parameters, which is an operator, elements of which may occur as parameters of the Lie superalgebra.

67. To keep the computations as compact as possible we will suppose that any relation R to be solved is a sum of generators and commutators of two generators. Taking into account the points raised above we can deduce the following strategy for finding a solution of a relation:

1. If the relation contains at least one commutator whose coefficient does not depend on a parameter, choose one and solve for it.

2. If the relation contains commutators, but all possessing coefficients depending on parameters, do not solve the relation.

3. If the relation does not contain commutators, but at least one generator whose coefficient does not depend on a parameter, choose one and solve for it.

4. If the relation does not contain commutators and all generators have coefficients depending on parameters, try solve the relation by solving the set of coefficients regarded as a linear set of equations in an appropriate set of parameters.

These tasks can most conveniently be performed by using the procedures *operator_coeff*, for finding all generators with their corresponding coefficients, and *solvable_kernels* from the TOOLS package. The procedure *operator_coeff* has already been used and described before.

The call *solvable_kernels(exprn, k_oplist, c_oplist)* will return an algebraic list of kernels x from operators occuring on *k_oplist*, such that x occurs linearly in *exprn* and the coefficient of x does not depend on any operator occuring on *c_oplist*. From this it is clear that *solvable_kernels* can be fruitfully used in step 1, 2 and 4.

The process described above will be performed by the procedure *relation_analysis*. It returns either the kernel for which the relation is solved or '*unsolvable* or '*nested_commutator* if the relation for whatever reason is not solvable or contains nested commutators.

define *zero_list* ≡ '(*list* 0);
lisp operator *relation_analysis*;
lisp procedure *relation_analysis(relation, bracketname)*;
begin scalar *generatorname, parameters, kernel_list, solvable_kernels,*
    test, kernel, optimal_kernel, coefficient, clear_list;

check_if_bracketname_is_a_liebracket_in ("RELATION_ANALYSIS:");
generatorname←get(bracketname,generatorname);
parameters←get(bracketname,parameters)
kernel_list←operator_coeff(relation,generatorname);
return
    if *kernel_list = zero_list* **then** 0
    else
        if *independent_part_of kernel_list* ≠ 0 **then** ⟨Solve *relation* for a commutator 68⟩
        else⟨ Solve *relation* for a generator or parameters 70 ⟩
end$

————————————— End RWEB section—————————————

It suffices to compute only Jacobi identities using all triples of generators of the algebra. But direct useful information may be derived only, if the identity does not contain 'nested' Lie products to get one of the cases indicated above. If for instance the Jacobi identity using x_i, x_j, x_k is computed and (e.g.) $[x_i, x_j] = [x_1, x_2]$, then normally a term $[x_k, [x_1, x_2]]$ will occur, giving no relation between generators or 'not nested' Lie products.

This means that one should avoid these cases (needless computations). We solved this problem as follows: in the vector structure containing all the (known) Lie products of generators not only the value of a product is stored but some extra information too (as a dotted pair). Namely the information if a product is a sum of generators or not. It is even a little bit more sophisticated using the same trick of REDUCE of avoiding unnecessary resimplifications of already computed algebraic expressions:

Algebraic expressions are stored in REDUCE as follows: (!*SQ . standardquotient . !*sqvar!*) ,a 'sq-prefixform', and the value of !*sqvar!* is most of the time the list '(T)'. The internal algebraic simplification procedures check the last element of such a sq-prefixform (e.g. by CADDDR('a sqprefvorm')), and if it is 'T' no resimplifications is necessary. If now an operation is executed, which may make resimplifications necessary in the future, the 'T' in the !*sqvar!* is replaced by RPLACA(!*sqvar!*,NIL) meaning that in all (existing) sq-prefixforms the last element has changed to 'NIL'! Furtheron !*sqvar!* gets imidiately a new value '(T).

We use the same mechanism to indicate the case, that a linear dependence of generators (via evaluation of a Jacobi identity) has occured, meaning, that already computed Lie products give rise to new relations to be analyzed. This is sometimes deeply recursive. This behavior is implemented in our analysis of Jacobi identities and the simplification functions for Lie algebras.

Another important aspect in this range of application is that in our applications all the Lie algebras were equipped with a multigrading. If you are trying to find an algebra element of a 'given' multigrade, the number of possibilities to construct or find such an element decreases enormously. The multigrading is a reliable guide in the computations! The package, which was roughly described on the previous pages does contain several suitable functions.

- A function which delivers all generators of a given multigrade of the algebra as an algebraic list.

- A function which delivers all Lie products of a given multigrade as an algebraic list.

- Introducing new generators, its multigrade and its 'history' is saved in the system.

- The function which analyzes all (suitable) Jacobi identities records the multigrade of newly found Lie products. These are saved in a 'vector' structure too.

5 Conclusions

The message we tried to bring over may be repeated here again.

- Very often a special field of applications needs its own programming, though useful packages exist in all of the 'Algebra Systems'.

- For efficient programming in 'Symbolic Algebra' good knowledge of the used system is necessary. For us it is of greatest importance, that the source code of REDUCE is generally available contrary to systems where one buys only the 'Executable System'.

- The programming of functions to your own needs should nevertheless be constructed by using techniques of 'Software Ingeneering'. Our experience with a 'WEB'-system, literate programming, fits to some degree into this philosophy.

6 Bibliography

[Vinogradov] Krasilshchik, I.S, Lychagin, V.V., Vinogradov, A.M., Geometry of jetspaces and nonlinear partial differential equations, Advanced Studies in Contemporary Mathematics, Vol. 1 (1986).

[Estabrook] Wahlquist, H.D. and Estabrook, F.B.; Prolongation structures of nonlinear evolution equations I, J. Math. Phys. 16 (1975),1-7.

[Roelofs] Roelofs, G.H.M, Hijligenberg, H. van de, Prolongation structures for supersymmetric equations, Journal of Physics A, Math. and General, Vol 23 (1990).

[Kac] Kac, V.G., Infinite Dimensional Lie Algebras, Boston, Basel, Stuttgart, Birkhaeuser, 1983.

[Gabber] Gabber, O., Kac, V.G., On defining relations of Certain Infinitedimensional Lie Algebras. Bull. Am. Math. Soc., Vol 5, No. 2, Sep. 1981, 185-189.

[Knuth] Knuth, D.E., The WEB System of Structured Documentation, Stanford Computer Science Report STAN-CS-980, Dept. of Computer Science, Stanford University, Stanford, CA, Sep 1983.

[Knuth] Knuth, D.E., Literate Programming. The Computer Journal, Vol. 28, No. 2, 1984.

[Krommes] Krommes, J.A., The WEB System of Structured Software Design and Documentation for FORTRAN, Ratfor, and C. Technical Report, Princeton University, Princeton, NJ, Nov. 1989. Available via anonymous ftp from: ss01.pppl.gov in directory /pub/fweb.

Computer Algebra Approaches to Enzyme Kinetics

J.P. Bennett, J.H. Davenport, M.C. Dewar & D.L. Fisher
School of Mathematical Sciences
University of Bath, United Kingdom

M. Grinfeld
Department of Engineering Mathematics
University of Bristol, United Kingdom

H.M. Sauro
Department of Animal Genetics
University of Edinburgh, United Kingdom

Abstract

The rate law governing the kinetics of a single enzyme mediated reaction may be derived relatively easily by hand given knowledge of the enzyme's mechanism. Such rate laws are typically non-linear in the concentrations of metabolites involved. When a number of enzymes interact the composite rate law for the complete system involves the simultaneous solution of the individual enzyme rate laws. We show how computer algebra can be used to solve this previously intractable problem, using the method of Gröbner Bases. We present an experimental example where kinetic parameters for individual enzymes are measured by making observations of a multi-enzyme system, and fitting these data to the rate law for the complete system.

1 Introduction

The mathematics of single enzyme kinetics are fairly well-understood: for example the reversible reaction which converts a substrate S to product P mediated by enzyme E, would typically have a reaction mechanism that can be written as

$$S + E \rightleftharpoons E^* \rightleftharpoons P + E$$

(often abbreviated to $S \overset{E}{\rightleftharpoons} P$ in what follows). This would have the Michaelis-Menton rate law[1]:

$$v = \frac{V_{\text{max}_f}[S]/K_{M_S} - V_{\text{max}_r}[P]/K_{M_P}}{1 + [S]/K_{M_S} + [P]/K_{M_P}} = \frac{\frac{V_{\text{max}_f}}{K_{M_S}}\left([S] - \frac{[P]}{K_{\text{eq}}}\right)}{1 + [S]/K_{M_S} + [P]/K_{M_P}}$$

where

---

[1]Biochemical notation is somewhat confusing to the mathematician. To a mathematician, $k[X]$ signifies the ring of polynomials in the variable X over the field k, whereas to a biochemist it signifies the product of a constant k and the concentration of substance X. Other than in reaction schemes, we will use capital letters to stand for the *concentrations* themselves. Note that, in reaction schemes, the notation $A + B$ means "substance A and substance B".

v is the net rate of formation of substance P, i.e. $d[P]/dt$ or $-d[S]/dt$;

$V_{\max_f}$ is the maximum possible forward velocity (and depends on the total concentration of the enzymes $[E] + [E^*]$);

$V_{\max_r}$ is the maximum possible reverse velocity (also depending on the total concentration of the enzymes $[E + E^*]$);

K_{M_S} is the Michaelis constant for S — that concentration of S at which the reaction proceeds at half the velocity $V_{\max_f}$ in the absence of P and

K_{M_P} is the Michaelis constant for P — that concentration of P. at which the reaction proceeds in reverse at half the velocity $V_{\max_r}$ in the absence of S;

K_{eq} is the equilibrium constant for the reaction — $[P]/[S]$ when $v = 0$.

These kinetic parameters are not absolutely constant. They depend on experimental conditions such as temperature and pH, as well as many factors less well-understood. In particular, *in vitro* experiments on an isolated enzyme may well give different results from what actually takes place in a living organism (*in vivo*), where large numbers of enzymes are interacting.

One factor distinguishing biochemistry is the difficulty of obtaining accurate values of these constants. They vary widely in magnitude, ten orders of magnitude not being unknown, and the literature is inconsistent about the experimental conditions under which they are measured. This complicates all attempts at numerical fitting of experimental data to these constants.

Let us consider the two enzyme system:

$$S \overset{E_1}{\rightleftharpoons} X \overset{E_2}{\rightleftharpoons} P,$$

where X is the product of the first reaction and substrate to the second. Each reaction's kinetics will satisfy a Michaelis-Menton rate law. However, the intermediate X may well be hard to measure (if it is not, then we can treat the two reactions separately). However typically S and P are metabolites whose concentrations do not vary greatly (known as *pool metabolites* in the cell). For example glucose levels are well buffered in the cell to prevent them changing much. Under these circumstances we can make the *steady state assumption* that the concentration of X, which we cannot measure, is nevertheless assumed to be constant.

In this case, where the rates of the two reactions must thus be equal, it is possible to eliminate (the concentration of) X from the two Michaelis-Menton equations, and find that v satisfies a quadratic equation (i.e. one of total degree two) in S and P, whose coefficients involve the eight parameters $V_{\max_f}^{(1)}$, $V_{\max_r}^{(1)}$, $K_{M_S}^{(1)}$, $K_{M_X}^{(1)}$, $V_{\max_f}^{(2)}$, $V_{\max_r}^{(2)}$, $K_{M_X}^{(2)}$ and $K_{M_P}^{(2)}$ [Sies 1982]. This analysis is pretty tedious, and, as far as the authors are aware, has not previously been extended to more complex chains. Indeed in his seminal paper introducing the concept of Metabolic Control Analysis no less an authority than Henrik Kacser declares that in general such systems of simultaneous non-linear equations are not generally soluble [Kacser & Burns 1973].

2 A Three-Enzyme Problem

In 1988 [Bennett *et al.* 1988], we applied the method of Gröbner bases [Buchberger 1985] to perform the same calculation for a three-enzyme chain, i.e.

$$S \overset{E_1}{\rightleftharpoons} X_1 \overset{E_2}{\rightleftharpoons} X_2 \overset{E_3}{\rightleftharpoons} P,$$

where the constants for each enzyme E_i are: the forward and reverse Michaelis constants $k_f^{(i)}$ and $k_r^{(i)}$, the equilibrium constant $k_{eq}^{(i)}$ and the maximum forward velocity $v_{max}^{(i)}$.

The Reduce [Hearn 1987] program to perform this elimination is trivial:

```
v1    := (v1max/k1mf)*(s-x1/k1eq)/(1+s/k1mf+x1/k1mr) $
v2    := (v2max/k2mf)*(x1-x2/k2eq)/(1+x1/k2mf+x2/k2mr) $
v3    := (v3max/k3mf)*(x2-p/k3eq)/(1+x2/k3mf+p/k3mr) $
bas   := groebner({num(v-v1),num(v-v2),num(v-v3)},{x1,x2,v}) $
```

The total time required was four seconds (on a Sun 4/280). In the formulae produced, the first two elements of bas, i.e. the expressions for X_1 and X_2 in terms of v, S and P would together occupy a whole page. The last element of bas, the equation for v, is a cubic whose coefficients are given below.

Factorised coefficient of v^0

$$-\left(\left(\left(v_{max}^{(2)}k_r^{(2)}k_f^{(3)}k_{eq}^{(3)}+k_f^{(2)}k_{eq}^{(2)}v_{max}^{(3)}k_r^{(3)}\right)k_f^{(1)}k_{eq}^{(1)}+v_{max}^{(1)}k_r^{(1)}k_f^{(2)}k_{eq}^{(2)}k_f^{(3)}k_{eq}^{(3)}\right)P-\left(\left(v_{max}^{(2)}+v_{max}^{(3)}\right)\times\right.$$

$$k_r^{(1)}k_{eq}^{(2)}-v_{max}^{(2)}k_f^{(3)}-k_f^{(2)}k_{eq}^{(2)}v_{max}^{(3)}\right)k_f^{(1)}k_{eq}^{(1)}k_{eq}^{(2)}k_{eq}^{(3)}k_r^{(3)}-\left(\left(k_r^{(2)}-k_f^{(3)}\right)k_f^{(2)}+Sk_{eq}^{(1)}k_r^{(2)}\right)v_{max}^{(1)}k_r^{(1)}\times$$

$$k_{eq}^{(2)}k_{eq}^{(3)}k_r^{(2)}-\left(v_{max}^{(2)}+v_{max}^{(3)}\right)Sk_{eq}^{(1)}k_r^{(1)}k_{eq}^{(2)}k_r^{(2)}k_{eq}^{(3)}k_r^{(3)}\right)$$

Factorised coefficient of v^1

$$-\left(\left(\left(v_{max}^{(2)}k_r^{(2)}k_f^{(3)}k_{eq}^{(3)}+k_f^{(2)}k_{eq}^{(2)}v_{max}^{(3)}k_r^{(3)}\right)v_{max}^{(1)}k_r^{(1)}+k_f^{(1)}k_{eq}^{(1)}v_{max}^{(2)}k_r^{(2)}v_{max}^{(3)}k_r^{(3)}\right)P+\left(\left(v_{max}^{(2)}k_f^{(3)}+\right.\right.\right.$$

$$k_f^{(2)}k_{eq}^{(2)}v_{max}^{(3)}\right)+\left(v_{max}^{(2)}+v_{max}^{(3)}\right)Sk_{eq}^{(1)}k_{eq}^{(2)}\right)v_{max}^{(1)}k_r^{(1)}k_r^{(2)}k_{eq}^{(3)}k_r^{(3)}+k_f^{(1)}k_{eq}^{(1)}k_r^{(1)}v_{max}^{(2)}k_{eq}^{(2)}k_r^{(2)}v_{max}^{(3)}k_{eq}^{(3)}k_r^{(3)}+$$

$$Sk_{eq}^{(1)}k_r^{(1)}v_{max}^{(2)}k_{eq}^{(2)}k_r^{(2)}v_{max}^{(3)}k_{eq}^{(3)}k_r^{(3)}\right)$$

Factorised coefficient of v^2

$$\frac{-v_{max}^{(1)}k_r^{(1)}v_{max}^{(2)}k_r^{(2)}v_{max}^{(3)}k_r^{(3)}\left(P-Sk_{eq}^{(1)}k_{eq}^{(2)}k_{eq}^{(3)}\right)}{k_{eq}^{(1)}k_{eq}^{(2)}k_{eq}^{(3)}\left(\left(\left(k_r^{(2)}-k_f^{(3)}\right)k_f^{(2)}-k_r^{(1)}k_r^{(2)}\right)k_f^{(1)}k_r^{(3)}-Pk_f^{(1)}k_f^{(2)}k_f^{(3)}-Sk_r^{(1)}k_r^{(2)}k_r^{(3)}\right)}$$

Factorised coefficient of v^3

$$\left(k_{eq}^{(1)}k_{eq}^{(2)}k_{eq}^{(3)}\left(\left(\left(k_r^{(2)}-k_f^{(3)}\right)k_f^{(2)}-k_r^{(1)}k_r^{(2)}\right)k_f^{(1)}k_r^{(3)}-Pk_f^{(1)}k_f^{(2)}k_f^{(3)}-Sk_r^{(1)}k_r^{(2)}k_r^{(3)}\right)\right)$$

The expressions were factorised by Reduce (using the command on factorize), and converted to TEX using a package written by one of the authors. This clearly proves that computer algebra can solve problems which are too difficult to do by hand.

3 Utility of the Solution

The sceptic naturally asks: "what use are such complicated formulae"? Other than using them as wall-paper, human beings do not want to look at expressions of this size. One major use of such formulae is numerical computation: Reduce, like many algebra systems, can convert such expressions automatically into Fortran expressions that can be incorporated into numerical programs.

There are other uses. We must mention the work of Lecourtier & Raksanyi [1985], who show that computer algebra systems can be used to test the structural properties of compartmental models. If the parameters involved are not structurally identifiable, then no amount of experimentation and numerical computation is going to deduce the correct values.

There are also interesting questions raised by the form of these equations. A three-enzyme chain in steady state generates a cubic equation for v, and we conjecture that an n-enzyme state will generate an equation of degree n. All systems analysed so far have only had one physically valid solution (i.e. with v real and all concentrations real and positive). Although for linear enzyme chains it would be surprising to find otherwise we might expect to find that in cyclic biochemical pathways there may be more than one physically realistic steady state. If these were not stable in time we have the possibility of oscillatory behaviour, and a basis for biological clocks.

4 A Specific Example

During the past year we have performed experiments to test the suitability of this work in looking at complete enzymes systems. We have worked with the *in vitro* system of fumarase, malate dehydrogenase (MDH) and aspartate aminotransferase (AAT). In the body this is part of the pathway involved in protein biosynthesis, and converts fumarate to aspartate and α-ketoglutarate. For practical experimental purposes we have chosen to run it in the reverse direction.

$$\text{aspartate} + \alpha\text{-ketoglutarate} \quad \overset{v_1}{\underset{\text{AAT}}{\rightleftharpoons}} \quad \text{glutamate} + \text{oxaloacetate}$$

$$\text{oxaloacetate} + \text{NADH} \quad \overset{v_2}{\underset{\text{MDH}}{\rightleftharpoons}} \quad \text{malate} + \text{NAD}^+$$

$$\text{malate} \quad \overset{v_3}{\underset{\text{fumarase}}{\rightleftharpoons}} \quad \text{fumarate}$$

We run the reaction in a closed vessel, starting off with just aspartate, α-ketoglutarate and NADH. The progress is followed by UV-spectroscopy to monitor the breakdown of NADH. This data is logged automatically every few seconds by an IBM PC connected by an analogue to digital converter. The precise experimental details are described elsewhere [Fisher 1990].

This work is in progress, and to date we have concentrated solely on the two enzyme system of aspartate aminotransferase and malate dehydrogenase. We make our quasi-steady state assumption that $v_1 = v_2$. Our observed flux v is the same as v_2. The rate laws are somewhat complex:

$$v_1 = \cfrac{\cfrac{V_{\max_{1,f}}[\text{asp}][\alpha\text{-kg}]}{K_{I_{1,\text{asp}}}K_{M_{1,\text{akg}}}} - \cfrac{V_{\max_{1,r}}[\text{oaa}][\text{glu}]}{K_{I_{1,\text{oaa}}}K_{M_{1,\text{glu}}}}}{\begin{array}{l}\cfrac{[\text{asp}]}{K_{I_{1,\text{asp}}}} + \cfrac{K_{M_{1,\text{asp}}}[\alpha\text{-kg}]}{K_{I_{1,\text{asp}}}\cdot K_{M_{1,\alpha\text{-kg}}}} + \cfrac{[\text{oaa}]}{K_{I_{1,\text{oaa}}}} + \\[2ex] \cfrac{K_{M_{1,\text{oaa}}}[\text{glu}]}{K_{I_{1,\text{oaa}}}K_{M_{1,\text{glu}}}} + \cfrac{[\text{asp}][\alpha\text{-kg}]}{K_{I_{1,\text{asp}}}K_{M_{1,\alpha\text{-kg}}}} + \cfrac{[\text{asp}][\text{oaa}]}{K_{I_{1,\text{asp}}}K_{I_{1,\text{oaa}}}} + \cfrac{K_{M_{1,\text{asp}}}[\alpha\text{-kg}][\text{glu}]}{K_{I_{1,\text{asp}}}K_{M_{1,\alpha\text{-kg}}}K_{I_{1,\text{glu}}}} + \\[2ex] \cfrac{[\text{oaa}][\text{glu}]}{K_{I_{1,\text{oaa}}}K_{M_{1,\text{glu}}}}\end{array}}$$

$$v_2 = \cfrac{\cfrac{V_{\max_{2,f}}[\text{NADH}][\text{oaa}]}{K_{I_{2,\text{NADH}}}K_{M_{2,\text{oaa}}}} - \cfrac{V_{\max_{2,r}}[\text{mal}][\text{NAD}^+]}{K_{M_{2,\text{mal}}}K_{I_{2,\text{NAD}^+}}}}{\begin{array}{l}1 + \cfrac{[\text{NADH}]}{K_{I_{2,\text{NADH}}}} + \cfrac{K_{M_{2,\text{NADH}}}[\text{oaa}]}{K_{I_{2,\text{NADH}}}K_{M_{2,\text{oaa}}}} + \cfrac{K_{M_{2,\text{NAD}^+}}\cdot\text{mal}}{(K_{M_{2,\text{mal}}}\cdot K_{I_{2,\text{NAD}^+}})} + \cfrac{[\text{NAD}^+]}{K_{I_{2,\text{NAD}^+}}} + \\[2ex] \cfrac{[\text{NADH}][\text{oaa}]}{K_{I_{2,\text{NADH}}}K_{M_{2,\text{oaa}}}} + \cfrac{K_{M_{2,\text{NAD}^+}}[\text{NADH}][\text{mal}]}{K_{I_{2,\text{NADH}}}K_{M_{2,\text{mal}}}K_{I_{2,\text{NAD}^+}}} + \cfrac{K_{M_{2,\text{NADH}}}[\text{oaa}][\text{NAD}^+]}{K_{I_{2,\text{NADH}}}K_{M_{2,\text{oaa}}}K_{I_{2,\text{NAD}^+}}} + \\[2ex] \cfrac{[\text{mal}][\text{NAD}^+]}{K_{M_{2,\text{mal}}}K_{I_{2,\text{NAD}^+}}} + \cfrac{[\text{NADH}][\text{oaa}][\text{mal}]}{K_{I_{2,\text{NADH}}}K_{M_{2,\text{oaa}}}K_{I_{2,\text{mal}}}} + \cfrac{[\text{oaa}][\text{mal}][\text{NAD}^+]}{K_{I_{2,\text{oaa}}}K_{M_{2,\text{mal}}}K_{I_{2,\text{NAD}^+}}}\end{array}}$$

The K_I's are *inhibition constants*, and represent the metabolite's tendency to bind the enzyme at the wrong stage, inhibiting its activity.

In addition since the system is closed we have five *conservation relationships* on the metabolite concentrations:

$$
\begin{aligned}
[\text{asp}] - [\alpha\text{-kg}] &= [\text{asp}_0] - [\alpha\text{-kg}_0] \\
[\text{asp}] + [\text{glu}] &= [\text{asp}_0] + [\text{glu}_0] \\
[\text{mal}] - [\text{NAD}^+] &= [\text{mal}_0] - [\text{NAD}^+_0] \\
[\text{NADH}] + [\text{NAD}^+] &= [\text{NADH}_0] + [\text{NAD}^+_0] \\
[\text{asp}] + [\text{oaa}] + [\text{mal}] &= [\text{asp}_0] + [\text{oaa}_0] + [\text{mal}_0]
\end{aligned}
$$

The last of these in particular is not obvious, and the determination of such conservation rules is itself a subject for automation. Programs such as SCAMP and iMAP can solve these problems [Sauro & Fell 1990]

This now gives us seven simultaneous equations, from which we must eliminate six unknown concentrations and obtain a formula for the flux through the complete system.

With such complex constituent rate laws, this is not a trivial problem. Unlike most Gröbner basis problems the complexity is not in the degree of the polynomials involved, but in the coefficients. We can only achieve a result by a three step elimination and substitution. The resulting rate law is a cubic, with coefficients of many thousands of terms. Fortunately the REDUCE computer algebra system can produce these directly as a FORTRAN program.

4.1 Fitting the Data

Even with several thousand data points, it is impossible to fit all nineteen parameters. The four $V_{\max}$'s can be ignored, since their values were used in the first place to assay enzyme activity. That leaves eight parameters for MDH and seven for AAT.

Our rate law relates v (i.e. $-d\text{NADH}/dt$) to NADH, while our experimental data is for NADH and t. We need to integrate the rate law to fit to the data. We essentially have an overconstrained

Parameter	Published Value (Molar)	Experimental Value (Molar)
$K_{I_2,oaa}$	5.0×10^{-4}	$3.6 \pm 0.7 \times 10^{-4}$
$K_{I_2,NADH}$	5.2×10^{-6}	$6.7 \pm 0.5 \times 10^{-6}$
$K_{I_2,mal}$	1.7×10^{-1}	$1.7 \pm 0.0 \times 10^{-1}$
K_{I_2,NAD^+}	1.0×10^{-3}	$8.4 \pm 4.0 \times 10^{-3}$

Table 1: Inhibition constants for malate dehydrogenase

multi-point boundary problem. One approach would be to take estimates of the kinetic parameters, and numerically integrate from the starting point, determining the residuals as NADH concentration from this integration less experimental NADH concentration at the various time points. Least squares minimisation can then be used to home in on the best parameter estimates. However this is very sensitive to choice of starting conditions for the integration. Instead we take each experimental point in turn as starting condition, and integrate to the next point. This technique, based on that of [Bock 1981] is much more robust, and using standard NAG Fortran Library routines [NAG LTD. 1990] we can obtain parameter estimates. If we try to fit all 15 parameters at once minimisation is not successful—any minimum is lost in experimental noise. However if we assume some parameters hold their published values, then we can fit the remaining parameters with some degree of success.

4.1.1 Error estimation

To date we have only used the statistical technique of bootstrap to estimate the error behaviour. Typical results are for the inhibition constants of MDH (table 1).

However since we have analytic forms of the rate law we can instead use differentiation to get a feel for the error behaviour. For a given parameter, k_i, we wish to know how a small relative change in k_i will influence v. In other words

$$lim\frac{\Delta v/v}{\Delta k_i/k_i} = \frac{\partial \ln v}{\partial \ln k_i}$$

We do not know the experimental error in v directly, but this can be estimated from the known errors in NADH and t. We can then relate the error in k_i to this, and so decide which kinetic parameters can be estimated with low error by this technique. Conversely we can adjust experimental design to minimise errors.

4.2 Problems

A number of problems have arisen during this work. The full analysis and development of techniques is still in progress.

4.2.1 Gröbner Basis calculation

Calculation of Gröbner Bases even for the two enzyme system is computationally difficult, because of the complexity of the coefficients. The algorithms need modifying to handle this particular case effectively.

4.2.2 FORTRAN rounding errors

Although the rate law as calculated by REDUCE is correct, in putting it out in FORTRAN rounding errors are introduced, due to the limitations of double precision floating point. For example we get rather different results depending on whether calculations are done with concentrations expressed in molar or millimolar units. This is currently a topic of computer algebra research. However one approach would be to have rate law evaluation (as needed by the numerical integrator) done in REDUCE. This would degrade performance, but eliminate accuracy problems.

4.2.3 The steady-state assumption

It is not always obvious that the steady-state assumption is valid. For example in our closed system the initial substrate and final product do change in concentration during the reaction. *A posteriori* verification must be used to check that the assumption $v_1 = v_2$ is correct. In this example *post hoc* analysis shows that the assumption holds to a close enough approximation, since the substrates are present in reasonable excess. However even in biological systems where substrates and products are known to be held constant in pools there could be oscillations in intermediate concentrations, particularly where cycles are involved, and this must be checked for.

5 Conclusions

The steady state assumption is very attractive, since it renders the mathematics of complex systems tractable, and the biochemistry observable. But, in the manner in which it is stated, it is unverifiable. We believe that our methodology of

1. computing the necessary equations (algebraically) under the steady-state assumption;

2. using numerical methods to fit these to experimental data to obtain approximate values of the parameters;

3. using numerical integration of the original differential equations (i.e. without the steady state assumption) and least-squares fitting of the results to the experimental data to refine these values;

4. performing an *a posteriori* verification of the approximate validity of the steady state assumption;

provides a way of incorporating the steady-state methodology without being completely reliant on an unproven assumption. The computationally intensive numerical integrations at step 3 are not performed until after approximate values of the parameters have been determined, and so many fewer iterations are necessary.

The authors are grateful to the Science and Engineering Research Council of the United Kingdom for their support under grant GR/F/66726.

References

[Bennett *et al.* 1988] J. P. Bennett, J. H. Davenport, and H. M. Sauro. Solution of some equations in biochemistry. Technical Report 88-12, University of Bath, June 1988.

[Bock 1981] H. G. Bock. Numerical treatment of inverse problems in chemical reaction kinetics. In K. H. Ebert, P. Deuflhard, and W. Jäger, editors, *Modelling of chemical reaction systems*, pages 102–125. Springer-Verlag, 1981.

[Buchberger 1985] Buchberger,B., Gröbner bases, an algorithmic method in polynomial ideal theory. *Recent Trends in Multi-Dimensional System Theory (ed. N.K. Bose)* Reidel, 1985, pp. 184-232.

[Fisher 1990] D. L. Fisher. Applications of computer algebra to enzyme analysis. Technical Report 90-32, University of Bath, February 1990.

[Hearn 1987] A. H. Hearn. *The REDUCE User's Guide*. The RAND Corporation, 1987.

[Kacser & Burns 1973] Kacser,H. & Burns,J.A., The Control of Flux. *Symposium of the Society for Experimental Biology*, **27**, 1973, pages 65–104.

[Lecourtier & Raksanyi 1985] Lecourtier,Y. & Raksanyi,A., Algebraic Manipulation Routines for Testing Structural Properties. In *IFAC Identification and System Parameter Estimation*, 1985, pages 543–549.

[NAG LTD. 1990] NAG LTD. *The NAG Fortran Library Manual — Mark 14*, 1990.

[Sauro & Fell 1990] Sauro,H.M. & Fell,D.A., Simulators and Analysers—A description of two computer programs, SCAMP and *i*MAP. *Proceedings of BioThermoKinetics 1990*, to be published by Intercept, Andover, UK, 1991.

[Sies 1982] Sies,H. (ed.), *Metabolic Compartmentation* (Chapter 1). Academic Press, London, 1982.

CALCUL FORMEL
POUR LA "POURSUITE SINGULIERE" DE TRAJECTOIRES

M.A. FAYAZ, Laboratoire des Signaux et Systèmes, CNRS-ESE,
Plateau de Moulon, 91192 Gif-sur-Yvette Cedex, France

Résumé Dans ce papier nous présentons un programme en REDUCE qui permet de calculer le *rang de singularité* $r(x_s, y_d)$ pour une trajectoire $y_d \in C^\infty(\mathbb{R})$ et un point singulier x_s donnés. On a démontré [1], [2] que ce rang fournissait *une condition nécessaire* pour la poursuite de la trajectoire y_d à travers le point singulier x_s et permettait de quantifier le maximum de régularité du couple $(\bar{u}_d, \bar{x}_d)$, solution du problème de poursuite. Dans [3], nous montrons que si $r(x_s, y_d)$ est infini et le "saut de l'index relatif" au point singulier est fini, alors il existe une commande C^∞ permettant le passage à travers le point singulier.

Abstract In this paper we present a REDUCE programme for computing the *rank of singularity* $r(x_s, y_d)$ for a given output trajectory $y_d \in C^\infty(\mathbb{R})$ and a singular point x_s. In [1] and [2] it is shown that this rank provides *a necessary condition* for tracking y_d through x_s and quantifies the maximum regularity of $(\bar{u}_d, \bar{x}_d)$, solution to the tracking problem. In [3], we show that if $r(x_s, y_d)$ is infinite and the "jump of the relative degree" at the singular point is finite, then there exists, locally, a C^∞ control driving the system through the singular point.

I. INTRODUCTION

L'analyse des systèmes dynamiques dont la complexité ne cesse de croître, nécessite de plus en plus, l'établissement d'une synergie entre des théories mathématiques et des outils informatiques adéquats. La théorie de contrôle des systèmes a souvent recours à la manipulation d'expressions symboliques et algébriques. Dans ce contexte, les langages formels apparaissent comme des outils indispensables. Les travaux de Jacob et Oussous [4], Lamnabhi [5], Claude et Dufresne [6], Marino et Cesaro [7], Barbot [9], et Diop [8], entre autres, mettent en exergue l'efficacité de ces langages dans le cadre de l'analyse et la commande des systèmes non linéaires. Ces langages trouvent aussi leur intérêt dans d'autres domaines comme la mécanique, la robotique et l'optique.

Les caractéristiques fondamentales de ces langages sont principalement : la manipulation des polynômes, des fractions rationnelles, des fonctions élémentaires, les matrices de ces éléments ainsi que la dérivation et l'intégration formelle [11], [12], [13].

Dans cet article, nous présentons un programme en REDUCE qui permet de calculer le rang de singularité $r(x_s, y_d)$ pour une trajectoire $y_d \in C^\infty(\mathbb{R})$ et un point singulier x_s donnés. On a démontré [1], [2] que ce rang fournissait une condition nécessaire pour la poursuite de la trajectoire y_d à travers le point singulier x_s et permettait de quantifier le maximum de régularité du couple $(\bar{u}_d, \bar{x}_d)$, solution au problème de poursuite. En calculant $r(x_s, y_d)$, nous obtenons aussi les conditions initiales pour des systèmes différentiels singuliers d'ordre k tels qu'ils sont définis dans [1] et [2]. Leur résolution fournit une paire admissible $(\bar{u}_d, \bar{x}_d)$ de commande et de trajectoire d'état pour la poursuite de y_d à travers x_s telle que $u_d \in C^k$ et $x_d \in C^{k+1}$. Dans la partie II, nous rapellerons brièvement les définitions et résultats théoriques obtenus dans [2]. La partie III est consacrée au programme et à quelques exemples d'application.

II. RAPPELS

Considérons un système MIMO, carré, affine en la commande de la forme :

$$\dot{x} = f(x) + \sum_{i=1}^{m} g_i(x) u_i \qquad\qquad (*)$$
$$y = h(x), \qquad x(0) = x_0, x \in M, \qquad y \in \mathbb{R}^m$$

Soit $y_d \in C^\infty(\mathbb{R})$ la fonction à poursuivre.

Définition 1

x_s est un point singulier pour la poursuite de trajectoires si:

a) det $(D_\alpha(x_s)) = 0$

b) il n'existe pas de voisinage de x_s, $\vartheta(x_s)$, tel que det $(D_\alpha(x)) = 0$
   pour tout $x \in \vartheta(x_s)$

$D_\alpha(x)$ étant la matrice de découplage et α l'index relatif du système $(*)$ ou de son extension dynamique.

Définition 2

On dit que y_d peut être poursuivie à travers le point singulier x_s, si il existe une paire admissible $(\bar{u}_d, \bar{x}_d)$, définie sur un intervalle$[0, t_s + \varepsilon)$, telle que $\bar{u}_d = u_d$ et $\bar{x}_d = x_d$ s u r $[0, t_s)$, et $\bar{x}_d(t_s) = x_s$. Dans la suite nous supposerons $x_s = x_0$ et $t_s = 0$.

Définition 3

Etant donnés un point singulier x_0 et une fonction $y_d \in C^\infty(\mathbb{R})$, on appelle $r(x_s, y_d)$ le rang de singularité pour la trajectoire y_d et le point x_0, le plus grand entier tel qu' il existe $\vec{v} = (v_0, v_1, ..., v_r) \in \mathbb{R}^{r+1}$ vérifiant les équations:

$$y_d^{(\alpha+k)} = C_{k+\alpha}(x_0, v_0, ..., v_{k-1}) + D_\alpha(x_0)v_k \quad \text{avec} \quad 0 \le k \le r \quad \text{and} \quad v_i = u^{(i)}(0)$$

Si ces équations sont satisfaites pour tout $k \ge 0$ alors $r = \infty$.

Si ces équations sont satisfaites pour $0 \le k \le \alpha + r_0$ alors $r = r_0$.

Si ces équations ne sont pas satisfaites pour une valeur de k, telle que $0 \le k < \alpha$ alors r n'est pas défini.

III. CALCUL DU RANG DE SINGULARITÉ

Avant de montrer comment on peut obtenir systématiquement le rang de singularité $r(x_0, y_d)$, nous allons détailler les principales étapes intermédiaires nécessaires à ce calcul.

Les notations suivantes seront utilisées:

NBETAT : Nombre d'etats

NBENT : Nombre d'entrées

NBSOR : Nombre de sorties

NBDER : Nombre de derivées (choisi $\ge$ NBETAT)

IDXR(I) : Index relatif de la sortie I

MDEC : Matrice de découplage

RSIN(I) : Rang de singularité de la sortie I

U(I,J) : Derivée d'ordre J de l'entrée I

YP(I,J) : Derivée d'ordre J de la sortie I

XP(I) : $\dot{x}_i$

III.1. Calcul de l'index relatif :

Le calcul de l'index relatif se fait par l'alternance des deux operations suivantes:

a) Le calcul pour chacune des sorties du système des derivées de Lie successives par rapport aux champs de vecteurs f et g_i . En REDUCE la dérivée de Lie d'une fonction $h(x)$ par rapport à champs de vecteurs $f(x)$ peut être exprimer par:

LIFH:=(FOR I:=1:NBETAT SUM DF(H, X(I))*F(I));

b) Test de dependance de $y_j^{(k)}$ par rapport aux entrées. En REDUCE:

DEP :=(FOR I:=1:NBENT SUM DF(Y(J,K), U(I,0)));
IF DEP $\neq$ 0 THEN INDXR(J) :=k
ELSE <<k:=k+1; GOTO a >>

III.2. Calcul de la matrice de découplage :

Les entrées de la matrice de decouplage sont données par:

MDEC(I,J):=DF(Y(I,IDXR(I)),U(J,0));

III.3. Calcul des points singuliers :

Les points singuliers sont les racines de la matrice de découplage du système ou de son extension dynamique.

PTSIN := FACTORIZE (DET(MDEC));

III.4. Calcul de $r(x_0, y_d)$:

Le calcul de $r(x_0, y_d)$ comprend les étapes suivantes:
- Choix d'un point singulier.
- Evaluation des $y_j^{(k)}$ ($1 \leq j \leq$ NBSOR , k $\leq$ NBDER) au point singulier x_0.
- Evaluation des $y_{d_j}^{(k)}$ (t) ($1 \leq j \leq$ NBSOR , k $\leq$ NBDER) à l'instant t_0.
- Test de cohérence entre $y_j^{(k)}$ et $y_{d_j}^{(k)}$.

EXEMPLES

Ci-dessous nous donnons trois exemples qui résument les principales situations apparaissant lors du calcul le rang de singularité. Ces exemples reflètent bien le dialogue interactif et les resultats fournis par le programme.

Exemple 1

Le système :

$$\dot{x}_1 = x_1 u_1$$
$$\dot{x}_2 = x_1 x_2 + x_3 u_1$$
$$\dot{x}_3 = x_2 u_2$$
$$y_1 = x_1$$
$$y_2 = x_2$$

La procédure :

```
NB DE VAR D'ETAT =?
 3;
NB D'ENTREES =?
2;
NB DE SORTIES =?
 2;
ENTRER LES EQ D'ETAT
XP(1):=
X(1)*U(1,0);
XP(2):=
X(1)*X(2)+X(3)*U(1,0);
XP(3):=
X(2)*U(2,O);
ENTRER LES SORTIES
Y(1,0):=
X(1);
Y(2,0):=
X(2);
ENTRER LE NB DE DERIVATION (>=NBRETAT)
4;
L'INDEX RELATIF:
IDXR(1)=1
IDXR(2)=1
MATRICE DE DECOUPLAGE:
MDEC(1,1)=X(1)
MDEC(1,2)=0
MDEC(2,1)=X(3)
MDEC(2,2)=0
```

LA 2ième COLONNE DE LA MATRICE DE DECOUPLAGE EST IDENTIQUEMENT NULLE. L'INDEX RELATIF N'EST PAS *BIEN* DEFINI.IL FAUT ETUDIER LA POSSIBILITE D'UNE EXTENSION DYNAMIQUE

Cet exemple met en exergue une situation propre au systèmes multi-entrées multi-sorties. En effet, les premières derivées des deux sorties $y_1^{(1)}$ et $y_2^{(1)}$ sont affectées par u_1 mais pas par u_2. Dans ce cas, si le rang de la matrice de découplage du système est constant dans un voisinage $\vartheta(x_0)$ d'un point x_0, alors en ajoutant une dymamique sur une des entrées du système, on peut obtenir un système ayant un index relatif bien défini au point x_0. Cette opération porte le nom d'algorithme d'extension dynamique. Pour plus de détails à ce sujet on peut consulter par exemple le livre d' Isidori [10].

Exemple 2

Le système :

$$\dot{x}_1 = x_1 + x_2 u_1$$
$$\dot{x}_2 = x_2 x_3$$
$$\dot{x}_3 = x_1 u_2$$
$$y_1 = x_1$$
$$y_2 = x_2$$

La procédure :

```
NB DE VAR D'ETAT =?
 3;
NB D'ENTREES =?
2;
NB DE SORTIES =?
2;
ENTRER LES EQ D'ETAT
XP(1):=
X(1)+X(2)*U(1,0);
XP(2):=
X(2)*X(3);
XP(3):=
X(1)*U(2,0);
ENTRER LES SORTIES
Y(1,0):=
X(1);
Y(2,0):=
X(2);
ENTRER LE NB DE DERIVATION (>=NBRETAT)
4;
L'INDEX RELATIF:
IDXR(1)=1
IDXR(2)=2
MATRICE DE DECOUPLAGE:
MDEC(1,1)=X(2)
MDEC(1,2)=0
MDEC(2,1)=0
MDEC(2,2)=X(2)*X(1)
L'ENSEMBLE DES POINTS SINGULIERS = {X(2),X(2),X(1)}
ENTRER LE POINT SINGULIER DE VOTRE CHOIX
XS(1):=
```

0;
XS(2):=
1;
XS(3):=
1;
ENTRER LA FONCTION A POURSUIVRE
YT(1,0):=
TT**5;
YT(2,0):=
SIN(TT);
ENTRER LE TEMPS Ts
Ts:=
0;
RS(1)=0
RS(2)=INDEFINI
LE RANG DE SINGULARITE N'EST PAS DÉFINI. LA TRAJECTOIRE Y_D
NE PEUT PAS ETRE POURSUIVIE À TRAVERS CE POINT SINGULIER .

Exemple 3

Le système :

$$\dot{x}_1 = u$$
$$\dot{x}_2 = x_1^2$$
$$y = x_2$$

La procédure :

NB DE VAR D'ETAT =?
2;
NB D'ENTREES =?
1;
NB DE SORTIES =?
1;
ENTRER LES EQ D'ETAT
XP(1):=
U(1,0);
XP(2):=
X(1)**2;
ENTRER LES SORTIES
Y(1,0):=
X(2);
ENTRER LE NB DE DERIVATION (>=NBRETAT)
7;
L'INDEX RELATIF:
IDXR(1)=2
MATRICE DE DECOUPLAGE:
DEC(1,1)=2*X(1)

```
L'ENSEMBLE DES POINTS SINGULIERS = {X(1)}
ENTRER LE POINT SINGULIER DE VOTRE CHOIX
XS(1):=
0;
XS(2):=
0;
ENTRER LA FONCTION A POURSUIVRE
YT(1,0):=
TT**3
ENTRER LE TEMPS Ts
Ts:=
 0;
RS(1) >= 5
RANG DE SINGULARITE >=5
```

Pour ce système le rang de singularité est ≥5. Si il existe une paire de solution, permettant la poursuite à travers le point singulier, alors $\bar{u}_d$ sera au moins C^5 et $\bar{x}_d$ aux moins C^6. Cette limitation à 5 est due au fait que l' ordre de dérivation à été fixé à 7, alors que théoriquement on peut montrer que $r = \infty$. Nous fixons le nombre de dérivations à priori pour éviter les boucles infinies.

IV. CONCLUSION

Cette présentaion a eu pour but de montrer qu'ici encore les langages formels offrent des possibilités pour l'implémentation des algorithmes structurés, manipulant des expressions algébriques et symboliques.

Notre algorithme permet un test rapide, pour savoir si une trajecctoire donnée remplit les conditions nécessaires de passage à travers un point singulier. Il nous permet également de connaître le maximum de régularité de la commande $\bar{u}_d$ et de la trajectoire d'etat $\bar{x}_d$. Il permet en particulier de décider si on doit avoir recours à l'approche de la commande discontinue [14], ou à des methodes approchées [15].

Références

[1] P. E. Crouch I. S. Ighneiwa and F. Lamnabhi-Lagarrigue, *On the singular output tracking problem.* A paraître dans Mathematical theory of Signals and Systems.

[2] M.A. Fayaz and F. Lamnabhi-Lagarrigue, *On the singular output tracking for nonlinear multivariable systems,* in "Analysis and optimisation of systems ", Lect. Notes in Control Inform., Springer-Verlag, Berlin, 1990. N° 144, PP. 611-619.

[3] M.A. Fayaz, *Solving the problem of singular output-tracking for a class of nonlinear systems and output trajectories*, Preprint.

[4] G. Jacob et N. Oussous, *Un logiciel calculant la réalisation minimale des systèmes analytiques de série generatrice finie*, Publication N° 123, 1988, LIFL USTLFA.

[5] M. Lamnabhi, *Functional analysis of nonlinear circuits: a generating power series approach* . IEE Proceeding, vol. 133, N° 5, October 1986.

[6] D. Claude et P. Dufresne, *A MAXIMA Application to nonlinear systems decoupling* in "Computer Algebra", EUROCAM'82, Lect. Notes in Computer Science, Springer-Verlag, 1982, 144, PP. 284-301.

[7] R. Marino, G. Cesaro, *Nonlinear Control Theory and Symbolic Manipulation* , in "Mathematical theory of Networks and Systems", Lect. Notes in Control and Inform., Springer-Verlag, 1984, N° 58, PP. 726-740.

[8] S. Diop, *Elimination in control theory* . Math. Control Signals and Systems 4, 1991, PP. 17-32.

[9] J. P. Barbot, *A computer aided design for Sampling nonlinear systems* , in "Applied Algebra, Algebraic Algorithm and Error Correcting Codes", Lec.Notes in Comp. Sc. 357, T. Mora ed., 1989.

[10] A. Isidori, *Nonlinear Control Systems: An Introduction,* 2nd Edition, Springer-Verlag, Communication and Control Engineering Series.

[11] J. Davenport , Y. Siret, E. Tournier,*Calcul formel, Système et algorithmes de manipulation algébrique.* "Etudes et Recherches en Informatique", Ed.MASSON.

[12] G. Rayna, REDUCE, *Software for Algebraic Computation,* Springer-Verlag.

[13] A.C. Hearn, REDUCE, User's Manual, Version 3.3.The Rand Corporation, Santa Monica, 1988.

[14] M. Fliess, P. Chantre S. Abu El Ata, A. Coic, *Discontinuous predictive control, inversion and singularities: Application to a heat exchanger,* in"Analysis and optimisation of systems", Lect. Notes in Control Inform., Springer-Verlag, Berlin, 1990. N° 144, PP. 851-857.

[15] J.W. Grizzle, M.D. Di Benedetto, *Approximation by regular input-output maps*, preprint 90 .

USING SYMBOLIC CALCULUS FOR
SINGULARLY PERTURBED NONLINEAR SYSTEMS

by

J.P. BARBOT and N. PANTALOS

Laboratoire des Signaux et Systèmes
Plateau du Moulon
91192 Gif-sur-Yvette
FRANCE

ABSTRACT

The formal Gröbner formula for the problem of the inversion of analytic functions, is used in order to compute the quasi-steady-state model for singularly perturbed nonlinear systems. For this, the use of symbolic calculus is of paramount importance and a computer program, in REDUCE symbolic language, is proposed.

1. INTRODUCTION

The great expanding of various applications during the two last decades, according to many theoretical results using singular perturbation methods, makes these methods to be considered as an extremely useful and powerful tool for modelling and control linear processes, generally of high dimension [1]. Recently, singular perturbation techniques have been introduced in a nonlinear context and some important results have already appeared in the continuous-time case [2], [3], [4].

In both linear and nonlinear cases, the above methods deal with systems which inherently contain slow and fast dynamic phenomena. In general, the dynamical behaviour of such systems is characterized by a rapid attraction of the fast system's states to a slow integral manifold. Contrary to the linear case, where the exact slow integral manifold can be obtained by solving a Riccati-type equation, in the nonlinear case, the computation even of the zero order approximation of this manifold requires heavy manipulations. Frequently in many applications, only zero approximation is used and leads to the the so-called quasi-steady-state model. Due to the fact that this model describes essentially the slowest (average) part of the system's dynamics and is of lower dimension than the initial one, it is often used as the basic control design model. Therefore, from a theoretical and a practical point of view, computing the quasi-steady-state model is of paramount importance.

In the present paper, a computer program (using the REDUCE symbolic language) is proposed, in order to compute the quasi-steady-state model for nonlinear systems expressed in the standard singularly perturbed state form. This program is based on the Gröbner formula [5], [6]. The solution is expressed in terms of formal Lie derivatives exponentials and computed using symbolic calculus [7].

An outline of the paper is as follows. Section 2 reviews briefly the integral manifold approach for singularly perturbed nonlinear systems. Section 3 presents the Gröbner formula; a proof is also given in the case of the inversion of a family of analytic functions. Finally, section 4 collects two examples representing a nonlinear system with fast actuators [8] and a three-dimensional nonlinear autonomous system from chemistry [9] and points out some difficulties which may result from physical singularities.

2. THE INTEGRAL MANIFOLD APPROACH

Given a nonlinear singulary perturbed system

$$\dot{x} = f(x,z,u,\varepsilon) \qquad x \in R^n, \ z \in R^m, \ u \in R^p, \ \varepsilon \text{ is a small positive parameter} \tag{2.1}$$

(Σ)

$$\varepsilon \dot{z} = g(x,z,u,\varepsilon) \tag{2.2}$$

where the functions f and g are assumed to be sufficiently many times continuously differentiable on their arguments, it is well-known that the solution of (Σ) consists of a slow-steady-state and a fast boundary layer. This two-time scale behaviour of the solution, as trajectory in R^{n+m}, may be observed geometrically using the integral manifold approach. The fundamental property of the integral manifold is that once the fast system's states reach this manifold, they remain on this thereafter.

Let M_ε denote the above n-dimensional slow manifold defined by

$$M_\varepsilon = \{ (x,z) \colon z = \phi \ (x, \ u, \ \varepsilon) \ \} \tag{2.3}$$

Taking the derivative of z along the trajectories (2.1) and (2.2), and assuming that the system's input is only x-dependent, we obtain the so-called manifold condition:

$$\varepsilon \left\{ \frac{\partial\phi(x,u,\varepsilon)}{\partial x} + \frac{\partial\phi(x,u,\varepsilon)}{\partial u} \frac{\partial u}{\partial x} \right\} f(x,\phi,u,\varepsilon) = g(x,\phi,u,\varepsilon) \tag{2.4}$$

which $\phi \ (x, \ u, \ \varepsilon)$ must satisfy for all x of interest and $\varepsilon \in [0, \ \varepsilon^*]$, where ε^* is a positive constant. We note that the assumption that the system's input is only x-dependent is not restrictive for two reasons. Firstly, the control law is frequently designed on the reduced order model which is independent of the state z; secondly, we have already supposed that the slow manifold is attractive which implies that z is driven to this manifold without control.

The approximation procedure of the solution $\phi \ (x, \ u, \ \varepsilon)$ of (2.4) starts by substituting into the manifold condition (2.4) the power series expansion of $\phi \ (x, \ u, \ \varepsilon)$ as:

$$\phi \ (x, \ u, \ \varepsilon) = \sum_{i=0}^{\infty} \varepsilon^i \ \phi_i(x,u) \tag{2.5}$$

and calculating $\phi_i(x,u)$ by equating terms in like power of ε. For this, we have to suppose that functions f and g can be also expanded in powers of ε. For zero and first order approximations, we verify easily the following expressions:

for $\varepsilon=0$: $0 = g(x,\phi_0(x,u),u,0)$ $\qquad\qquad\qquad\qquad$ (2.6.a)

and   for $\varepsilon^2 = 0$:

$$\{ \frac{\partial\phi_0(x,u)}{\partial x} + \frac{\partial\phi_0(x,u)}{\partial u}\frac{\partial u}{\partial x} \} f(x,\phi_0,u,\varepsilon) - \frac{\partial g(x,\phi_0,u,\varepsilon)}{\partial\varepsilon}/z=\phi_0,\varepsilon=0 =$$

$$\{ \frac{\partial(g(x,z,u,0))}{\partial z}/z=\phi_0 \} \ \phi_1 \qquad\qquad\qquad\qquad (2.6.b)$$

If only the zero order approximation is considered, which means that $z = \phi_0(x,u)$, then substitution of this value of z into (2.1) results in the so-called *quasi-steady state model*. We note that it is equivalent to neglect completely all the fast system's dynamics in (Σ), by setting $\varepsilon=0$ in (2.2) and to solve the algebraic equation: $0 = g(x,z,u,0)$ in z.

In order to compute explicitelly the solution of (2.6.a), we use the Gröbner formula, presented in [5] in a slightly more general mathematical context.

3. THE GRÖBNER FORMULA

Consider the following vectorial equation

$$y = h(w) \qquad\qquad \text{with } y\in R^k \text{ and } w\in \dot{R}^k$$

where h is an analytic function. It is assumed that $(\frac{\partial h}{\partial w})_{|w=w_0}$ is invertible for any w_0 belonging in a neighbourhood of the solution.

Theorem [5] The local inverse function ϕ, solution of the above equation, is given by the following Gröbner formula :

$$w = \phi(y,w_0) = e^{S(y,w_0,\cdot)} \text{ Id }_{|w=w_0}$$

$$\text{with } \quad S(y,w_0,\cdot) = \sum_{i=1}^{k} (y^i - h(w_0)^i)L_{D_i}(.) \triangleq L_{D(y-h(w_0))}(.)$$

D_i denotes the ith column of the inverse of the Jacobian matrix of the function h:

$$D = (J_h)^{-1}|_{w_0} \triangleq (\frac{\partial h}{\partial w})^{-1}|_{w=w_0}$$

and the y^i and h^i denotes the i-component of the vectors y and h respectively; finally, Id denotes the identity function.

The above solution is expressed in terms of Lie series which converge absolutely and uniformly in a neighbourhood of the solution. The assumption about the invertibility of the Jacobien matrix of the function h at $w=w_0$, is exactly the required assumption in the implicit function theorem. Finally, we note that the initial value w_0 fixes implicitly the validity domain of the solution.

Proof The function $\phi(y,w_0)$ will be the local inverse function of h in the neighbourhood of $y_0 = h(w_0)$ if and only if the Taylor's series expansions (about $y=h(w_0)$) of the functions $\phi(y,w_0)$ and $h^{-1}(y)$ coincide.

This means that :

$$\frac{\partial^i \phi(y,w_0)}{\partial^i y}\Big|_{y=h(w_0)} = \frac{\partial^i h^{-1}(y)}{\partial^i y}\Big|_{y=h(w_0)} \qquad \forall\ i \in N \qquad (**)$$

The result follows by induction. In fact :

a - for i = 0 we get

$$\phi(y,w_0)\big|_{y=h(w_0)} = e^0\, Id_{|w=w_0} = w_0 = h^{-1}(y_0)$$

b - for i = 1 we have

$$\frac{\partial\,\phi(y,w_0)}{\partial y}\Big|_{y=h(w_0)} = \frac{\partial\,e^{S(y,w_0,.)}\,Id\,\big|_{w=w_0}}{\partial y}\Big|_{y=h(w_0)}$$

$$= \frac{\partial\,S(y,w_0,.)}{\partial y}\Big|_{y=h(w_0)}\,e^{S(y,w_0,.)}\,Id\,\big|_{w=w_0}$$

$$= L_D e^{S(y,w_0,.)}\,Id\,\big|_{w=w_0} = L_D\,e^0\,Id_{|w=w_0}$$

$$= D_{|w=w_0} = [\,J_h|_{w=w_0}]^{-1} = \frac{\partial\,h^{-1}}{\partial y}\big|_{y=h(w_0)}$$

c - we assume that (**) holds at order p, which means that

$$L_D^p\,e^{S(y,w_0,.)}\,Id\,\big|_{w=w_0} \triangleq \frac{\partial^p\,\phi(y,w_0)}{\partial^p y}\big|_{y=h(w_0)} = \frac{\partial^p h^{-1}}{\partial^p y}\big|_{y=h(w_0)}$$

d - finally, at order p+1, we get

$$\frac{\partial^{p+1}\,\phi(y,w_0)}{\partial^{p+1}y}\big|_{y=h(w_0)} = L_D^{p+1}\,e^{S(y,w_0,.)}\,Id\,\big|_{w=w_0}$$

$$= L_D L_D^p\,e^{S(y,w_0,.)}\,Id\,\big|_{w=w_0}$$

$$= L_D\frac{\partial^p\,\phi(y,w_0)}{\partial^p y}\big|_{y=h(w_0)}$$

$$= \frac{\partial}{\partial w}\left(\frac{\partial^p\,\phi(y,w_0)}{\partial^p y}\big|_{y=h(w_0)}\right)|_{w_0}\frac{\partial\,w}{\partial y}\big|_{y=h(w_0)}$$

$$= \frac{\partial}{\partial w}\left(\frac{\partial^p h^{-1}}{\partial^p y}\big|_{y=h(w_0)}\right)|_{w_0}\frac{\partial\,w}{\partial y}\big|_{y=h(w_0)}$$

$$= \frac{\partial^{p+1} h^{-1}}{\partial^{p+1}y}\big|_{y=h(w_0)}$$

and the proof is completed.

In order to solve (2.6.a), applying the above formula with y=0 , one takes:

$$\phi_0 = e^{S(y,z_0,.)} \, Id \, |_{z=z_0} \tag{3.1}$$

where $S(y,z_0,.) = \sum_{i=1}^{m} - g^i \, L_{D_i} \, (.)$, and g^i denotes the i-th component of the vector $g(x,z,u,0)$

evalued at $x=x_0$, $z=z_0$ and $u=u_0$. D_i denotes the i-th colomn of the matrix $\{J_g\}^{-1}|_{z=z_0}$. More explicitely, formula (3.1) gives:

$$\phi_0 = z_0 + \sum_{i=1}^{m} - g^i \, L_{D_i} \, (\, Id \,) \, |_{z=z_0} + \frac{1}{2!} \sum_{i=1}^{m} \sum_{j=1}^{m} g^i \, g^j \, L_{D_j} \, (L_{D_i} \, (Id)) |_{z=z_0} + ...$$

On the other hand, computing ϕ_0 permits to obtain, if desired, a better approximation of the solution ϕ of (2.4), that is $\phi = \phi_0 + \varepsilon \phi_1$, where the correction term ϕ_1 is given from (2.6.b).

In the following, we exhibit some structural-type informations concerning with the proposed program. This program is shared into three parts. The two first parts include a collection of data acquisition and auxiliary procedures used in the program. The third part deals with computation of the solutions ϕ_0 and ϕ_1 (see (3.1) and (2.6.b) respectively). This part is organized as follow:

- calculus of the Jacobien matrix J_g
- test about the invertibility of J_g and calculus (if possible) of the inverse matrix of J_g,

  denoted by D.
- calculus of the zero-order approximation ϕ_0 (see (3.1)) using the procedures
    LIEDERIV and GROBNERFORM, which are given below:
- calculus of ϕ_1 according to (2.6.b).

PROCEDURE LIEDERIV (function1, function2, dimz)

% It deals with Lie derivation generated by function1 and operating on function2

```
        begin
                scalar result;
                result := for i:=1 : dimz sum (df (function1, z(i))*function2(i));
                return result;
        end;
```

PROCEDURE GROBNERFORM (function1, function2, dimz, *order*)

% It deals with Gröbner formula computation
% *order* takes the values 1,2,..., according to the desired order of the expansion of ϕ_0

```
begin
        scalar solution;
        solution := function1;
        deri (0):= function1;
        factorielle : = 1;
        for i:=1 : order do
                begin
                        factorielle:= factorielle * i;
                        deri(i):= liederiv(deri(i-1),function2,dim_z);
                        solution:= solution + deri(i)/factorielle;
                end;
        return solution;
end;
```

4. EXAMPLES

4.1 The fast actuator system

Consider the following nonlinear analytic system with fast actuators [8], described by

$$\dot{x} = \sin x + (x^2 + 1) z \qquad -\frac{\pi}{2} < x < \frac{\pi}{2}$$

$$\varepsilon\dot{z} = -x - \tan z - \varepsilon z + u \qquad -\frac{\pi}{2} < z < \frac{\pi}{2}$$

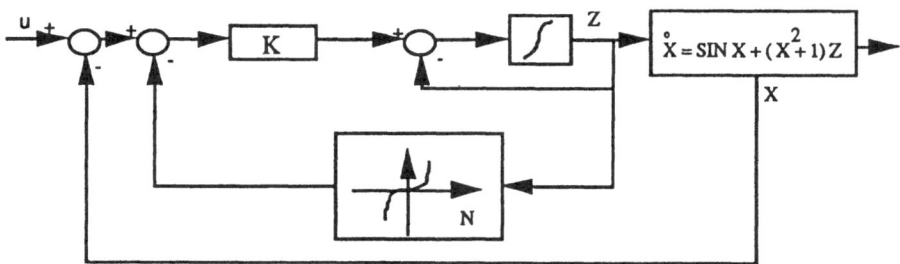

Figure 1: The fast actuator system.

Neglecting fast dynamics by setting $\varepsilon = 0$ in (4.1.2), results in the algebraic equation :

$$0 = -x - \tan z + u$$

with solution :

$$z = \phi_0 (x,u) = \arctan (u-x)$$

In this case, the proposed program gives the formal series expansion of the above solution, where the order of its expansion is chosen by the user. In the following, we give zero and first order approximations (according to (2.6.b)); the order of the expansion of the solutions ϕ_0, ϕ_1 is chosen to be equal to 10:

$$\phi_0 = \frac{1}{315} \{ 35* (u-x)^9 - 45* (u-x)^7 + 63* (u-x)^5 - 105* (u-x)^3 + 315* (u-x) \}$$

$$\phi_1 = -\{(u-x)^8 - (u-x)^6 + (u-x)^4 - (u-x)^2 + 1\} * \{ \frac{\partial u}{\partial x} - 1 \} * \{ x^2 * \phi_0 + \sin x + \phi_0 \} * \cos^2(\phi_0)$$

4.2 The Belousov-Zhabotinsky reaction

Consider the following three-dimensional autonomous system from chemistry describing in particular an approximated kinetic model of the Belousov-Zhabotinsky reaction [9].

$$\dot{x}_1 = \alpha(z) \, x_1 - \lambda(z) \, x_2 + a(z)$$

$$\dot{x}_2 = \beta(z) \, x_2 - \mu(z) \, x_1 + b(z)$$

$$\varepsilon \, \dot{z} = - x_1 - \frac{3}{2} (\frac{z^3}{3} - z) \triangleq g(x_1, z)$$

Setting $\varepsilon = 0$, leads to the reduced-order system, which defines a two-dimensional flow on the following manifold:

$$M_0 = \{ (x_1, x_2, z): \ 0 = - x_1 - \frac{3}{2} (\frac{z^3}{3} - z) \}$$

The projection of M_0 on the (x_1, z)-plane is sketched in Figure 1. This is divised into three parts according to the sign of $\dfrac{\partial g(x_1, z)}{\partial z}$, which defines the stability (or instability) domains. In particular, we verify that part II is instable (repellor) contrary to parts I and III which are stable (attractors).

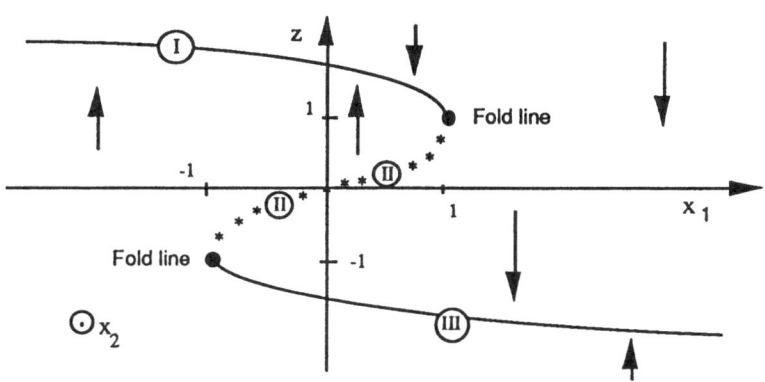

<u>Figure 2</u>: Projection of M_0 on the (x_1, z)-plane.

The fold curves C^- and C^+ of M_0 are defined by

$$\frac{\partial g(x_1, z)}{\partial z} = \frac{3}{2}(z^2 - 1) = 0$$

which implies that C^- and C^+ are straight lines perpendicular to (x_1, z)-plane at the points $(-1, -1)$ and $(1, 1)$. Obviously, along these curves, which represent the physical system's singularities, Gröbner's formula is not valid.

Outside C^- and C^+, infinite Gröbner formula expansion gives the exact solution of the equation $g(x_1, z) = 0$, according to the initial value of the state z. For example, if $z_0 = 2$, the proposed program provides the exact solution which will be valid only for all $x_1 \in]-\infty, 1[$.

On the other hand, troncate the above expansion, results in the restriction of the validity domain of the solution to a vinicity of $(x_{1,0}, z_0)$, where

$$x_{1,0} = \frac{3}{2}(\frac{z_0^3}{3} - z_0)$$

In the following, we give the solutions obtained using the proposed program, corresponding to parts I, II and III, where the order δ of the approximation of the solution is equal to 5. We verify that for a fixed value of x_1, the above solutions provide three different values for z. In what follows, we take $\alpha(z) = a(z) = -10$ and $\lambda(z) = 10$. Moreover, since $g(x_1, z)$ is independant of the state x_2, the solutions ϕ_0 and ϕ_1 (but not ϕ_2) will be independant of the polynoms $\beta(z)$, $\mu(z)$ and $b(z)$.

for $z_0 = -2$: $\phi_0(x_1) = -2(2288\, x_1^5 + 17920\, x_1^4 + 69212\, x_1^3 + 201728\, x_1^2 + 787501\, x_1 - 4143616)/4782969$

with $\phi_0(0) = 1.732$

for $z_0 = 0$: $\phi_0(x_1) = 2 (16 x_1^4 + 36 x_1^2 + 243) x_1 / 729$

with $\phi_0(0) = 0$

for $z_0 = 2$: $\phi_0(x_1) = - 2 (2288 x_1^5 - 17920 x_1^4 + 69212 x_1^3 - 201728 x_1^2 + 787501 x_1 + 4143616) / 4782969$

with $\phi_0(0) = - 1.732$.

Finally, looking for a better approximation of the exact slow manifold, we observe that x_2 appears in the expression of ϕ_1, which for $z_0 = 2$ and $\delta = 5$ is given by:

$$\phi_1 = - 40 (x_2 + x_1 +1) (11440 x_1^4 + 71680 x_1^3 + 207636 x_1^2 + 403456 x_1 + 787501) / (14348907 (\phi_0^2 - 1))$$

Geometrically, this is interpreted as an inclination about x_2-axis which may change the structural properties of the slow reduced-order system. In consequence, the expression of ϕ_1 may be rendered in some practical cases, necessary. Figure 3, illustrates the contribution of ϕ_1 relative to the zero order solution ϕ_0, for $\varepsilon = 0.1$.

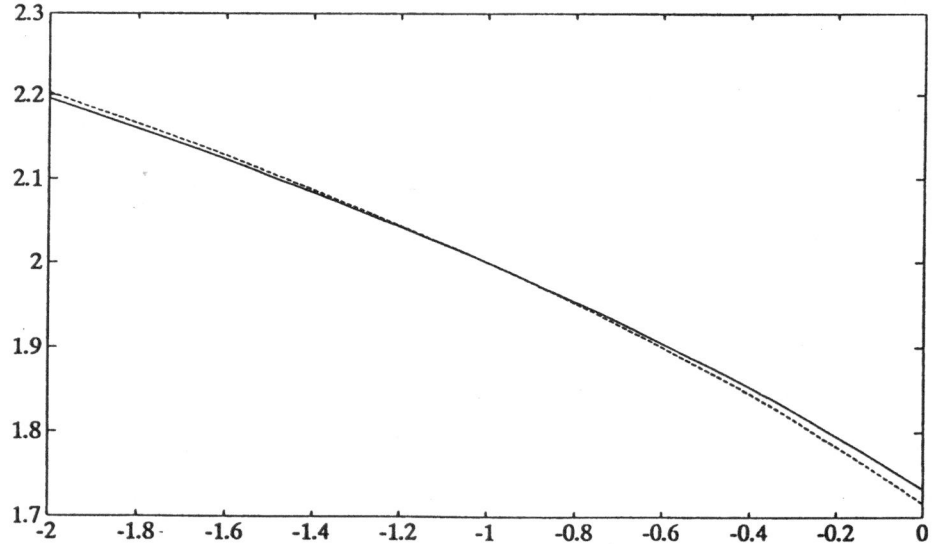

<u>Figure 3</u>. Contribution of ϕ_1. Continuous and dotted lines represent respectively, zero (ϕ_0) and first-order ($\phi_0 + \varepsilon\phi_1$) approximated solutions both projected into the (x_1,z)-plane at $x_2=2$.

Acknowledgements

The authors appreciate many fruitful discussions and suggestions with Professors D. Normand-Cyrot and S. Monaco.

REFERENCES

[1] P.Kokotovic, H.K.Khalil, J.O'Reilly, Singular Perturbation Methods in Control Analysis and Design, London, Academic Press, 1986.

[2] R.Marino, P.V.Kokotovic, A geometric approach to composite control of two-time-scale systems, Proceeding of 25th Conference on Decision and Control, Athens, Greece, 1986.

[3] M.W.Spong, K.Khorasani, P.V.Kokotovic, An Integral Manifold approach to the Feedback Control of Flexible Joint Robots, IEEE Journal of Robotics and Automation, Vol. RA-3, N° 4, 1987.

[4] P.M.Sharkey, J O'Reilly, Composite control of non-linear singularly perturbed systems : a geometric approach. Int.J.Control,Vol 48, N° 6, 1988.

[5] W.Gröbner, Serie di Lie e loro applicazioni, Cremonese, Rome, 1973.

[6] S.Monaco, D.Normand-Cyrot, Nonlinear decoupling in discrete-time, IFAC Symposium on Nonlinear Control Systems Design, Capri, Italy, 1989.

[7] J.P.Barbot, A Computer-aided Design for Sampling a Nonlinear Analytic System, in Lecture Notes in Computer Science, N° 357, 1988, Springer-Verlag.

[8] J-P Su, J-G Hsieh, Composite feedback control for a class of nonlinear singularly perturbed systems with fast actuators, Int. J. Control, vol. 52, N° 3, 1990.

[9] M. Canalis, Illustrations et etude de systèmes lents-rapides dans R^3, Thèse à l'Université de Nice, 1984.

[10] A.C.Hearn, REDUCE user's manual, The Rand Corporation, Santa Monica, 1984.

CALCUL D'APPROXIMATIONS DE LA SOLUTION D'UN SYSTEME NON LINEAIRE UTILISANT LE LOGICIEL SCRATCHPAD

Alexis MARTIN
Laboratoire des Signaux et Systèmes, CNRS
Supelec
91192 Gif-sur-Yvette Cedex, France

1.   Introduction

L'analyse des systèmes peut se définir comme l'étude de la réponse d'un système non linéaire en boucle ouverte à une entrée u(t). Cette étude est équivalente mathématiquement à la recherche de solutions d'une équation différentielle ordinaire, c'est à dire en pratique à rechercher une suite de fonctions convergeant vers la solution. L'expression de ces fonctions se complique à chaque itération, si bien qu'une analyse générale exacte de la solution s'avère très souvent difficile. On préfère donc limiter le champ de l'étude et particulariser les questions. Si par exemple le système est un circuit électronique non linéaire soumis à une tension d'entrée u(t) = sinωt, on peut se demander ce que vaut une harmonique particulière de la sortie en fonction des divers paramètres du modèle.

Des méthodes numériques de recherche de solutions ont été implémentées dans divers logiciels, comme SPICE, pour les circuits électroniques, ou NEPTUNIX pour des systèmes quelconques. Ces techniques s'appuient sur des algorithmes de point fixe comme Newton-Raphson ou sur des développements en série et ne fournissent qu'un tableau de valeurs numériques. Il est notamment impossible de comprendre la variation du résultat obtenu en fonction des divers paramètres du modèle sans recalculer un tableau de points pour chaque valeur de ces paramètres. Par conséquent toute analyse structurelle globale à l'aide de ces méthodes numériques est difficile.

Jusqu'à présent les méthodes non numériques ont été peu utilisées pour résoudre ce type de problème. Citons toutefois les travaux de [1]. Un des objectifs de notre étude est de proposer une autre approche combinant à la fois des techniques nouvelles de calcul comme la combinatoire et des outils nouveaux pour la mise en oeuvre informatique, à savoir l'utilisation du logiciel SCRATCHPAD. Le résultat obtenu ne sera plus un tableau de points, mais un ensemble d'expressions mathématiques simples où les paramètres du système apparaîtront explicitement. Ceci permettra a posteriori une analyse de l'influence de ces paramètres sur telle ou telle variable du système.

Nous allons tout d'abord définir quelques notions générales utiles pour notre étude. Un système de contrôle *non linéaire* à *représentation d'état*, linéaire par rapport au contrôle, peut se décrire dans un voisinage de x(0) par un système d'équations différentielles non linéaires sous la forme suivante :

$$(SNL) \begin{cases} \dot{x} = f(x) + \sum_{i=1}^{m} u_i g_i(x) \\ \\ y = h(x) \qquad\qquad\qquad x(0) = x_0 \end{cases}$$

où x est une fonction du temps, définie de $\mathbb{R}$ dans $\mathbb{R}^n$, u_i, i = 1, ..., m de $\mathbb{R}$ dans $\mathbb{R}$ sont les m entrées possibles, et g_i, f et h sont des fonctions analytiques de $\mathbb{R}$ dans $\mathbb{R}^n$, définies dans un voisinage de x(0).

Lorsque les fonctions u_i, g_i, f et h sont fixées, on sait que la solution (x(t), y(t)) de ce système existe sous certaines conditions dans un voisinage de x(0) et qu'elle est unique pour une condition initiale x(0) fixée, voir Gröbner[4]. Dans un premier temps, on recherche une *solution fonctionnelle*, c'est à dire que l'on ne fixe pas les fonctions u_i à priori, i = 1, ..., m . Lorsque l'on remplace ensuite les fonctions u_i par leurs expressions respectives, on obtient alors la solution déjà calculée dans le cas général d'une EDO quelconque.

Nous construisons donc une suite d'*approximants fonctionnels* de la solution, c'est-à-dire une suite de fonctionnelles des u_i convergeant vers la solution fonctionnelle de l'équation. Il existe un grand nombre de suites de fonctionnelles convergeant vers la solution, mais seules présentent de l'intérêt pour notre étude les suites de fonctionnelles qui sont aisément calculables et qui convergent rapidement vers la solution : après quelques itérations, les termes successifs de la suite doivent avoir un comportement qualitativement proche de la fonctionnelle solution de l'EDO. Notre choix de suite fonctionnelle est une suite obtenue à partir des *noyaux de Volterra* , qui constituent une approximation de type Taylor dans l' espace fonctionnel normé engendré par les fonctions u_i, voir Fliess et al.[3]. Pour un grand nombre de circuits électroniques non linéaires on peut vérifier par simulation que cette suite constitue bien un approximant fonctionnel convergeant rapidement vers la solution [1].

En notant w_i le i-ème noyau de Volterra et y(t) la solution du système non linéaire (SNL), la formule générale qui définit de façon unique les noyaux de Volterra associés à la sortie y(t) s'écrit :

$$y(t) = w_0(t) + \sum_{i=1}^{m} \int_0^t w_1(t-\tau_1) u_i(\tau_1) d\tau_1 +$$

$$\sum_{k\geq 2} \sum_{i_1, i_2, ..., i_k = 1}^{m} \int_0^t \int_0^{\tau_k} ... \int_0^{\tau_2} w_k(t, \tau_k, \tau_{k-1}, ..., \tau_1) u_{i_k}(\tau_k) ... u_{i_1}(\tau_1) d\tau_k ... d\tau_1$$

Un *système bilinéaire* est un système non linéaire à représentation d'états défini par l'équation suivante :

$$\text{(BLN)} \begin{cases} \dot{x} = Ax + \sum_{i=1}^{m} u_i \, A_i x \\ \\ y = Cx \qquad\qquad\qquad x(0) = x_0 \end{cases}$$

L'intérêt des sytèmes bilinéaires provient des nombreux résultats qui ont pu être obtenus pour cette classe particulière de systèmes non linéaires. Par exemple on peut aisément calculer les expressions des noyaux de Volterra successifs en fonction des matrices définissant le système.

En utilisant une méthode fondée sur la linéarisation de Carleman, Brockett[2] a montré que sous certaines hypothèses on peut associer à un système non linéaire une suite de systèmes bilinéaires tel que le n-ième système bilinéaire construit a les mêmes n premiers noyaux de Volterra que le système non linéaire considéré. Ainsi les noyaux de Volterra du système non linéaire peuvent se calculer grâce à l'approximant bilinéaire.

A l'approche analytique classique que nous avons rappelée jusqu'ici, [2], [6], on peut substituer une approche de type syntaxique, utilisant le vocabulaire des automates finis [5], ou bien une méthode combinatoire utilisant des arborescences [8] ou encore une technique algébrique fondée sur les séries génératrices en variables non commutatives [3]. Ces divers points de vue sont en fait équivalents à certains égards [5], [7]. Par exemple, dans [7], la véritable nature de certaines séries génératrices est éclaircie en construisant une suite d'approximants combinatoires. Cette suite d'approximants est en fait identique à celle construite par Brockett. Dans la suite de notre exposé, nous utiliserons conjointement l'approche analytique et l'approche algébrique avec quelques incursions dans la théorie combinatoire.

2. Algorithme de construction des approximants

2.1. Rappels [3]

Soient $u_1(t), u_2(t), \dots u_m(t)$ les entrées du système et soit un ensemble fini, appelé alphabet, $X = \{z_0, z_1, \dots z_m\}$. On note X^* l'ensemble des mots engendrés par X. On peut ainsi résumer l'approche algébrique introduite par Fliess: On considère la lettre z_0 comme un opérateur codant l'intégration par rapport au temps et z_i, $i = 1, \dots, m$, comme des opérateurs codant l'intégration par rapport au temps après une multiplication par $u_i(t)$. Ainsi à tout mot de X^* on peut faire correspondre une intégrale itérée, notée $I^t\{w\}$, qui peut être définie récursivement comme suit :

$$I^t\{\emptyset\} = 1 \text{ et pour } w = z_\alpha v \in X^*, \quad I^t\{w\} = \int_0^t d\tau \, I^\tau\{v\} \text{ si } \alpha = 0$$

et $\quad \mathbf{I^t}\{w\} = \displaystyle\int_0^t u_i(\tau)\,d\tau\, \mathbf{I^\tau}\{v\}$ $\quad$ si $\quad \alpha = i$.

Considérons maintenant le système de contrôle non linéaire suivant :

$$(\text{SNL}) \quad \begin{cases} \dot{x} = f(x) + \displaystyle\sum_{i=1}^m u_i g_i(x) \\[2mm] y = h(x) \qquad\qquad\qquad x(0) = x_0 \end{cases}$$

La solution du système $y(t)$ peut s'écrire sous la forme suivante :

$$y(t) = y(0) + \sum_{v \geq 0} \; \sum_{j_0,\,\dots,\,j_v = 0}^m f_{j_v} \dots f_{j_0} \cdot h(x) \Big|_{x = x_0} \mathbf{I^t}\{z_{j_0} \dots z_{j_v}\}$$

en posant $f_0 = f$, $f_i = g_i$, ou plus simplement :

$$y(t) = \sum_{d \in X^*} n(d)\, \mathbf{I^t}\{w(d)\}$$

où $\mathbf{I^t}\{w(d)\}$ représente l'intégrale itérée associée au mot $w(d)$ et $n(d)$ le coefficient associé à cette intégrale. Ce développement fonctionnel des fonctions u_i s'appelle *développement fonctionnel de Fliess* de la solution du système. La série formelle associée à y est une *série génératrice en variables non commutatives* qui s'écrit :

$$s = y(0) + \sum_{v \geq 0} \; \sum_{j_0,\,\dots,\,j_v = 0}^m f_{j_v} \dots f_{j_0} \cdot h(x) \Big|_{x = x_0} z_{j_0} \dots z_{j_v}$$

On peut de même associer une série génératrice à toute variable d'état du système. Dans ces séries, si l'on étudie par exemple le cas où l'alphabet est réduit à deux éléments, on peut regrouper les mots de la manière suivante :

$$\{\varnothing,\ z_0,\ z_0^2,\ z_0^3, \dots z_0^n, \dots\}, \ \{z_1,\ z_0 z_1,\ z_0^2 z_1,\ z_0^3 z_1, \dots,\ z_0^m z_1 z_0^n, \dots\},$$
$$\{z_1^2,\ z_0 z_1^2,\ z_0^2 z_1^2,\ z_0^3 z_1^2, \dots,\ z_0^m z_1 z_0^n z_1 z_0^p, \dots\} \ , \ \dots.$$

Ce regroupement de mots induit un regroupement de termes dans les séries génératrices à deux lettres, et on peut vérifier que le i-ème regroupement de termes correspond au i-ème noyau de Volterra de la variable $x(t)$ (ou $y(t)$) associée à cette série génératrice. Ce regroupement de termes nous permet de calculer la série génératrice sous la forme d'une somme :

$$g = g^1 + g^2 + \dots g^n + \dots$$

où g^n est l'ensemble des mots contenant n occurences de la lettre z_1, correspondant au n-ième noyau de Volterra. Ce procédé est généralisable aux séries génératrices fondées sur un alphabet fini quelconque.

2.2. Description de l' algorithme

Après cette esquisse du cadre théorique dans lequel nous nous plaçons, nous allons maintenant décrire notre algorithme de résolution, l'implémentation sur ordinateur en langage formel SCRATCHPAD sera détaillée par la suite.

Nous allons présenter ici un exemple simple, avec une seule entrée u(t), deux variables d'état x_1 et x_2 et une sortie $y = x_1$. Nous supposerons de plus que la non-linéarité du système libre est quadratique, cependant notre algorithme s'applique sans difficultés théoriques majeurs à des systèmes généraux (SNL). La seule hypothèse importante que nous ferons est que le système est à un point d'équilibre à t = 0, autrement dit, le noyau de Volterra d'ordre 0 associé à x_1 ou x_2 est nul. Les équations du système ne peuvent donc comporter de constante. De plus, il est toujours possible de se ramener à des conditions initiales nulles par un changement de variable $x = x - x_0$.

Dans les cas pratiques envisagés, on peut trouver une base qui diagonalise la matrice représentant la partie linéaire, en effet la matrice de la partie linéaire admet de façon générique n valeurs propres distinctes. Si la matrice est seulement trigonalisable sous la forme réduite de Jordan, l'algorithme peut encore être appliqué dans ce cas, au prix de quelques complications supplémentaires dans les relations de récurrence que nous allons établir. Notons que dans la base de vecteurs propres ainsi définie, la non-linéarité quadratique se présente alors sous la forme la plus générale possible.

On pose donc :

$$(\text{SNL2}) \quad \begin{cases} \dot{x}_1 = \lambda_1 x_1 + a_{20}^1 x_1^2 + a_{11}^1 x_1 x_2 + a_{02}^1 x_2^2 + u\, a_1 \\ \dot{x}_2 = \lambda_2 x_2 + a_{20}^2 x_1^2 + a_{11}^2 x_1 x_2 + a_{02}^2 x_2^2 + u\, a_2 \\ y = x_1 \end{cases}$$

$x_1(0) = 0, \; x_2(0) = 0$

Cette diagonalisation de la partie linéaire permet une très grande simplification des calculs ultérieurs : au lieu de résoudre un système linéaire d'équations on applique simplement une relation de récurrence, ce qui est infiniment moins coûteux en temps de calcul : sans diagonalisation, les calculs sont trop importants pour être réellement effectués. Dans la base de vecteurs propres, le nombre de termes à manipuler est inférieur au nombre de termes à manipuler dans une base quelconque, où la matrice de la partie linéaire du système n'est pas diagonale. Ainsi seules les véritables caractéristiques du système sont conservées en évitant la redondance de l'information à traiter.

Nous allons maintenant appliquer les notions théoriques exposées dans les paragraphes précédents : nous avons vu qu' il existait des approximants bilinéaires pour une large classe de systèmes non linéaires. On pose $x_{ij} = x_1^i \, x_2^j$ et l'équation différentielle associée à x_{ij} s'obtient en remplaçant les expressions de $\dot{x}_1$ et $\dot{x}_2$ par les équations de (SNL2) et en supprimant les termes de degré homogène en x_2 et en x_1 strictement supérieur à n, de manière à obtenir un système bilinéaire. Cet approximant bilinéaire est tel que les n premiers noyaux de Volterra associés aux solutions x_{ij} de ce système sont identiques aux n premiers noyaux de Volterra associés aux solutions x_{ij} du système (SNL2). Explicitons les équations de ces approximants bilinéaires pour notre exemple :

$$
\text{BLN(n)}
\begin{cases}
\text{si } i+j = n \quad \dot{x}_{ij} = (i\,\lambda_1 + j\,\lambda_2)x_{ij} + i\,u\,a_1 x_{i-1j} + j\,u\,a_2 x_{ij-1} \\[4pt]
\text{si } i+j < n \quad \dot{x}_{ij} = (i\,\lambda_1 + j\,\lambda_2)x_{ij} + i\,u\,a_1 x_{i-1j} + j\,u\,a_2 x_{ij-1} + \\[2pt]
\qquad\qquad\qquad (i\,a_{20}^1 + j\,a_{11}^2)\,x_{i+1j} + (i\,a_{11}^1 + j\,a_{02}^2)\,x_{ij+1} + \\[2pt]
\qquad\qquad\qquad i\,a_{02}^1\,x_{i-1j+2} + j\,a_{20}^2\,x_{i+2j-1} \\[6pt]
y = x_{10}
\end{cases}
$$

Approximer la solution de (SNL2) consiste à considérer la somme de ses projections sur ses noyaux de Volterra successifs. Si on définit la variable x_{ij}^n comme étant la projection de la variable d'état x_{ij} de BLN(n) sur son noyau de Volterra d'ordre n, on approxime x_{ij} solution de (SNL2) par la somme des x_{ij}^k, k variant de 1 à n (Rappelons que $x_{ij}^0 = 0$). Nous allons nous placer à présent dans le contexte des séries génératrices pour la suite de nos calculs. Nous considérons l'alphabet à 2 lettres $\{z_0, z_1\}$ et nous intégrons le système différentiel ci-dessus en remplaçant les opérateurs intégraux ainsi obtenus par les lettres z_0 ou z_1. On obtient ainsi un nouveau jeu de relations. On note g_{ij} la série génératrice associée à x_{ij}, et g_{ij}^n la série génératrice associée à x_{ij}^n, série génératrice constituée par l'ensemble des mots de g_{ij} contenant n occurences de la lettre z_1 (voir les regroupements de termes évoqués dans la partie précédente). L'approximation de g_{ij} est alors la somme des g_{ij}^k, k variant de 1 à n.

Remarquons que si l'on considère 2 approximants bilinéaires d'ordre n et m, avec $n < m$, alors les n premiers noyaux de Volterra de ces 2 approximants sont identiques. x_{ij}^n et g_{ij}^n sont donc indépendants du bilinéaire BLN(m) considéré, pourvu que m soit supérieur à n. A l'aide du système BLN(n), on obtient ainsi les relations de récurrence :

$$\text{si } i+j = n \qquad g_{ij}^n = (1 - (i\,\lambda_1 + j\,\lambda_2)z_0)^{-1}\,(i\,a_1\,z_1\,g_{i-1j}^{n-1} + j\,a_2\,z_1\,g_{ij-1}^{n-1})$$

$$\text{si } i+j < n \qquad g_{ij}^n = (1 - (i\,\lambda_1 + j\,\lambda_2)z_0)^{-1}\,(\,i\,a_1\,z_1\,g_{i-1j}^{n-1} + j\,a_2\,z_1\,g_{ij-1}^{n-1} + $$

$$(i\, a_{20}^1 + j\, a_{11}^2)\, z_0\, g_{i+1j}^n + (i\, a_{11}^1 + j\, a_{02}^2)\, z_0\, g_{ij+1}^n +$$

$$i\, a_{02}^1\, z_0\, g_{i-1j+2}^n + j\, a_{20}^2\, z_0\, g_{i+2j-1}^n)$$

$$g = g_{10}^n$$

Ces relations sont obtenues de la manière suivante : on considère le nouveau jeu de relations obtenues avec les séries génératrices à partir du système BLN(n). On ne conserve que les termes contenant exactement n occurences de z_1, ce qui revient à projeter BLN(n) sur son n-ème noyau de Volterra. Les n–1 premiers noyaux de Volterra de BLN(n) nous sont déjà connus par BLN(1), BLN(2), ... BLN(n–1). Les noyaux de Volterra de BLN(n) d'ordre supérieur à n n'ont à priori pas de rapport avec les noyaux de Volterra d'ordre supérieur à n de (SNL2). L'intérêt principal de ces relations est donc de nous permettre de calculer les noyaux de Volterra de (SNL2) à tout ordre.

Pour calculer les g_{ij}^n on procède de la manière suivante : on calcule d' abord g_{01}^1 et g_{01}^1, puis g_{20}^2, g_{11}^2, g_{02}^2, g_{10}^2 et g_{01}^2, etc ... De façon plus générale, on va dans le sens des n croissants, et à chaque étape n on calcule toutes les séries génératrices associées aux solutions de BLN(n) : (x_{01}, x_{10}, x_{20}, ..., x_{ij}, ... x_{0n}), avec i+j < n, en commençant par les termes g_{ij} tels que i+j = n, puis i+j = n-1 et ainsi de suite. Ce calcul nous donne un approximant fonctionnel. Pour obtenir la solution finale pour une entrée u donnée, on substitue à z_1 la transformée de Borel-Laplace de u, moyennant un mélange[3], ce qui ne demande que peu de temps de calcul si l'entrée u est par exemple une somme d'exponentielles.

Pour n fixé, n = 3, 4, 5, ces relations nous ont permis de construire un graphe, ou automate, donnant le coefficient associé à un produit donné de fractions ou de variables non commutatives. Toujours en utilisant la relation de récurrence, on peut déterminer quels sont les chemins de ces graphes à choisir pour obtenir le pôle désiré, et répondre ainsi à la question posée en introduction : si l'entrée du système est u(t) = sinωt, la sortie en régime permanent possède une composante sur sin3ωt, par exemple, composante que l'on peut exprimer en fonction des paramètres du système. Ici le résultat est une somme de 38 fractions rationnelles à structure simple, ce qui reste raisonnable. En effet, en suivant chacun des chemins du graphe des relations à l'ordre 3, on obtiendra un coefficient fonction des paramètres du système, la somme pondérée des coefficients associés aux chemins étant précisément le coefficient de sortie.

Ce genre d'approche est typique de l'esprit de la combinatoire [7], même si nous n'avons pas utilisé les outils de la combinatoire pour développer notre algorithme. Comme nous allons le voir par la suite, le calcul des g_{ij}^n devient fastidieux pour n > 5, mais grâce à ces graphes et à Scratchpad, où les structures de données sont très hiérarchisées, il est possible de se restreindre au calcul des seuls termes de g_{ij}^n nous intéressant vraiment. En dépit donc de ces limitations, le calcul formel permet la compréhension des propriétés structurelles des EDO, ce qui constitue un progrès par rapport aux méthodes numériques.

3. Utilisation du logiciel SCRATCHPAD

3.1. Description du programme

En SCRATCHPAD on doit tout d'abord créer les structures mathématiques que l'on désire manipuler, en spécifiant leur représentation interne et en donnant des fonctions de base sur ces structures. Une fois ces structures construites, on peut alors écrire l'application souhaitée.

On souhaite définir un domaine d'objets mélangeant à la fois des paramètres a_1, a_2,... issues du système (SNL2) et des séries génératrices. On procède en plusieurs étapes : On crée tout d'abord un domaine POLE, dont les éléments appartiennent à un alphabet, comme z_0, z_1, z_2, ... codées en interne par les entiers 0, 1, 2,... ou des pôles comme $(1 - \lambda z_0)^{-1}$ codé en interne par λ et représenté par (λz_0) *. On peut alors créer un domaine de polynômes non commutatifs : plus précisément, on définit un anneau de polynômes non commutatifs dont les indéterminées appartiennent au domaine POLE et dont les coefficients sont des polynômes habituels à coefficients entiers et à indéterminées quelconques, en particulier a_1, a_2, a_{20}^1, a_{11}^1, a_{02}^1, a_{20}^2, a_{11}^2, a_{02}^2. Le domaine des polynômes non commutatifs généraux a été conçu par Petitot[9], nous n'y avons apporté que quelques modifications pour notre application particulière.

Les relations de récurrence liant les g_{ij}^n sont implémentées dans la fonction **voltg**. voltg(i,j,n) calcule g_{ij}^n, le résultat étant un polynôme du domaine défini ci-dessus. Dans cet exemple, voltg(1, 0, 3) est une somme de 38 termes et voltg(1, 0, 4) une somme de 492 termes. Une simulation avec un modèle de transistor bipolaire nous montre cependant qu'il faut aller au moins jusqu'à l'ordre 7 pour obtenir une bonne approximation de la sortie, lorsque l'on prend l'entrée égale à u(t) = sinωt. Dans ce cas, plutôt que de chercher à calculer g_{10}^7, qui contiendrait un nombre trop important de termes, nous ne calculons qu'une partie du résultat, en faisant préciser à l'utilisateur quels sont les coefficients qui l'intéressent parmi l'ensemble des termes de g_{10}^7. De manière plus générale on ne calcule qu'un coefficient particulier de g_{ij}^n. C'est l'objet de la fonction **calcmon** qui prend deux arguments, un monôme et un triplet i,j,n, et qui renvoie le coefficient voulu

```
voltg(i,j,n) ==  -- les valeurs de voltg(1,0,1) et voltg(0,1,1) sont données

    (n=1) and (i=1) => a1 * ([g1]::L P I::POLE::XP) * z1
    (n=1) and (j=1) => a2 * ([g2]::L P I::POLE::XP) * z1

    i + j = n =>       -- si i + j = n
         ([i*g1 + j*g2]::L P I::POLE::XP) * (i*a1*z1_
            *voltg(i-1,j,n-1)  +  j*a2*z1*voltg(i,j-1,n-1))

    i + j < n =>       -- on teste le cas i + j < n
         ([i*g1 + j*g2]::L P I::POLE::XP) *
         (     i*a1*z1  *  voltg(i-1,j,n-1) _
            +  j*a2*z1  *  voltg(i,j-1,n-1) _
            +  (i*a120 + j*a211) * z0 * voltg(i+1,j,n) _
            +  (i*a110 + j*a202) * z0 * voltg(i,j+1,n) _
            +  i*a102  * z0 * voltg(i-1,j+2,n) _
            +  j*a220  * z0 * voltg(i+2,j-1,n)   )
```

```
calcmon(word,i,j,n)==

      -- coefpar contient le resultat desire
      -- coefg1 est le coefficient entier de g1
      -- coefg2 est le coefficient entier de g2
      --npole est l'exposant du pole dans le mot

      while ((i+j) < n) repeat

      -- on decompose le mot, produit de poles en une liste
      -- a chaque iteration, on lit le pole courant
      -- et on met a jour coefg1, coefg2, et npole
      -- Puis on effectue les tests suivants :

  (npole = 1) and (coefg1 = (i - 1)) and (coefg2 = j) =>
      coefpar := (i * a1) * coefpar
      n := n - 1
      j := j - 1

  (npole = 1) and (coefg1 = i) and (coefg2 = j - 1)) =>
      coefpar := (j * a2) * coefpar
      n := n - 1
      j := j - 1

  (npole = 0) and (coefg1 = (i - 1)) and (coefg2 = j + 2)) =>
      coefpar := (i * a102) * coefpar
      i := i - 1
      j := j + 2

  (npole = 0) and (coefg1 = i) and (coefg2 = (j + 1)) =>
      coefpar := ((i * a111) + (j * a202)) * coefpar
      j := j + 1

  (npole = 0) and (coefg1 = (i + 1)) and (coefg2 = j) =>
      coefpar := coefpar * ((i * a120) + (j * a211))
      i := i + 1

  (npole = 0) and (coefg1 = (i + 2)) and (coefg2 = j - 1)) =>
      coefpar := (j * a220) * coefpar
      i := i + 2
      j := j - 1

  while (#lfac > 1) repeat

      (coefg1 = (i - 1)) and (coefg2 = j) =>
          coefpar := i * a1 * coefpar
          i := i - 1
          n := n - 1

      (coefg1 = i - 1)) and (coefg2 = j) =>
          coefpar := j * a2 * coefpar
          j := j - 1
          n := n - 1

i = 1 => a1 * coefpar   --valeur finale du coefficient formel
j = 1 => a2 * coefpar   --valeur finale du coefficient formel
```

3.2. Applications

```
lambda (1)  ->voltg(1,0,2)$VOLTSER
```

$$2\, a_{20}^1\, (a_1)^2\, (g_1 z_0)^*\, z_0\, (2 g_1 z_0)^*\, z_1\, (g_1 z_0)^*\, z_1$$

$$+\, 2\, a_{02}^1\, (a_2)^2\, (g_1 z_0)^*\, z_0\, (2 g_2 z_0)^*\, z_1\, (g_2 z_0)^*\, z_1$$

$$+\, a_{11}^1\, a_1\, a_2\, (g_1 z_0)^*\, z_0\, ((g_1 + g_2) z_0)^*\, z_1\, (g_1 z_0)^*\, z_1$$

$$+\, a_{11}^1\, a_1\, a_2\, (g_1 z_0)^*\, z_0\, ((g_1 + g_2) z_0)^*\, z_1\, (g_2 z_0)^*\, z_1$$

```
lambda (2) -> word := [g1]*0*[2*g1]*1*[g1]*1
```

$$(g_1 z_0)^* z_0 (2g_1 z_0)^* z_1 (g_1 z_0)^* z_1$$

```
lambda (3) ->calcmon(word,1,0,2)$CALCMON
```

$$2\, a_{20}^1\, (a_1)^2$$

```
lambda (4)     ->word:=[g1]*0*[2*g1]*0*[g2+2*g1]*0*[2*g2+2*g1]*0*
    [2*g2+3*g1]*1*[2*g2+2*g1]*1*[g2+2*g1]*1*[g1+g2]*1*[g1]*1
```

$$(g_1 z_0)^* z_0 (2g_1 z_0)^* z_0 ((g_2+2g_1)z_0)^* z_0 ((2g_2+3g_1)z_0)^* z_0 ((2g_2+3g_1)z_0)^* z_1$$

$$((2g_2+2g_1)z_0)^* z_1 ((g_2+2g_1)z_0)^* z_1 ((g_2+g_1)z_0)^* z_1 (g_1 z_0)^* z_1$$

```
lambda (5)     ->calcmon(word,1,0,5)$CALCMON
```

$$((96\,(a_{11}^1)^2 + 48\, a_{02}^2\, a_{11}^1)(a_{20}^1)^2 + (96\,(a_{11}^1)^2 + 48\, a_{02}^2\, a_{11}^1)\, a_{11}^2\, a_{02}^1)\, a_1$$

L'auteur remercie Madame Françoise Lamnabhi-Lagarrigue pour son écoute attentive, et son aide patiente et précieuse.

BIBLIOGRAPHIE :

[1] S. BACCAR, G. SEIGNIER, F. LAMNABHI-LAGARRIGUE, "Utilisation du calcul formel pour la modélisation et la simulation de circuits électroniques faiblement non linéaires", à paraître dans Annales de Télécommunications.

[2] R.W.BROCKETT, "Volterra series and geometric control theory", Automatica, 12, pp. 167-176, 1976.

[3] M.FLIESS, M. LAMNABHI and F. LAMNABHI-LAGARRIGUE, "An algebraic approach to nonlinear functional expansions", IEEE Trans. Circuits and Syst., 30, pp. 554-570, 1982.

[4] W. GRÖBNER, "Differentialgleichungen, I: Gewöhnliche Differentialgleichungen", in Mathematik für Physiker Vol. 6, B.I. Wissenschaftsverlag.

[5] C.HESPEL and G.JACOB, "Approximations bilinéaires tronquées: Carleman, Volterra finies, Padé, Automate géométrique et structurel", in "Algebraic Computation in Control", (Eds. Jacob and Lamnabhi-Lagarrigue), Lect. Notes Contr. Inf. Sc., 1991.

[6] A. J. KRENER, "Bilinear and nonlinear realizations of input-output maps", SIAM J. Control, 13, 1975, pp. 827-833.

[7] F.LAMNABHI-LAGARRIGUE, P.LEROUX and X.G.VIENNOT, "Combinatorial approximations of Volterra series by bilinear systems", in "Analysis of controlled dynamical systems", (Eds.Bonnard, Bride, Gauthier and Kupka), Birkhaüser, Boston, 1991.

[8] P.LEROUX and X.G.VIENNOT, "A combinatorial approach to nonlinear functional expansions", 27th Conf. on Decision and Control, Austin, Texas, 1988, pp.1314-1319. To appear in Theoretical Computer Science.

[9] M.PETITOT, "Polynômes non commutatifs sous Scratchpad", mai 1990, Rapport de DEA, LIFL, USTL.

Scratchpad implementation of the local minimal realization of dynamic systems*

N.E. Oussous and M.Petitot

Laboratoire d'Informatique Fondamentale de Lille
U.A. 369 du C.N.R.S., Université de Lille I,
59655 Villeneuve d'Ascq Cedex. FRANCE.
e−mail: oussous@lifl.lifl.fr   petitot@lifl.lifl.fr

Abstract : In this paper we give, an algorithm that computes the local minimal analytical realization of finite generating power series. In the same time, this algorithm proves that any finite generating series has a polynomial realization: observation and vector fields components are commutative polynomials.
That algorithm is implemented in the computer algebra system *Scratchpad.*

1   Introduction

We study the local minimal realization. Fliess [2] and C.Reutenauer [11] have shown the existence and uniqueness of local and minimal realization of noncommutative power series. And their algorithm allows to recursively compute the components of the analytical expansion of the *observation* and of the involved *vector fields*.

The present paper gives an account in this way, in the restrictive case of finite generating series.

First, we solve completely, with an algorithm, in *Scratchpad,* the minimal realization problem, with exact explicit (and not only recursive) computation of the observation and of the vector fields.

Secondly, the given algorithm is a proof of the following fact: *" any finite generating series admits a minimal realization in which the observation, and the vector field components, are commutative polynomials".* This fact appeared as a conjecture in the preceding *Macsyma* implementation [5, 8].

The use of the system *Scratchpad* is motivated by two essential facts: the software developpement in *Scratchpad* requests a good knowledge of the mathematical structures

---

*This work is partially supported by IBM France and by the PRC Mathématiques et Informatique.

and then, we can have a structured solutions as near as possible to the theoretical proofs. The second fact is that the system *Scratchpad* allows a good control of the internal representation of the manipulated objects.

2   The realization problem

Locally, the realization problem can be expressed as follows [2]:

> Let an Input/Output behaviour given by its generating power series, is there a differential system like (Σ) which has the same generating power series? In positive case, describe it.

2.1   Nonlinear dynamic systems

We consider a system of the following form:

$$(\Sigma) \begin{cases} \dot{q}(t) = \sum_{i=0}^{m-1} u_i(t) Y_i(q) \quad \text{with} \ \ u_0(t) \equiv 1, \\ y(t) \ = \ h(q(t)). \end{cases}$$

where

q belongs to a *connected $\mathbb{R}$-analytical variety* Q,
the Y_i's are *analytical vector fields*,
h, is a *$\mathbb{R}$-analytical function* called *observation*,
Y_i and h are defined in a neighbourhood of the given *initial state* $q(0)$,
the inputs $u_1, ..., u_{m-1}$ are *real and piecewise continuous*.
The vector $u = (u_0, u_1, \ldots, u_{m-1})$ will be called the input function.

Each *input u_i*, is associated with a *letter x_i* $(0 \leq i \leq m - 1)$. The set of all letters, $X = \{x_0, x_1, ..., x_{m-1}\}$, will be called the *command alphabet*. Let X^* be the free-monoid generated by X.

For each word $w \in X^*$, we denote by Y_w the differential operator defined as follows:

$$\begin{cases} Y_\epsilon \ = \ Identity, \quad \epsilon \text{ is the empty word}, \\ Y_{ux_i} \ = \ Y_u \circ Y_i, \end{cases} \tag{1}$$

where the Y_i's are vector fields given by (5). The *action* of differential operator Y_w over analytic function f defined on the variety Q, is denoted by $Y_w \circ f$.

For small enough time and inputs, the output $y(t)$ of the system (Σ) is given by the *Peano-Baker-Formula* [2, 11], called also the *Fliess fundamental formula*:

$$y(t) \ = \ \sum_{w \in X^*} (Y_w \circ h)_{|_{q(0)}} \int_0^t \delta_u w. \tag{2}$$

where $|_{q(0)}$ means evaluation in $q(0)$, and $\int_0^t \delta_u w$ is the *iterated integral* defined recursively as follows:

$$\begin{cases} \int_0^t \delta_u \varepsilon &= 1, \\ \int_0^t \delta_u(v x_i) &= \int_0^t (\int_0^\tau \delta_u v) u_i(\tau) d\tau. \end{cases} \tag{3}$$

In particular, since $u_0(\tau) \equiv 1$, $\int_0^t \delta_u x_0 = \int_0^t u_0(\tau) d\tau = t$.

2.2 Generating power series and vector fields

The Input/Output behaviour of the system (Σ) is completely defined by its *generating power series* G which is the noncommutative power series in the variables $x_0, x_1, ..., x_{m-1}$ (see paragraph 4.1), given by the following formula [2]

$$G = \sum_{w \in X^*} (Y_w \circ h)_{|_{q(0)}} w. \tag{4}$$

The variety Q is of dimension N. Let $z_1, z_2, ..., z_N$ be a system of *local coordinates*. The $\mathbb{R}$-algebras of formal series and polynomials in *commutative variables* $z_1, z_2, ..., z_N$ are denoted respectively by $\mathbb{R}[\![z_1, z_2, ..., z_N]\!]$ and $\mathbb{R}[z_1, z_2, ..., z_N]$. The *vector fields* Y_i used above can be written:

$$Y_i = \sum_{k=1}^N \theta_i^k(z) \frac{\partial}{\partial z_k}, \qquad \theta_i^k(z) \in \mathbb{R}[z_1, z_2, ..., z_N]. \tag{5}$$

Thus, let f be an analytic function defined over the variety Q, then the action of the vector fields Y_i over the function f can be written:

$$Y_i \circ f = \sum_{k=1}^N \theta_i^k(z) \frac{\partial f}{\partial z_k}. \tag{6}$$

In particular, for a local coordinate z_i considered as an analytic function, we have

$$Y_i \circ z_j = \theta_i^j(z). \tag{7}$$

2.3 The observation and vector fields encoding

Let us call $\mathcal{O}_{q(0)}$ the algebra of real analytic functions defined in a neighbourhood of $q(0)$. According to Fliess [2], we can associate with each system (Σ) a map $\sigma : \mathcal{O}_{n(0)} \longmapsto \mathbb{R}\langle\langle X \rangle\rangle$, defined by:

$$\forall h \in \mathcal{O}_{q(0)}, \qquad \sigma(h) = \sum_{w \in X^*} (Y_w \circ h)_{|_{q(0)}} w.$$

Therefore, $\sigma(h)$ is the generating power series of the system (Σ) whose the observation is h. The map σ is a morphism which translates the algebra structure of $\mathcal{O}_{q(0)}$ in the formal power series algebra (with Shuffle product) structure.

For any power series S, we note $x_i \triangleleft S$ the "left remainder" of S by the letter x_i (see the section 4 below).

Let h an analytic function and Y_i a vector field:

$$
\begin{aligned}
\sigma(Y_i \circ h) &= \sum_{w \in X^*} (Y_w \circ (Y_i \circ h))|_{q(0)} w \\
&= \sum_{w \in X^*} \langle \sigma(h)|wx_i \rangle w \\
&= \sum_{w \in X^*} \langle x_i \triangleleft \sigma(h)|w \rangle w \\
&= x_i \triangleleft (\sigma h)
\end{aligned}
$$

We see that with the vector field Y_i is associated the operator, left remainder, '$x_i \triangleleft$' by the letter x_i. If we denote by Z_i the image by σ of the local coordinate z_i, then

$$x_i \triangleleft Z_j = \sigma(Y_i \circ z_j) = \sigma(\theta_i^j(z)).$$

Thus, we have a mean to compute the vector fields components.

3 Realization of nonlinear dynamic systems

3.1 Differentially produced power series

Definition 3.1 [2] *The formal power series $G \in \mathbb{R}\langle\langle X \rangle\rangle$ is differentially produced if and only if, there exist:*

1. *an integer $r \in \mathbb{N}$,*

2. *an homomorphism $\mathcal{Y}$ from X^* into differential operator algebra over $\mathbb{R}[z_1, ..., z_r]$ such that $\forall x_i \in X, Y_i = \mathcal{Y}(x_i)$ is a vector field,*

3. *a commutative power series $h \in \mathbb{R}[z_1, ..., z_r]$,*

such that:

$$\forall w \in X^*, \quad \langle G|w \rangle = (\mathcal{Y}(w) \circ h)|_0, \tag{8}$$

where $|_0$ means evaluation in $z_1 = ... = z_r = 0$.

The couple $(\mathcal{Y}, h)$ is called *differential representation* of G, of dimension r.

From (4) and (8), it is obvious that:

G is the generating power series of a system like (Σ) if and only if G is produced differentially.

3.2   The realization algorithm schema

Let G be a series of Lie–Hankel rank equal to r. Let $\mathcal{A}(G)$ be the set of Lie polynomials that annihilate G :

$$\mathcal{A}(G) = \{ \, P \in \mathcal{L}ie\langle X\rangle \mid G \triangleright P = 0 \, \}$$

$\mathcal{A}(G)$ is clearly a Lie–sub–algebra of $\mathcal{L}ie\langle X\rangle$.

We set:

$$\mathcal{V}(\mathcal{A}(G)) = \{ \, Q \in \mathbb{R}\langle\!\langle X\rangle\!\rangle \mid Q \triangleright \mathcal{A}(G) = 0 \, \}.$$

This set is the abstract differential variety generated by G.

According to Reutenauer [11], the realization consists to

1. Find some Lie polynomials $P_1, ..., P_r$ that form a basis of $\mathcal{L}ie\langle X\rangle$ modulo $\mathcal{A}(G)$, i.e. such that $(G \triangleright P_1, ..., G \triangleright P_r)$ is a basis of the vector space $G \triangleright \mathcal{L}ie\langle X\rangle$, (for the right remainder action $\triangleright$, see section 4).

2. Find r *formal power series* Z_j without constant term such that:

$$\begin{cases} \langle Z_j \triangleright P_i | \varepsilon \rangle = \delta_{ij} & \text{for } i \leq r, \\ Z_j \in \mathcal{V}(\mathcal{A}(G)). \end{cases} \tag{9}$$

   observe that $\langle Z_j \triangleright P_i | \varepsilon \rangle = \langle Z_j | P_i \rangle$. The Z_j will be the locale coordinate.

3. Express G, which encodes the observation h [5], as a commutative *series* on the Z_j for the shuffle product:

$$G = \sum_\alpha c_\alpha Z^{\sqcup\!\sqcup\, \alpha}.$$

4. Compute the series $x_i \triangleleft Z_j$, which encode the components $\theta_i^j(z) = Y_i \circ z_j$ of the vector fields Y_i, and express them as commutative series on the Z_j for the shuffle product.

5. Translate (directly) these expressions as giving a differential production of G.

The steps 3 and 4 are not effective in the following sense: the analytical series expanssions are obtained only via recursive equations allowing successive computations of the coefficients.

3.3   Effective computation

Here, we restrict the problem to *finite generating series* (polynomial), in order to obtain a complete computer algorithm. This is achieved by a particular choice of the local coordinates. We are not satisfied with the condition (9) in the item 2. We impose in addition the following strongly triangular condition:

Traingular condition: we choose locale coordinates Z_j and Lie–polynomials P_i satisfying:

$$\begin{aligned} Z_j \triangleright P_i &= 0 \quad \text{if } j > i, \\ Z_i \triangleright P_i &= 1. \end{aligned} \tag{10}$$

Indeed, Z_j and P_i satisfy the condition (9) in the item 2, and the matrix $(Z_j \triangleright P_i)_{1 \le i,j \le r}$ is upper triangular. This condition allows us to use a sample algorithm [7] to express any noncommutative polynomial, belonging in $\mathcal{V}(\mathcal{A}(G))$, as a linear combination of shuffles of the Z_j's. That algorithm is based on the idea that each polynomials can be rebuilt from its successive derivatives. Note that right remainder by a Lie–polynomials is a derivative for the shuffle product.

To construct the Z_j's, we start with the Lie–Hankel matrix whose rows are indexed by Lie–polynomials (in lexicographic by length order) and columns are indexed by the words (in lexicographic by length order). We apply the *"Gaussian Elimination"* method from right to left (begining with the longest words) and we set

$$Z_j = w_j \triangleleft G - \langle G | w_j \rangle, \quad 1 \le j \le r.$$

where the w_j's are the *"caracteristic words"* which index the columns whose pivot exists. The operations applied over rows are transcribed on the Lie–polynomials P_i. After this operation, we obtain another Lie–polynomial set which can be renamed also $(P_i)_{1 \le i \le r}$.

Before to present the algorithm tools, we give the results of a computation of our *Scratchpad algorithm:*

Example 3.1 *Let $G = 4x_2^4 - x_0 x_1^4 + 3x_2 x_1^3$. The Lie-rank of G is equal to 6. We obtain the following local coordinate system:*

$$\left\{ \begin{aligned} Z_0 &= 3x_2 x_1^3 + 4x_2^4 - x_0 x_1^4, \\ Z_1 &= 3x_2 x_1^2 - x_0 x_1^3, \\ Z_2 &= 3x_2 x_1 - x_0 x_1^2, \\ Z_3 &= 3x_2 - x_0 x_1, \\ Z_4 &= 4x_2, \\ Z_5 &= -x_0 \end{aligned} \right.$$

The observation:

$$h = z_0.$$

The vector fields:

$$\left\{ \begin{aligned} Y_0 &= -\frac{\partial}{\partial z_5} \\ Y_1 &= z_1 \frac{\partial}{\partial z_0} + z_2 \frac{\partial}{\partial z_1} + z_3 \frac{\partial}{\partial z_2} + z_5 \frac{\partial}{\partial z_3} \\ Y_2 &= \frac{1}{96} z_4^3 \frac{\partial}{\partial z_0} + 3 \frac{\partial}{\partial z_3} + 4 \frac{\partial}{\partial z_4} \end{aligned} \right.$$

The computation was be done in 13 secondes.

4 Tools

Under the above paragraphs, it's clear that formal power series are the main tool in this study. We use also Lyndon basis of the free Lie–algebra. In this section, we remind the essential definitions on these tools.

4.1 Noncommutative formal power series

Let $X = \{x_0, x_1, ..., x_{m-1}\}$ be a finite set of *letters* called *alphabet*. We denote by X^* the free monoid generated by X. The empty word is denoted by ε^1. We denote by $|w|$ the *length* of the word w.

A formal power series S is a *mapping* from X^* into $\mathbb{R}$ which associates each word w with its *coefficient* $S(w)$ denoted by $\langle S|w\rangle$. S will also be expressed by the formal sum:

$$S = \sum_{w \in X^*} \langle S|w\rangle w.$$

The set of all formal power series is denoted by $\mathbb{R}\langle\langle X\rangle\rangle$ and is an algebra for the *Cauchy product*[2].

The *support* of the formal power series S is the language:

$$\text{Supp}(S) = \{ w \in X^* \mid \langle S|w\rangle \neq 0 \}.$$

Any power series without constant term is called *proper*.

A *polynomial* is a formal power series that has a *finite support*. The set of all polynomials is a sub–algebra of $\mathbb{R}\langle\langle X\rangle\rangle$ denoted by $\mathbb{R}\langle X\rangle$. The *degree* of a polynomial P is defined as follows:

$$\deg(P) = \begin{cases} \sup\{|w| \; with \; w \in supp(P)\}, & if \quad P \neq 0, \\ -\infty, & if \quad P = 0. \end{cases}$$

We denote $\mathcal{L}ie\langle X\rangle$ the *free–Lie–algebra* generated by X in which the *Lie–brackets* are defined by

$$[P, Q] = PQ - QP \qquad P, Q \in \mathcal{L}ie\langle X\rangle.$$

Many operations may be defined over the words and power series. In particular, we can define the two following operations.

4.1.1 Shuffle product

The shuffle product of two words [1, 3] is defined recursively as follows

$$\begin{cases} \forall w \in X^*, \quad w \shuffle \varepsilon = \varepsilon \shuffle w = w, \\ \forall u, v \in X^*, \; \forall x, y \in X, \; (xu) \shuffle (yv) = x(u \shuffle (yv)) + y((xu) \shuffle v). \end{cases}$$

---

[1]The empty word is implemented by the multiplicative identity element 1.
[2]The Cauchy product is the extension of the concatenation product defined over words.

This product is commutative and associative, and can be extended to formal power series by setting for S and T belonging in $\mathbb{R}\langle\langle X \rangle\rangle$:

$$S \uplus T = \sum_{u,v \in X^*} \langle S|u\rangle\langle T|v\rangle u \uplus v.$$

4.1.2 Remainder's computation

Definition 4.1 [3] *Let $w \in X^*$ be a word. We set, for $x \in X$,*

$$x \triangleleft w = \left\{ \begin{array}{ll} v & if \quad w = vx, \\ 0 & otherwise. \end{array} \right. \qquad \left(resp. \; w \triangleright x = \left\{ \begin{array}{ll} v & if \quad w = xv,, \\ 0 & otherwise. \end{array} \right. \right)$$

$x \triangleleft w$ (resp. $w \triangleright x$) is called left (resp. right) remainder of the word w by the letter x.

Property 4.1 $\forall w \in X^*, \forall x, y \in X$, *we have:*

$$(xy) \triangleleft w = x \triangleleft (y \triangleleft w) \qquad (resp. \; w \triangleright (xy) = (w \triangleright x) \triangleright y).$$
$$(x \triangleleft w) \triangleright y = x \triangleleft (w \triangleright y).$$

Definition 4.2 *Let $S \in \mathbb{R}\langle\langle X \rangle\rangle$ be a formal power series. We set, for $u \in X^*$,*

$$u \triangleleft S = \sum_{w \in X^*} \langle S|wu\rangle w \qquad (resp. \; S \triangleright u = \sum_{w \in X^*} \langle S|uw\rangle w);$$

which will be called the left (resp. right) remainder of the power series S by the word u.

In other words, we have, for $w \in X^*$,

$$\langle u \triangleleft S|w\rangle = \langle S|wu\rangle, \quad and \quad \langle S \triangleright u|w\rangle = \langle S|uw\rangle.$$

Example 4.1 *Let $X = \{x_0, x_1\}$, $S = x_0 x_1 x_0 + 2x_1 x_0$, and $u = x_1 x_0$, then $u \triangleleft S = x_0 + 2$ and $S \triangleright u = 2$.*

4.1.3 Lie–Hankel matrix

Definition 4.3 [4, 11] *Let $S \in \mathbb{R}\langle\langle X \rangle\rangle$. We define the Lie–Hankel matrix associated with S as an infinite array, denoted by $\mathcal{LH}_S$, whose the lines are indexed by some totally ordered basis of $\mathcal{L}ie\langle X \rangle$ and the columns are indexed by X^* (sorted for lexicographic by length order) such that:*

$$\mathcal{LH}_S(P_i, w) = \langle S|P_i w\rangle = \langle S \triangleright P_i|w\rangle.$$

Definition 4.4 [11] *The Lie-Rank of the power series S, denoted by $\mathcal{LR}_S$, is the rank of the Lie-Hankel matrix $\mathcal{LH}_S$, if it is finite and $+\infty$ elsewhere.*

---

[3]A *Macsyma* implementation of the remainder computation is given in [10].

4.2   Lyndon words and Lyndon basis

Let $(X = \{x_0, x_1, \ldots x_{m-1}\}, <)$ be a totally ordered alphabet. We define the lexicographical order over X^* as follows :

$\forall\, u, v \in X^*, \quad u < v \quad$ if and only if,

$$\begin{cases} \text{either} & \exists w \neq \varepsilon \text{ such that } uw = v, \\ \text{or} & \exists x, y, z \in X^* \text{ and } a, b \in X \text{ such that } u = xay, \ v = xbz \text{ and } a < b. \end{cases}$$

Definition 4.5 [6] *Let $w \in X^*$. w is a Lyndon word if it is strictly smaller than any of its proper right factors. Let L denotes the set of Lyndon words over X.*

Example 4.2 $X = \{x_0, x_1\}, \quad x_0, \ x_1, \ x_0x_1, \ x_0^2x_1, \ x_0x_1^2, \ \ldots \quad$ *are Lyndon words.*

Property 4.2 *Let $w \in L \backslash X$ and m its longest proper right factor in L. If $w = lm$, then $l \in L$ and $l < lm < m$. The couple $\sigma(w) = (l, m)$ is called the standard factorisation of w.*

Lemma 4.1 [4][6] *Let $w \in L \setminus X$ and $\sigma(w) = (l, m)$ its standard factorization, and let n a Lyndon word such that $w < n$. Then, the pair (w, n) is the standard factorization of wn if and only if $n \leq m$.*

Each Lyndon word b is associated with a Lie polynomial P_b, defined recursively as follows:

$$\begin{cases} P_x = x \quad for \quad x \in X, \\ P_b = [P_l, P_m], \quad for \ b \in L \backslash X, \ such \ that \ \sigma(b) = (l, m). \end{cases} \tag{11}$$

So we get the *Lyndon basis* (called also Chen–Fox–Lyndon basis) ([6, 12]). We can produce Lyndon words and Lyndon basis by enumerating them in *lexicographic by length order*.

5   Architecture of the implementation in Scratchpad

In *Scratchpad*, the programming has the philosophy that the other systems haven't. Indeed, to have some general tools, which may be efficients and reusable, we must define a good architecture of the software before the programming.

We list the useful *Scratchpad* objects for the realization of Polynomial Dynamic Systems (*i.e.* whose generating power series are finite).

---

[4]This lemma is used to develop, in *Scratchpad*, an algorithm to generate Lyndon words and Lyndon basis up to a given degree.

5.1 List of the used abbreviations

```
OFMON1    : OrderedFreeMonoid1(S:OrderedSet)

XDPOLY    : XDistribuedPolynomial(vl:OrderedSet, R:Ring)

XRPOLY    : XRecursivePolynomial(vl:OrderedSet, R:Ring)

MAGMA     : Magma(VarSet:OrderedSet)

LWORD     : LyndonWord(VarSet:OrderedSet)

VECFIELD  : VectorField(VarSet:OrderedSet,

                              R:PartialDifferentialRing VarSet)

STDBASIS  : StandardBasis(VarSet:OrderedSet, K:Field, cmp)

XTAYLOR   : XTaylorDevelopment(VarSet:OrderedSet, K: Field, Sy:Symbol,

                              XPOLY:XFreeAlgebra(VarSet,K))

DPS       : DynamicPolynomialSystem(CA:OrderedSet, Sy:Symbol, K:Field)
```

5.2 Semantic of the used objects

OFMON1 management of the words over any ordered alphabet. This alphabet can be infinite but the words must have finite length.

XDPOLY management of the noncommutative polynomials over any alphabet vl whose coefficients are in any ring R. These polynomials have a distributed representation.

XRPOLY For the users, this domain is the same as the above one but the polynomials have the recursive representation. This representation is of use in the expensive computations (in particular for shuffle product).

MAGMA management of brackted words over any ordered alphabet VarSet. These words are represented by trees.

LWORD management of Lyndon words. The implementation uses the MAGMA domain.

VECFIELD management of vector fields whose coefficients belong in any differential ring. The elementary partial derivatives are accomplished with respect to the variables in VarSet.

STDBASIS management of *standard basis* for polynomial vector spaces of finite dimension. With a fixed order over words (to compare two words, we use the function cmp given as a parameter), these standard basis are a canonic representation of vector spaces.

XTAYLOR this package is a generalisation of *TAYLOR formula*. Any element of the shuffle algebra of polynomials can be built from its successive *derivatives*.

DPS this domain is the implementation in minimal and analytical form of analytical systems whose *generating series are finite*. So, we get a *canonical representation* of these systems.

6 Conclusion

We have improved here a preceding *Macsyma* algorithm, by proving an underlying conjecture: *"any finite generating series has a completely polynomial realization"*. Other account have obtained.

Finally, working with *Scratchpad*, we keep us very close of the mathematical structures. We have developped some general tools which allows us to find some extensions for our study, in particular the realization of rational power series.

The study of the algorithm's complexity is in preparation and will be developped in another report.

References

[1] M.Fliess.– Fonctionnelles causales non linéaires et indéterminées non commutatives, *Bull. Soc. Math. France,* 109, p.3-40, 1981.

[2] M.Fliess.– Réalisation locale des systèmes non linéaires, algèbres de Lie filtrées transitives et séries génératrices, *Invent. Math.*71, p.521-537, 1983.

[3] G.Jacob and N.Oussous.– Sur un résultat de REE: séries de Lie et algèbres de mélange, Publication du LIFL Lille, IT–103, 1987.

[4] G.Jacob and N.Oussous.– Algebraic computation of analytical minimal realisation, *Publication du LIFL Lille,* IT–139, 1988.

[5] G.Jacob and N.Oussous.– Local and minimal realization of nonlinear dynamical systems and Lyndon words, *IFAC Symposium "Nonlinear Control Systems Design", Capri–Italy,* June 1989.

[6] M.Lothaire.– Combinatorics on words, *Reading, Massachusetts,* 1983.

[7] N.E.Oussous.– Macsyma Computation of Local Minimal Realization of Dynamical Systems of which Generating Power Series are Finite, *To appear in J. Symbolic Computation,* 1990.

[8] N.E.Oussous.– Computation, on Macsyma, of the minimal differential representation of noncommutative polynomials, *to appear in "Algebraic and Computing Traitment of Noncommutative power Series" (ed. G.Jacob and C.Reutcnauer), Theoret. Comput. Sci.,* 1991.

[9] N.E.Oussous et M.Petitot.– Polynômes non commutatifs : représentation et traitement par les systèmes de calcul formel, *Submited to "Formal Power Series and Algebraic Combinatorics", Bordeaux–France,* May 1991.

[10] N.E.Oussous.– Realization of Polynomial Dynamic Systems, *Deuxième Journées-Séminaire "Traitements Algébriques et Informatiques des Séries Formelles Non Commutatives", ENS de Paris–France,* Avril 1990.

[11] C.Reutenauer.– The local realisation of generating series of finite Lie rank, in: M.Fliess and M.Hazewinkel, eds., *Algebraic and Geometric Methods In Nonlinear Control Theory,* p. 33–43, 1986.

[12] G.Viennot.– *Algèbres de Lie libres et monoïdes libres,* (Lecture Notes In Mathematics, Springer–Verlag. 691,1978).

Annex

Source file for test

```
-- test of the DPS Package
-- February 7, 91
-- =============
++++++++++++++++++++++++++++++++++++++++++++++++++++++++++++++++++++++++++++++++
++ Purpose: From a noncommutative polynomial g, we compute minimal     ++
++          analytical dynamic system whose generating series is g.    ++
++++++++++++++++++++++++++++++++++++++++++++++++++++++++++++++++++++++++++++++++

)clear all
)load  OFMON1 LWORD PIVOT STDBASIS IV VECFIELD  XTAYLOR DPS

ca := IndexedVariable(x);                           -- Command alphabet
x0 :ca := var(0);
x1 :ca := var(1);
x2 :ca := var(2);


poly     := XRecursivePolynomial(ca,RN)        -- domain instanciation
sdyn     := DynamicPolynomialSystem(ca,'z,RN)

g:poly := 4*x2**4 - x0*x1**4 +3*x2*x1**3;
extend g

sd:sdyn := realisation(g)             --          realization algorithm

dimension sd                      -- dimension of the state space
Observation sd                    --          observation function
VectField(sd,x0)                  -- vector fields for each input
VectField(sd,x1)
VectField(sd,x2)
```

Result file

```
;ca := IndexedVariable(x);                          -- Command alphabet
                                                           Type: Void
                                          Time: .067 (IN) = .067 sec


;x0 :ca := var(0);

                                                           Type: Void
                                            Time: .1 (IN) = .1 sec
```

```
;x1 :ca := var(1);
```
                                                                    Type: Void
                                                        Time: .1 (IN) = .1 sec

```
;x2 :ca := var(2);
```
                                                                    Type: Void
                                            Time: .1 (IN) + .6 (EV) = .7 sec

```
;poly    := XRecursivePolynomial(ca,RN)        -- domain instanciation
```

   (5) XRecursivePolynomial(IndexedVariable x,RationalNumber)
                                                                  Type: Domain
                                                                  Time: 0 sec

```
;sdyn    := DynamicPolynomialSystem(ca,'z,RN)
```

   (6) DynamicPolynomialSystem(IndexedVariable x,z,RationalNumber)
                                                                  Type: Domain
                                            Time: .1 (IN) + .033 (OT) = .133 sec

```
;g:poly := 4*x2**4 - x0*x1**4 +3*x2*x1**3;
```
                                                                    Type: Void
                            Time: 1.7 (IN) + .067 (EV) + .067 (OT) = 1.833 sec

```
;extend g
```

$$(8) \quad 3x_2 x_1^3 + 4x_2^4 - x_0 x_1^4$$

            Type: XDistribuedPolynomial(IndexedVariable x,RationalNumber)
                                            Time: .1 (IN) + .083 (OT) = .183 sec

```
;sd:sdyn := realisation(g)              -- realization algorithm
```

   Lyndon words

   Gaussian Elimination over Lie-Hankel matrix

   Lie-Rank : 6

   (9)
   [varset= $[x_0, x_1, x_2]$, dim= 6,

```
vfields =
        d         d         d         d         d
  [- ----, z  ---- + z  ---- + z  ---- + z  ----,
    d z    1 d z     2 d z     3 d z     5 d z
      5         0         1         2         3


    1   3 d        d         d
    -- z  ---- + 3 ---- + 4 ----],
    96  4 d z      d z       d z
             0        3         4



  obs= z ,
       0

  serie= x x x x x (- 1) + x (x x x 3 + x x x 4)]
         0 1 1 1 1         2 1 1 1     2 2 2
```

Type: DynamicPolynomialSystem(IndexedVariable x,z,RationalNumber)
Time: .1 (IN) + 12.8 (EV) + .2 (OT) = 13.1 sec

;dimension sd   -- dimension of the state space

   (10) 6

Type: PositiveInteger
Time: .1 (EV) + .167 (OT) = .267 sec

;Observation sd   -- observation function

   (11) z
            0
Type: SparseMultivariatePolynomial(RationalNumber,IndexedVariable z)
Time: .5 (OT) = .5 sec

;VectField(sd,x0)   -- vector fields for each input

```
          d
  (12)  - ----
          d z
             5
```
Type: VectorField(IndexedVariable z,
       SparseMultivariatePolynomial(RationalNumber,IndexedVariable z))
Time: .017 (EV) + .15 (OT) = .167 sec

```
;VectField(sd,x1)


          d           d           d           d
  (13)  z ----  + z  ----  + z  ----  + z  ----
        1 d z     2 d z     3 d z     5 d z
              0         1         2         3
Type: VectorField(IndexedVariable z,
         SparseMultivariatePolynomial(RationalNumber,IndexedVariable z))
                                            Time: .1 (OT) = .1 sec


;VectField(sd,x2)


        1    3  d         d         d
  (14)  -- z  ----  + 3 ----  + 4 ----
        96 4  d z       d z       d z
                 0         3         4
Type: VectorField(IndexedVariable z,
         SparseMultivariatePolynomial(RationalNumber,IndexedVariable z))
                                Time: .1 (IN) + .083 (OT) = .183 sec
```

Poincaré's Linearization Method
Applied to the Design of Nonlinear Compensators

Arthur Krener, Mont Hubbard

Sinan Karahan, Andrew Phelps, Benoit Maag

Institute of Theoretical Dynamics

University of California

Davis, CA 95616 — U.S.A.

1. <u>Introduction</u> Over the past three years with funding from AFOSR, we have developed a methodology for the design of compensators for highly nonlinear plants. This project has led to the development of POINCARE, a MATLAB based package for the design of nonlinear feedback control laws and nonlinear state observers. POINCARE has generally proven to be a substantial improvement over standard linear designs when applied to systems with significant nonlinearities. This paper describes the software package, POINCARE, as well as related research on several important theoretical issues. The long term goal of our efforts is the development of methodologies for the design of compensators for plants exhibiting highly nonlinear behavior. These include high performance and V/STOL aircraft, high performance robots, advanced jet engines and nonlinear chemical processes. The existing methods of linear compensator design, the classical Bode plot, phase gain margins, etc., the semiclassical LQG approaches and the more modern H^{∞} and LQG/LTR approaches are not totally adequate to design controllers for plants which have

multiaxis, highly coupled nonlinearities, significant parameter variations and unmodeled dynamics, when these plants must perform well over a wide range of operating regimes. More sophisticated control methodologies are needed to obtain the full range of performance that such devices and processes are capable of achieving.

That is not to say that these linear methodologies are of no use when it comes to nonlinear design. On the contrary, they have proven highly successful in various linear applications and we believe that any broad and successful approach to nonlinear design must precede from this solid foundation and reduce to it when the plant is reasonably close to linear.

It is for this reason that our approach is based on linearizing the system model to the greatest extent possible by various coordinate changes, state feedback, input–output injection, etc. as available and then proceeding with tried and trustworthy linear methodologies on the linearized model. The resulting compensator is then transformed back to the original coordinates for implementation. The technique of H. Poincaré is a conceptually and numerically simple method for the term by term linearization of nonlinear dynamics. Its generalization to control problems forms the basis of our approach.

2. The Linearization Technique of Poincaré and Approximate Normal Forms of Control Systems.

Poincaré [Ar, G–H] considered the problem of linearizing an n–dimensional vector field i.e. an ordinary differential equation.

$$(2.1) \qquad \dot{x} = f(x)$$

around a critical point x^0, $f(x^0) = 0$ by a change of state coordinates

$$(2.2) \qquad z = \phi(x).$$

Rather than attack the problem in its entirety, he sought a term by term Taylor series solution. Suppose the critical point is $x^0 = 0$ and (2.1) can be expanded as

(2.3) $\dot{x} = Ax + f^{[2]}(x) + O(x)^3.$

when $f^{[2]}(x)$ is an n–dimensional vector field each component of which is a homogeneous polynomial of degree 2 in the coordinates $x_1, \ldots x_n$ and $O(x)^3$ denotes cubic and higher terms.

We seek a change of coordinates

(2.4) $z = x - \phi^{[2]}(x)$

where again $\phi^{[2]}$ is n–dimensional vector field of homogeneous polynomials of degree 2 which carries (2.3) into

(2.5) $\dot{z} = Az$

A straightforward calculation shows that (2.4) transforms (2.3) into

(2.6) $\dot{z} = Az + (f^{[2]}(x) - [Ax, \phi^{[2]}(x)]) + O(x)^3$

Hence (2.3) can be linearized to degree 2 iff the so–called Homological Equations [Ar]

(2.7) $f^{[2]}(x) = [Ax, \phi^{[2]}(x)]$

are satisfied. Notice that Lie bracketing by Ax is a linear mapping from quadratic vector fields $\phi^{[2]}(x)$ into quadratic vector fields $f^{[2]}(x)$. Therefore (2.7) is a square system of linear equations relating the $n^2(n+1)/2$ coefficients of the quadratic monomials making up the components of $\phi^{[2]}(x)$ to those of $f^{[2]}(x)$. Poincaré noted that if the eigenvalues of A are $\lambda_i; 1 \leq i \leq n$, then the eigenvalues of the linear operation of bracketing quadratic vector fields by Ax are $\lambda_i + \lambda_j - \lambda_k$ where $1 \leq i \leq j \leq n; 1 \leq k \leq n$. Hence (2.7) is solvable for any $f^{[2]}(x)$ iff the so-called nonresonance condition holds, i.e.

$\lambda_i + \lambda_j - \lambda_k \neq 0$, for $1 \leq i \leq j \leq n, 1 \leq k \leq n$.

If a system is linearized to degree $\rho - 1$, i.e.

$$(2.8) \qquad \dot{x} = Ax + f^{[\rho]}(x) + O(x)^{\rho+1}$$

then we can seek a degree ρ change of coordinates

$$(2.9) \qquad z = x - \phi^{[\rho]}(x)$$

carrying (2.8) into (2.5).

As before this is possible for all $f^{[\rho]}$ iff the homological equations

$$(2.10) \qquad f^{[\rho]}(x) = [Ax, \phi^{[\rho]}(x)]$$

are solvable. This is a larger linear system with eigenvalues $\lambda_{i_1} + ... + \lambda_{i_\rho} - \lambda_k$ for $1 \leq i_1 \leq i_2 \leq ... \leq i_\rho \leq n$ and $1 \leq k \leq n$, hence leads to additional nonresonance conditions.

Now consider a smooth nonlinear system of the form

$$(2.11a) \qquad \dot{x} = f(x) + g(x) u$$

$$(2.11b) \qquad y = h(x)$$

where u, x and y are m, n and p dimensional. For convenience we assume that control enters the dynamics linearly and does not appear explicitly in the output. These restrictions can easily by relaxed. Hunt–Su [H–S], Jakubczyck–Respondek [J–R] and others have considered the question of when such a system can be transformed by a change of state coordinates (2.2) into a system of the form

(2.12) $z = Az + B(\alpha(x) + \beta(x) u)$.

Such a system can be linearized by state feedback

(2.13a) $\alpha(x) + \beta(x) u = v$

or, equivalently,

(2.13b) $\alpha(\phi^{-1}(x)) + \beta(\phi^{-1}(z)) u = v$.

Such systems are said to be <u>feedback linearizable</u>. The above authors have derived the integrability conditions for the system of partial differential equations that must be satisfied by ϕ, α and β.

Practical applications of this approach predate the general theoretical development. It is at the core of many nonlinear control schemes such as the Total Automatic Flight Control System (TAFCOS) developed by G. Meyer and associates [M–C] at NASA Ames Research Center and the Resolved Acceleration approach to robot control of Luh, Walker and Paul [L–W–P]. Fortunately, for many mechanical systems the PDE's for ϕ, α and β are integrable and a solution is obvious. The solution is not unique. However, such systems are exceptional, for a generic system with $2m < n$, the PDE's are not integrable. Moreover, even if they are, a solution is not always easy to find.

By utilizing Poincaré's method we can bypass these difficulties at the cost of obtaining only approximate solutions to the PDE's irrespective of their integrability. Suppose that system dynamics (2.11) is expanded as follows

$$(2.14a) \qquad \dot{x} = Ax + Bu + f^{[2]}(x) + g^{[1]}(x)\, u + O(x,u)^3$$

and the output map is similarly expanded

$$(2.14b) \qquad y = Cx + h^{[2]}(x) + O(x)^3.$$

We consider the effect on (2.14) of quadratic changes of coordinates in the state space (2.4) and output space

$$(2.15) \qquad w = y - \gamma^{[2]}(y).$$

The resulting system is

$$(2.16a) \qquad z = Az + Bu$$
$$+ f^{[2]}(x) + g^{[1]}(x)u - [Ax + Bu, \phi^{[2]}(x)] + O(x,u)^3$$
$$(2.16b) \qquad w = Cz$$
$$+ h^{[2]}(x) + C\,\phi^{[2]}(x) - \gamma^{[2]}(y) + O(x)^3$$

By adding to and subtracting from (2.16a) the quadratic feedback expression

$$(2.17) \qquad u + \alpha^{[2]}(x) + \beta^{[1]}(x)\, u = v$$

we obtain the system

$$(2.17a) \qquad \dot{z} = Az + Bv + R_1^{[2]}(x,u) + O(x,u)^3$$
$$(2.17b) \qquad w = Cz + R_2^{[2]}(x) + O(x)^3$$

where the quadratic error terms are given by

(2.18a) $R_1^{[2]}(x,u) = f^{[2]}(x) + g^{[1]}(x)u$
$- [Ax + Bu, \phi^{[2]}(x)] - B(\alpha^{[2]}(x) + \beta^{[1]}(x)u)$

(2.18b) $R_2^{[2]}(x,y) = h^{[2]}(x) + C \phi^{[2]}(x) - \gamma^{[2]}(y) + O(x)^3.$

We refer to (2.18a) with $R_1^{[2]}(x,u) = 0$ as the controller homological equation of degree 2. It is a system of $n^2(n+1)/2 + m\, n^2$ linear equations in $n^2(n+1)/2 + m$ $n(n+1)/2 + m^2 n$ unknown coefficients of $\phi^{[2]}$, $\alpha^{[2]}$ and $\beta^{[1]}$. If there exists a solution to the controller homological equation of degree two then we say that the nonlinear system (2.14) is feedback linearizable to degree two.

It is easy to see that if a system is feedback linearizable then it is feedback linearizable to degree two. Given a system that is feedback linearizable to degree two, one can ask whether it is feedback linearizable to degree three, etc. This suggests an obvious question. If a real analytic system (2.11) is feedback linearizable to arbitrary degree, is it feedback linearizable? This may or may not be a hard question to resolve. The corresponding question for the Poincaré's problem is quite difficult, see Arnold [Ar]. From the point of view of design of nonlinear compensators, the question is somewhat irrelevant as it would be extremely impractical to solve the sequence of higher degree controller homological equations. The size of these equations grows exponentially in the degree.

Typically the controller homological equations of degree 2 are not solvable and so one is forced to seek an approximate solution, e.g. in the least square sense. Moreover, in many applications such as tracking, it is desirable not only to have $R_1^{[2]}(x,u)$ small but also $R_2^{[2]}(x)$. If we set the left sides of (2.18) to zero we obtain a system of $n^2(n + 1)/2$ $+ m^2 n + p\, n(n + 1)/2$ in $n^2(n + 1)/2 + m\, n(n + 1)/2 + m^2 n + p^2(p + 1)/2$ unknown coefficients of $\phi^{[2]}$, $\alpha^{[2]}$, $\beta^{[1]}$ and $\gamma^{[2]}$. Again, typically these equations do not have an exact solution and an approximate solution must be found instead.

The current version of POINCARE takes as data the linear parts A, B, C and the quadratic parts $f^{[2]}$, $g^{[1]}$, and $h^{[2]}$ if the system (2.14). If the desired goal is the design of a stabilizing state feedback control law, then, using the pole placement or LQR algorithm of A. Laub's Control Systems Toolbox for MATLAB, a stabilizing linear feedback gain F is found. Suppose the feedback

(2.19a) $u = F x + \bar{u}$

is chosen because of the desirable close loop stability of

(2.19b) $\dot{x} = \bar{A} x + B \bar{u}$

where

(2.19c) $\bar{A} = A + B F$

POINCARE allows one to simulate the linear feedback (2.19a) applied to the quadratic part of the system (2.14)

(2.20a) $\dot{\lambda} = \bar{A}x + \bar{f}^{[2]}(x) + B\bar{u} + g^{[1]}(x)\,\bar{u}$

where

(2.20b) $\bar{f}^{[2]}(x) = f^{[2]}(x) + g^{[1]}(x)\,Fx.$

If the stability of (2.20a) is satisfactory with $\bar{u} = 0$ over the range of the x of interest then one need not go any further. On the other hand if $\bar{f}^{[2]}$ has degraded the performance then one may seek an additional quadratic feedback to improve things. We find such a quadratic feedback by considering the effect of a quadratic coordinate change (2.4) and feedback

(2.21) $\bar{u} + \alpha^{[2]}(x) + \beta^{[1]}(x)\,\bar{u} = v$

on (2.20a).

(2.22a) $\dot{z} = \bar{A}z + Bv + \bar{R}^{[2]}(x, \bar{u}) + O(x, \bar{u})^3$

where

(2.22b)
$$\bar{R}^{[2]}(x, \bar{u}) = \bar{l}^{[2]}(x) + g^{[1]}(x)\, \bar{u}$$
$$- [\bar{A}\, x + B\, \bar{u}\,,\, \phi^{[2]}(x)] - B(\alpha^{[2]}(x) + \beta^{[1]}(x)\, \bar{u}).$$

If we assume that the open loop control $v = 0$ then $\bar{u} = O(x^2)$ and (2.22a,b) become

(2.22c)
$$\dot{z} = \bar{A}\, z + \bar{R}_1^{[2]}(x,0) + O(x)^3$$

(2.22d)
$$\bar{R}_1^{[2]}(x, 0) = \bar{l}^{[2]}(x) - [\bar{A}x\,,\, \phi^{[2]}] - B\, \alpha^{[2]}(x).$$

If we have chosen the linear feedback given F so that the eigenvalues of $\bar{A}$ are not resonant at degree two then there exist $\phi^{[2]}$ and $\alpha^{[2]}$ so that (2.22d) is zero. In fact there are $m\, n(n + 1/2$ linearly independent solutions parameterized by α. Which should be chosen? It is a trade off between the size of the quadratic feedback (2.21) as measured by $\alpha^{[2]}$ and the size of the quadratic change of coordinates (2.4) as measured by $\phi^{[2]}$. One approach is to make $\phi^{[2]}$ and $\alpha^{[2]}$ as small as possible in a least squares sense. This implies a choice of a metric on the space of coefficients of $\phi^{[2]}$ and $\alpha^{[2]}$.

An obvious choice is to take the standard inner product on the space of coefficients of $\phi^{[2]}$, $\alpha^{[2]}$. This is not as naive as it sounds because POINCARE has previously asked us for unit lengths of the components of x, u and y and has scaled the equation (2.14) accordingly. One can make more sophisticated choices as we shall describe later.

Given a choice of metric, POINCARE uses MATLAB's singular value decomposition (SVD) algorithm to find $\phi^{[2]}$ and $\alpha^{[2]}$ satisfying (2.22). It then uses the quadratic feedback

(2.23a)
$$\bar{u} = -\, \alpha^{[2]}(x)$$

to satisfy (2.14). The linear (2.19a) and quadratic (2.23a) feedback yield the total feedback.

(2.23b)
$$u = Fx - \alpha^{[2]}x.$$

Notice it does not depend on $\phi^{[2]}$. The closed loop dynamics in the original x coordinates has a quadratic component,

(2.24) $\qquad \dot{x} = \bar{A}\,x + \bar{f}^{[2]}(x) - B\,\alpha^{[2]}(x) + O(x)^3$

but in the transformed z coordinates (2.4) it does not,

(2.25) $\qquad \dot{z} = \bar{A}\,z + O(z^3).$

Because $\phi^{[2]}(x)$ and $\alpha^{[1]}(x)$ have been chosen to be small, one expects the performance of (2.24) to be similar to that of (2.25). The result should be an improvement over the system with only linear feedback,

(2.26) $\qquad \dot{x} = \bar{A}\,x + \bar{f}^{[2]}(x).$

To verify this, POINCARE simulates the response of three systems to a nonzero initial condition. It graphs the norm of the state $\|x(t)\|$ and the components of the output $y_i(t)$, $i = i,...,p$ of (2.14b) for the linear system with linear feedback,

(2.27) $\qquad \dot{x} = \bar{A}x,$

the quadratic system with linear feedback, (2.26) and the quadratic system with linear and quadratic feedback (2.24). The first is the linear ideal closed loop system, the second is the result of a standard design obtained by ignoring nonlinear terms and the third is the POINCARE approach. Generally in these simulations we have found that the performance of POINCARE system closely approximates the linear ideal, while the standard design is somewhat different and usually worse. By performance we mean things like settling times and basis of attraction. We present some simulations in Section 3.

The above problem is somewhat simplistic. Usually one wants to exercise open loop control on the stabilized system. For example one might desire the output $y(t)$ of the system (2.14) to track a reference signal $r(t)$. As before the first step is to do a linear design. One might try to use static state feedback and reference signal feedforward of the form

(2.28) $\qquad u = F\,x + \bar{u}$

to achieve the desired tracking for the linear part of (2.14). The input $\bar{u}(t)$ is a linear functional of the current and past values of the reference signal $\{\,r(s) : s \leq t\,\}$ typically obtained by passing the reference signal through an approximate right inverse of the linear system

(2.29a) $\dot{x} = \bar{A}\,x + B\,\bar{u}$

(2.29b) $y = C\,x$

We denote this dependency of $\bar{u}$ by $\bar{u}(t) = \bar{u}(t;r)$. We can use this as a basis for a nonlinear design via POINCARE.

We start by applying the feedback (2.28) to the nonlinear system (2.14) and obtain

(2.30a) $\dot{x} = \bar{A}\,x + B\,\bar{u} + \bar{f}^{[2]}(x) + \bar{g}^{[1]}(x)\bar{u} + O(x,\bar{u})^3$

(2.30b) $y = C\,x + h^{[2]}(x) + O(x)^3$

where

(2.30c) $\bar{g}^{[1]}(x) = g^{[1]}(x)\,G.$

We then apply the quadratic changes of coordinates and feedback (2.4), (2.15) and (2.21) to obtain

(2.31a) $\dot{z} = \bar{A}\,z + B\bar{u} + \bar{R}_1^{[2]}(x,\bar{u}) + O(x,\bar{u})^3$

(2.31b) $y = C\,z + \bar{R}_2^{[2]}(x,y) + O(x)^3$

where the quadratic residual $\bar{R}_1^{[2]}$ is as before (2.18a) after substitution of $\bar{A}$, B, $\bar{f}^{[2]}\,\bar{g}^{[1]}$ for A, B, $f^{[2]}$, $g^{[1]}$ and $\bar{R}_2^{[2]} = R_2^{[2]}$.

We wish to make the quadratic residuals in (2.31) as small as possible while at the same time using as keeping $\phi^{[2]}$, $\alpha^{[2]}$, $\beta^{[1]}$, $\gamma^{[2]}$ as small as possible. To make this precise we must define metrics on the space of coefficients of $(\bar{R}_1^{[2]}, \bar{R}_2^{[2]})$ and the space of coefficients of $(\phi^{[2]}, \alpha^{[2]}, \beta^{[1]}, \gamma^{[2]})$. An obvious choice is to use the standard metric on these spaces and again this is not a bad choice provided one has already scaled the equations (2.14). More sophisticated metrics can be obtained as follows. Choose metrics on the state, input and output spaces defined by positive definite matrices M_s, M_i, M_o, i.e.

$$\|x\|_s^2 = x^* M_s x$$
$$\|u\|_i^2 = u^* M_i u$$
$$\|y\|_o^2 = y^* M_o y$$

These could be the same metrics as those used in the LQR algorithm. Also choose a probability density $\rho(x,u)$ on the input cross state space. This density describes the expected operating regime; the likelihood that the system will be at state x with input u.

Then we define

(2.32a)
$$\left\| \begin{array}{c} R_1^{[2]} \\ R_2^{[2]} \end{array} \right\|^2 = \iint (\|R_1(x,u)\|_s^2 + \|R_2(x,C_x)\|_o) \, \rho(x,u) \, dx \, du$$

and

(2.32b)
$$\left\| \begin{array}{c} \phi^{[2]} \\ \alpha^{[2]} \\ \beta^{[2]} \\ \gamma^{[2]} \end{array} \right\|^2 = \iint \|\phi^{[2]}(x)\|_s^2 + \|\alpha^{[2]}(x) + \beta^{[1]}(x)u\|^2 + \|\gamma^{[2]}(x)\|_o^2) \, \rho(x,u) \, dx \, du$$

At present the only choices of metric available in POINCARE are the standard metrics on the spaces of coefficients and the metrics defined by (2.31a,b) when M_s, M_i and M_o are identity matrices and $\rho(x,u)$ is a Gaussian density with zero mean and unit covariance. These two choices are actually quite similar. We expect to add other options in the future.

POINCARE uses MATLAB's SVD algorithm to find $\phi^{[2]}$, $\alpha^{[2]}$, $\beta^{[1]}$, $\gamma^{[2]}$. This leads to the control law with linear part (2.28a) and quadratic part

(2.33a)
$$\bar{u} + \alpha^{[2]}(x) + \beta^{[1]}(x)\bar{u} = \bar{u}(t;r)^{[2]} \, \bar{u}(t;\gamma^{[2]}(r))$$

By neglecting cubic terms, we have the control law

(2.33b)
$$u = Fx + Gr$$
$$\quad - G(\alpha^{[2]}(x) + \beta^{[1]}(x)r + \bar{u}(t;\gamma^{[2]}(r))$$

and the closed loop system

(2.34a) $\dot{x} = \bar{A}x + Br$

$\qquad + \bar{l}^{[2]}(x) + \bar{g}^{[1]}(x)r - G(\alpha^{[2]}(x) + \beta^{[1]}(x)r + u(t;\gamma^{[2]}(r))$

(2.34b) $y = Cx + h^{[2]}(x)$.

This (2.34) should be compared with the ideal linear system (2.29) and the standard approach of applying the linear law (2.28) to the linear and quadratic parts of (2.14), i.e.,

(2.35a) $\dot{x} = \bar{A}x + B\bar{u}(t;r)$

$\qquad + \bar{l}^{[2]}(x) + \bar{g}^{[1]}(x)\bar{u}(t;r)$

(2.35b) $y = Cx + h^{[2]}(x)$.

Notice that the linear parts of all three systems are the same. POINCARE simulates and compares the behavior of the three systems (2.29), (2.35) and (2.34).

Of course in many situations full state observations are not possible and hence one must use a filter or observer to estimate the state from the past observations. If one were to ignore the nonlinearities of the system (2.14) one could design a linear observer of the form

(2.36a) $\dot{\hat{x}} = (A + KC)\hat{x} + Bu - Ky$

with linear error dynamics, $\tilde{x} = x - \hat{x}$, given by

(2.36b) $\dot{\tilde{x}} = (A + KC)\tilde{x}$.

If the linear part of (2.14) is observable then one can choose K to set the spectrum of $A + KC$ arbitrarily. One can also choose K by using a Kalman filtering formulation. MATLAB's Control Systems Toolbox has algorithms for either approach.

Of course if we use the linear observer (2.36a) to establish the state of the linear and quadratic part of (2.14) then the observer error dynamic has quadratic terms

(2.36c) $\dot{\tilde{x}} = (A + KC)\tilde{x}$

$\qquad + f^{[2]}(x) + g^{[1]}(x)u + Kh^{[2]}(x)$.

These extra terms may sufficiently degrade the performance of the observer (2.36a) so that it is unacceptable. POINCARE allows one to ameliorate the effect of the quadratic nonlinearities in the following fashion.

Consider once again the effect of quadratic changes of state (2.4) and output (2.15) coordinates on the quadratic part of (2.14). The result is (2.16). If we add and subtract an input–output injection term

$$\alpha^{[2]}(y) + \beta^{[1]}(y)u$$

to and from (2.16a) we obtain

(2.37a) $\qquad \dot{z} = Az + Bu + \alpha^{[2]}(y) + \beta^{[1]}(y)u + R_3^{[2]}(x,u,y) + O(x,u)^3$

(2.37b) $\qquad w = Cz + R_4^{[2]}(x,y) + O(x,u)^3$

where

(2.38a) $\qquad R_3^{[2]}(x,u,y) = f^{[2]}(x) - g^{[1]}(x)u - [Ax + Bu, \phi^{[2]}(x)] - \alpha^{[2]}(y) - \beta^{[1]}(y)u,$

(2.38b) $\qquad R_4^{[2]}(x,y) = h^{[2]}(x) + C\phi^{[2]}(x) - \gamma^{[2]}(y).$

We wish to emphasize that the current $\phi^{[2]}$, $\alpha^{[2]}$, $\beta^{[1]}$, $\gamma^{[2]}$ under discussion are different from those employed previously. In particular the dimensions and arguments of $\alpha^{[2]}$ and $\beta^{[1]}$ have changed. Previously $\alpha^{[2]}$ and $\beta^{[1]}$ were m × 1 and m × m valued functions of y. Later on, when it will be necessary to distinguish between these, we shall denote the previous ones with a subscript $_c$ as in $\phi_c^{[2]}$, $\alpha_c^{[2]}$, $\beta_c^{[1]}$, $\gamma_c^{[2]}$ since they are used for the design of control laws. The current ones shall be denoted with a subscript $_o$, as in $\phi_o^{[2]}$, $\alpha_o^{[2]}$, $\beta_o^{[1]}$, $\gamma_o^{[2]}$ since they will be used for the design of observers.

If we set $R_3^{[2]}$ and $R_4^{[2]}$ to zero then the system (22.37) is said to be in observer form of degree 2 and (2.38) are the observer homological equations of degree two.

Krener–Respondek [K–R], Bestle–Zeitz [B–Z] and others have looked at the question of when a nonlinear system can be transformed exactly into observer form by a smooth change of coordinates. For most systems, this requires checking a large set of integrability conditions and finding the change of coordinates by solving a set of PDE's. Phelps [P] has developed a method that greatly simplifies this, but it is still a formidable task.

Instead, we have approached the question term by term using the linearization technique of Poincaré.

We can design an observer for (2.37) with quadratic input–output injection

$$\text{(2.39a)} \qquad \dot{\hat{z}} = (A + KC)\hat{z} + Bu - Ky + \alpha^{[2]}(y) + \beta^{[1]}(y)u + K\gamma^{[2]}(y)$$

then the error $\tilde{z} = z - \hat{z}$ satisfies

$$\text{(2.39b)} \qquad \dot{\tilde{z}} = (A + KC)\tilde{z} + R_3^{[2]}(x,u,y) + KR_4^{[2]}(x,y) + O(x,\hat{z},u)^3.$$

POINCARE uses MATLAB's SVD algorithm to choose $\phi^{[2]}$, $\alpha^{[2]}$, $\beta^{[1]}$, $\gamma^{[2]}$ to minimize the quadratic part of (2.37b),

$$\text{(2.40)} \qquad \| R_3^{[2]} + KR_4^{[2]} \|^2$$

The metric used is either the standard one on the space of coefficients or

$$\| R_3^2 + KR_4^{[2]} \|^2 = \iint \|R_3^{[2]}(x,u,Cx) + KR_4^{[2]}(x,Cx)\|_s^2 \, \rho(x,u) \, dx \, du$$

where M_s is the identity and $\rho(x,u)$ is a Gaussian of zero mean and unit covariance. Notice that the linear observer gain K enters the quantity (2.40) to be minimized and hence influences the solution.

Of course, the error dynamics (2.39b) is nearly linear in the transformed coordinates. The implementation of the observer (2.39a) should be done in the original coordinates so we define $\hat{x}$ by

$$\text{(2.41a)} \qquad \dot{\hat{x}} = (A + KC)\hat{x} + Bu - Ky$$
$$+ [(A + KC)\hat{x} + Bu, \, \phi^{[2]}(\hat{x})]$$
$$+ \alpha^{[2]}(y) + \beta^{[1]}(y) + K\gamma^{[2]}(y) - \frac{\partial \phi^{[2]}}{\partial \hat{x}}(\hat{x}) \, Ky \, .$$

The error $\tilde{x} = x - \hat{x}$ dynamics is given by

$$\text{(2.41b)} \qquad \dot{\tilde{x}} = (A + KC)\tilde{x} + f^{[2]}(x) + g^{[1]}(x)u + Kh^{[2]}(x)$$
$$- [(A + KC)\hat{x} + Bu, \, \phi^{[2]}(x)] - \alpha^{[2]}(y) - \beta^{[1]}(y)u - K\gamma^{[2]}(y) + O(x,u)^3.$$

Then to $O(x,\hat{x})$, (2.39a) and (2.41a) are related by (2.41c) $\hat{z} = \hat{x} - \phi^{[2]}(\hat{x})$. Notice that the linear part of the observer (2.41a) is the same as the linear observer (2.36a). The extra quadratic terms in (2.41a) correct in part for the quadratic part of the original system (2.14). The linear part of the observer (2.41a) is stable by proper choice of K but the quadratic terms may destabilize it. Of course one might expect this since the quadratic terms of (2.14) are also destabilizing. But as a system with inputs u and y and output $\hat{x}$, (2.41a) it is not BIBO stable in contrast to the linear observer (2.36a).

POINCARE simulates three pairs of systems and observers to compare their performance. The first pair is a system consisting of the linear part of (2.14) and the linear observer (2.36a). This is the linear ideal and provides a benchmark for the other two. The second pair is the system consisting of the linear and quadratic parts of (2.14) and the linear observer (2.36a). This is a standard approach. The third is the system consisting of the linear and quadratic parts of (2.14) with the linear and quadratic observer (2.41a). This is the POINCARE approach. The three pairs of systems and observers are excited by initial condition errors and various kinds of inputs, u(t). The norms of the errors in x coordinates are graphed with respect to time. Generally one finds that the errors of the POINCARE approach are smaller and closer to those of the linear ideal than those of the standard approach, although this is not always the case. Simulations are given in the next section.

It is also possible to combine the two halves of POINCARE to obtain an input–output driven nonlinear compensator. The stable estimate of the nonlinear observer (2.41a) us fed into a linear and quadratic feedback control law. Once again the POINCARE approach can be simulated and compared with an ideal linear system with linear observer and linear state estimate feedback and a standard approach of a linear and quadratic system with linear observer and linear stable estimate feedback.

These three pairs of systems are as follows. The linear ideal is

(2.42a) $\dot{x} = Ax + Bu$

(2.42b) $\dot{\hat{x}} = (A + KC)\hat{x} + Bu - Ky$

(2.42c) $y = Cx$

(2.42d) $u = F\hat{x} + v$

The standard approach yields

(2.43a) $\dot{x} = Ax + Bu + f^{[2]}(x) + g^{[1]}(x)u$

(2.43b) $\dot{\hat{x}} = (A + KC)\hat{x} + Bu - Ky$

(2.43c) $y = Cx + h^{[2]}(x)$

(2.43d) $u = F\hat{x} + v$

The POINCARE approach yields

(2.44a) $\dot{x} = Ax + Bu + f^{[2]}(x) + g^{[1]}(x)u$

(2.44b) $\dot{\hat{x}} = (A + KC)\hat{x} + Bu - Ky$
$$+ [(A + KC)\hat{x} + Bu, \phi_o^{[2]}(\hat{x})]$$
$$+ \alpha_o^{[2]}(y) + \beta_o^{[1]}(y) + K\gamma_o^{[2]}(y) - \frac{\partial \phi_o^{[2]}(\hat{x})}{\partial \hat{x}} Ky$$

(2.44c) $y = Cx + h^{[2]}(x)$

(2.44d) $u = F\hat{x} + v - \alpha_c^{[2]}(\hat{x}) - \beta_c^{[1]}(\hat{x})v$

Equation (2.44d) should be plugged into (2.44b) before computing the bracket.

Simulations can be found in the next section.

3. Simulations
SESSION 1 Controller/Observer Example

_ QUADRATIC APPROXIMATION _
_ Controller - Observer _

----- What is your Macintosh running under? -----

    1) Finder
    2) Multifinder

 Select a menu number: 1
----- Enter mode -----

    1) Novice
    2) Expert

 Select a menu number: 1
----- M A I N   M E N U -----

    1) Help
    2) Enter nonlinear system (plant and output equations)
    3) Select the type of problem to be solved
    4) Simulate and plot results
    5) Quit

 Select a menu number: 1
_ QC is a menu driven script file for 2nd order approximate observers
_ for nonlinear control systems. Your control system should be in
_ the following format:
_
_ dx   [2]   [1]   3
_ -- = Ax + Bu + f (x) + g (x)u + O(x, u)
_ dt
_   [2]
_ y = Cx + h (x)

_
_ where A is the nxn plant matrix, B is the nxm input vector, f[2] is
_ the second degree part of the vector field, and g[1] is the first
_ degree part of the input vector. All of these terms should be
_ obtained by the user from the Taylor expansion of the control
_ system at the nominal operating point 0.
_   We compute the quadratic observer-controller parameters and plot
_ the system against linear and quadratic observer-controller pairs.
_ The corresponding function modules are documented on-line under
_ HELP QCRUN. For further details on the individual subroutines
_ and function programs called by the routines, use the help utility
_ in MATLAB.

___ Glossary of Q.C. Functions __

QCHELP introduces the quadratic approximation program.
QCSETUP inputs in the system functions F, G, f2, g1.
QCCOMP solves the homological equations for _, _ and _.
   _ QCSCALE rescales the system and control matrices.
   _ QCBIGMX sets up the linear equations in the coefficients.
   _ QCSVD solves these equations using the SVD algorithm.
QCFDBK gets the user-specified feedback.
QCGLOB sets up global symbolic variables for controller.
QCSIM controls the simulation subroutines.
QCINIT sets the initial conditions for the ODE's.
QCCNTRL creates the control M-file with several options.
QCSOLVE solves the ODE's for the system and observers.
QCPLOT plots the solutions with the following options:
   _ QCPHASE provides the phase plots;
   _ QCEXT extends the solutions;
   _ QCTIME provides the time plots.

QCRUN is this help display.

----- M A I N   M E N U -----

    1) Help
    2) Enter nonlinear system (plant and output equations)
    3) Select the type of problem to be solved
    4) Simulate and plot results
    5) Quit

Select a menu number: 2

----- Input one of the following: -----

    1) Enter data manually
    2) Enter filename for loading data

Select a menu number: 2

  Enter filename: example_1

Dimensions of the system:

[n, m] =
    3   1

Homogeneous order of system:
nchoose2 =
    6

  Linear plant matrix, A:
_____

   0   1   0
   0   0   1
  -3   -2   -1

  Eigenvalues of the open-loop plant:
_____

 -1.2757
  0.1378 + 1.5273i

0.1378 - 1.5273i

Second order part of the plant, f2:

1	0	-1	0	0	0
0	-1	0	1	0	1
0	0	0	0	1	0

Constant part of the input vector, B:

0
0
1

First order part of the input vector, g1:

1	0	0
0	0	0
0	1	0

Constant part of the output vector, C:

1   0   0

Second order part of the output vector, h2:

0   0   0   0   0   0

Scale factors of the states, x1 through x3:
xscale =
      1   1   1

Scale factor of the input, u1:
uscale =
      1

Scale factor of the output y1:
yscale =

1

----- M A I N   M E N U -----

    1) Help
    2) Enter nonlinear system (plant and output equations)
    3) Select the type of problem to be solved
    4) Simulate and plot results
    5) Quit

 Select a menu number: 3
----- Select the type of problem to be solved -----

    1) Quadratic Controller design (full observability)
    2) Quadratic Observer design
    3) Quadratic Controller-Quadratic Observer design
    4) Get more information on the above choices

 Select a menu number: 3
CLOSED LOOP FEEDBACK DESIGN FOR LINEAR PART
             OF THE PLANT

----- Input one of the following: -----

    1) Specify closed loop eigenvalues
    2) Design linear quadratic regulator with state
weighing
    3) Design linear quadratic regulator with output
weighing

 Select a menu number: 2

Quadratic regulator will minimize:

   —
   _ integral(x'Qx + u'Ru)dt

—
_ Enter 3x3 <sym. pos. semi-def.> matrix, Q:
—

```
___> _ [1 0 0;0 1 0;0 0 1]
_ Enter 1x1 <sym. pos. def.> matrix, R:
_
 ___> 1
```

Closed loop eigenvalues:
_____

-1.3294
-0.4975 + 1.4599i
-0.4975 - 1.4599i

Gain matrix F that places the poles of A + BF at above
_____

 -0.1623   -1.7015   -1.3244

Calculating scaled variables according to the scale
factors...

_
_   2
_ We construct a large linear system L X = B in the O(x,u)
_ coefficients of the homological equations. This system has
_
_   2   2
_ ROWS: n (n+1)/2 + mn
_
_   2   2
_ COLUMNS: n (n+1)/2 + mn(n+1)/2 + m n .
_
_ In general, the column rank is deficient and the solution is
_ overdetermined.
_ We use the SVD algorithm to get the "nearest" possible solution.

----- Please choose the method for minimization: -----

     1) Identity norm
     2) Norm weighed by a normal distribution

 Select a menu number: 2
Solving for the coordinate change and feedback...

Q u a d r a t i c C o n t r o l l e r P r o b l e m :

----- Please choose a menu item -----

1) Display coordinate change and feedback
2) Show remainder between the actual and closest linearizable system
3) Exit to main menu

  Select a menu number: 1
     [2]   [2]
   phi in second degree coordinate change z = x - phi_con(x):
   ~~~~~~

phi_con[2](1) = -0.0076646*x(1)*x(1) - 0.15515*x(1)*x(2) +
              0.67221*x(1)*x(3) - 0.023882*x(2)*x(2) +
              0.1054*x(2)*x(3) + 0.046875*x(3)*x(3)
phi_con[2](2) = -2.9808*x(1)*x(1) - 1.1353*x(1)*x(2)
              + 0.15515*x(1)*x(3) - 0.5626*x(2)*x(2) +
              0.016217*x(2)*x(3) - 0.043276*x(3)*x(3)
phi_con[2](3) = -0.49063*x(1)*x(1)-5.5871*x(1)*x(2) -
              1.2223*x(1)*x(3) - 2.1954*x(2)*x(2) -
              0.68737*x(2)*x(3) -0.7826*x(3)*x(3)

                                         [2]
   Second degree part of feedback: alpha_con
   ~~~~~~

alpha_con[2](1) = +2.8083*x(1)*x(1)+4.7978*x(1)*x(2)-
1.4355*x(1)*x(3)+7.6045*x(2)*x(2)+3.1339*x(2)*x(3)-
0.1493*x(3)*x(3)
                                       [1]
   First degree part of feedback: beta_con
   ~~~~~~

beta_con[1](1, 1) = +1.2223*x(1)+1.6874*x(2)+1.5652*x(3)

   Q u a d r a t i c C o n t r o l l e r P r o b l e m :

----- Please choose a menu item -----

    1) Display coordinate change and feedback
    2) Show remainder between the actual and closest
linearizable system
    3) Exit to main menu

Select a menu number: 2

$$[2]$$
Second degree part in the remainder: R(x)
~~~~~~

R[2](1) = -0.01731*x(1)*x(1)-0.15516*x(1)*x(3)-
0.01731*x(2)*x(2)-0.016222*x(2)*x(3)+0.069239*x(3)*x(3)
R[2](2) = 0
R[2](3) = 0

$$[1]$$
First degree part in the remainder: R (i'th col. of R[1]
multiplies u(i))
~~~~~~
~~~~~~~~~~~~~

R[1](1, 1) = +0.32779*x(1)-0.1054*x(2)-0.09375*x(3)
R[1](2, 1) = -0.15515*x(1)-0.016217*x(2)+0.086552*x(3)
R[1](3, 1) = 0

$$[2]$$
Second degree part in the remainder of the output: H(x)
~~~~~~

H[2](1) = -0.0076646*x(1)*x(1)

CLOSED LOOP FEEDBACK DESIGN FOR LINEAR PART OF THE
            O-B-S-E-R-V-E-R

----- Input one of the following: -----

      1) Specify observer eigenvalues
      2) Design linear Kalman filter

 Select a menu number: 2

Kalman filter will minimize the error covariance.

_
_ Enter 3x3 <sym. pos. semi-def.> driving noise covariance
matrix, Q:
_

___> _ [1 0 0;0 1 0;0 0 1]
_ Enter 1x1 <sym. pos. def.> observation noise covariance
matrix, R:

_

___> 1
  Observer eigenvalues:

  _____

  -1.5385
  -0.4085 + 1.5331i
  -0.4085 - 1.5331i

  Gain matrix K that places the poles of A + KC at above

  _____

  -1.3556
  -0.4188
   2.2571

S o l v i n g   f o r   t h e   Q u a d r a t i c   O b s e r v e r

----- Please choose the method for minimization: -----

      1) Identity norm
      2) Norm weighed by a normal distribution

 Select a menu number: 2

----- M A I N   M E N U -----

      1) Help
      2) Enter nonlinear system (plant and output equations)
      3) Select the type of problem to be solved
      4) Simulate and plot results
      5) Quit

 Select a menu number: 4
--Press any key--
Simulation with external disturbances:

Please define input u1

----- Disturbance input type -----

    1) Zero disturbance on the above input
    2) Impulse (not implemented yet)
    3) Step
    4) Sinusoid
    5) Random noise (not implemented yet)
    6) External data file (not implemented yet)

 Select a menu number: 4
 Enter sinusoid amplitude: 1
 Enter sinusoid frequency (*scaled* time): 6

_
_ Enter 3x1 vector of initial conditions, x10c: [0;0;0]
_
 ___>
Simulation will start from initial time t0c = 0.

 Enter final time (*scaled*), tfc: 12

_
_ Enter 3x1 Vector of observer initial conditions, x10o:
[0;0;0]
_

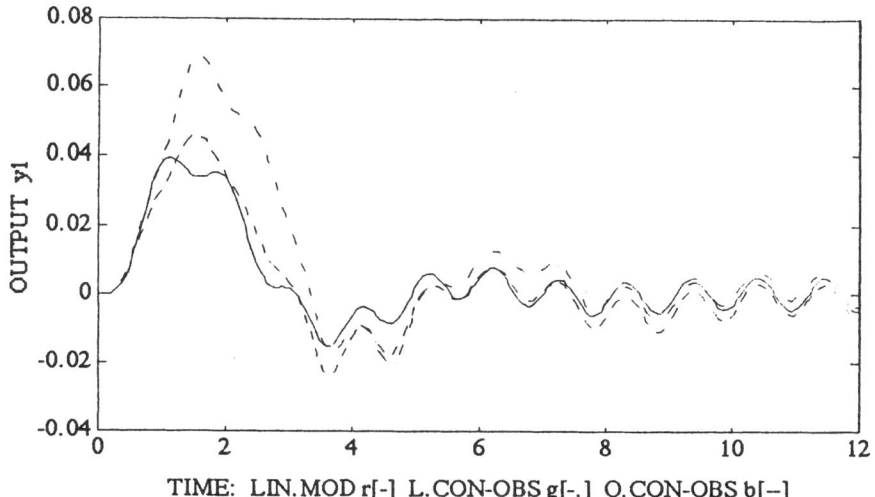

TIME: LIN.MOD r[-] L.CON-OBS g[-.] Q.CON-OBS b[--]

USING THE SAME SYSTEM AS ABOVE:

     1) Quadratic Controller design (full observability)
     2) Quadratic Observer design
     3) Quadratic Controller-Quadratic Observer design
     4) Get more information on the above choices

 Select a menu number: 1
CLOSED LOOP FEEDBACK DESIGN FOR LINEAR PART OF THE PLANT

----- Input one of the following: -----

     1) Specify closed loop eigenvalues
     2) Design linear quadratic regulator with state weighing
     3) Design linear quadratic regulator with output weighing

 Select a menu number: 2

Quadratic regulator will minimize:

    _ integral(x'Qx + u'Ru)dt

_ Enter 3x3 <sym. pos. semi-def.> matrix, Q:
_

___> [1 0 0;0 1 0;0 0 1]

_ Enter 1x1 <sym. pos. def.> matrix, R:

___> 1

Closed loop eigenvalues:
_____

-1.3294
-0.4975 + 1.4599i
-0.4975 - 1.4599i

Gain matrix F that places the poles of A + BF at above
_____

-0.1623   -1.7015   -1.3244

INITIAL CONDITION RESPONSE: X=[0.1 0.1 0.1]

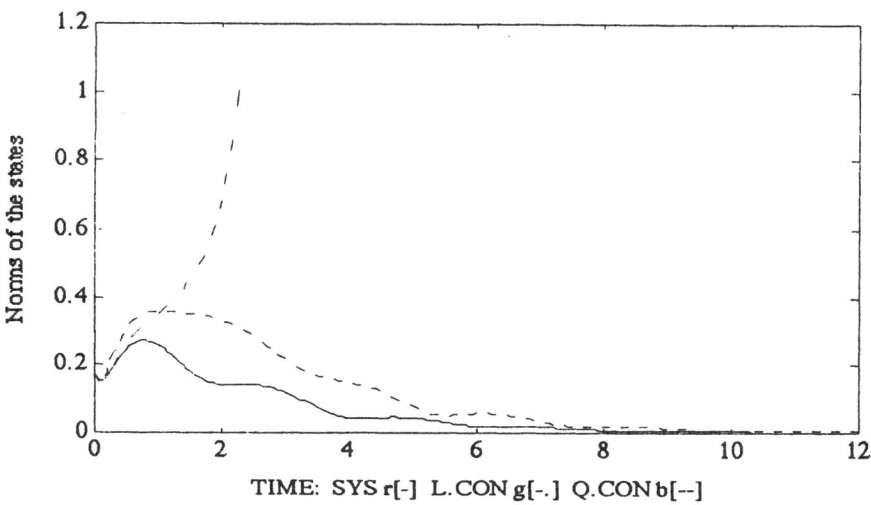

INITIAL CONDITION RESPONSE: X=[-0.1 -0.1 -0.1] (Negative of the above)

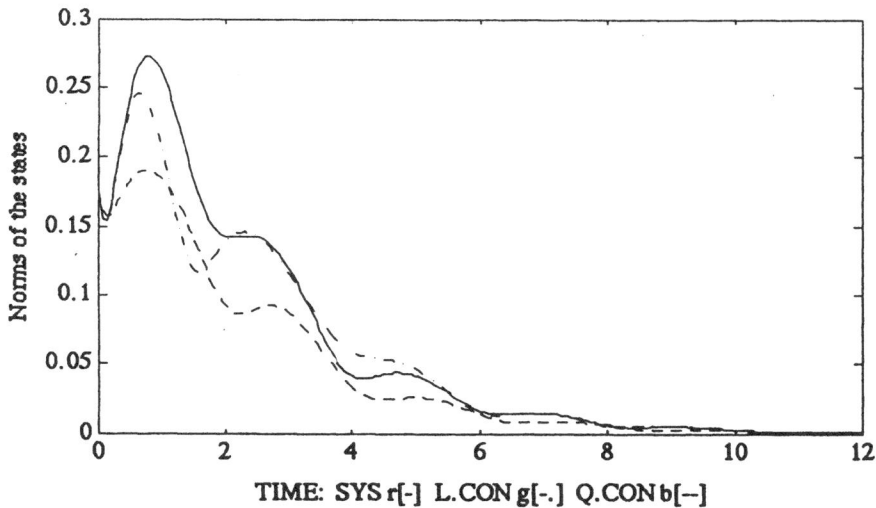

STEP RESPONSE: Step input magnitude is 0.1.

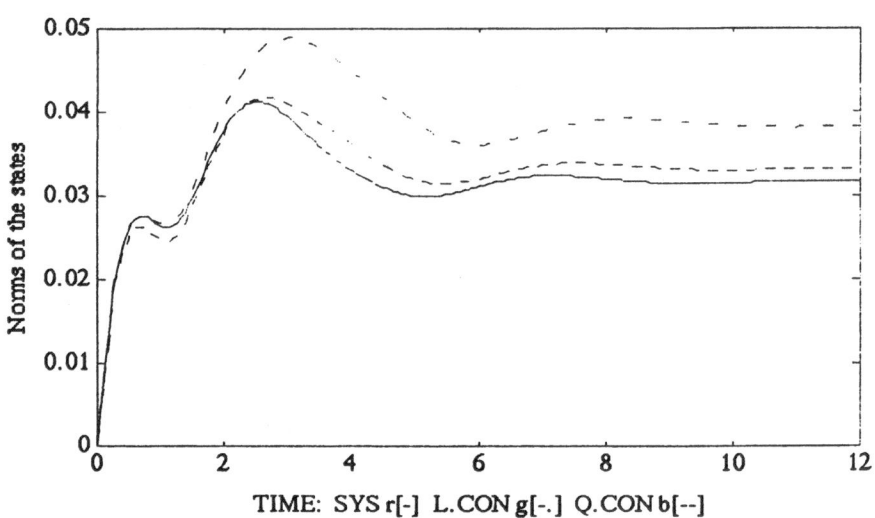

TIME: SYS r[-] L.CON g[-.] Q.CON b[--]

STEP RESPONSE: Step input magnitude is 0.2.

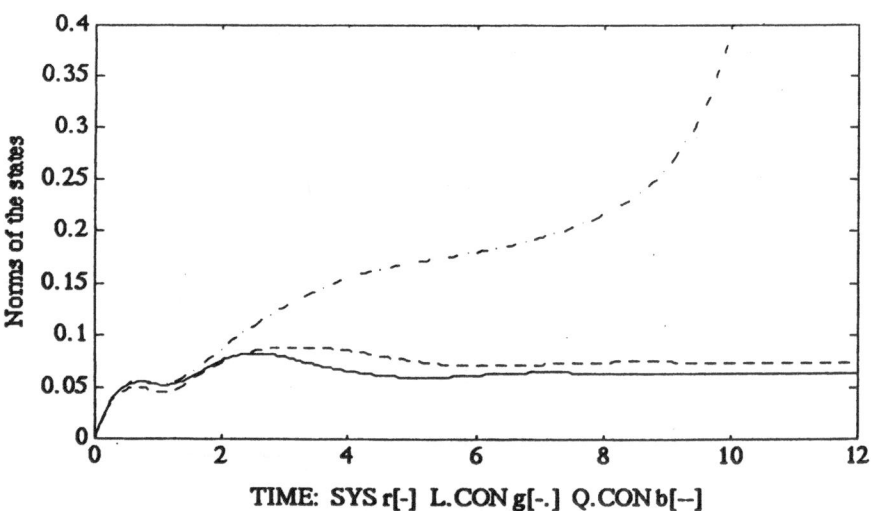

TIME: SYS r[-] L.CON g[-.] Q.CON b[--]

SESSION 2   OBSERVER EXAMPLE

Dimensions of the system:

[n, m] = 3   1

Homogeneous order of system: nchoose2 = 6

Linear plant matrix, A:

```
    0      1      0
    0      0      1
    0     -1      0
```

Eigenvalues of the open-loop plant:

```
    0
    0 + 1.0000i
    0 - 1.0000i
```

Second order part of the plant, f2:

```
    0      0      0      0      0      0
    0      0      0      0      0      0
    0     -5      0      0      3      0
```

Constant part of the input vector, B:

```
    0
    0
    1
```

First order part of the input vector, g1:

```
    0      0      0
    0      0      0
    1      0      0
```

Constant part of the output vector, C:

---

 1   0   0

Second order part of the output vector, h2:

---

 0   0   0   0   0   0

Scale factors of the states, x1 through x3:
xscale = 1   1   1

Scale factor of the input, u1:
uscale = 1

Scale factor of the output y1:
yscale = 1

----- M A I N   M E N U -----

    1) Help
    2) Enter nonlinear system (plant and output equations)
    3) Select the type of problem to be solved
    4) Simulate and plot results
    5) Quit

 Select a menu number: 3
----- Select the type of problem to be solved -----

    1) Quadratic Controller design (full observability)
    2) Quadratic Observer design
    3) Quadratic Controller-Quadratic Observer design
    4) Get more information on the above choices

 Select a menu number: 2
CLOSED LOOP FEEDBACK DESIGN FOR LINEAR PART OF THE
        O-B-S-E-R-V-E-R
----- Input one of the following: -----

    1) Specify observer eigenvalues
    2) Design linear Kalman filter

Select a menu number: 2

Kalman filter will minimize the error covariance.:
_
_ Enter 3x3 <sym. pos. semi-def.> driving noise covariance
matrix, Q:
_
  ___> _ [1 0 0;0 1 0;0 0 1]
_ Enter 1x1 <sym. pos. def.> observation noise covariance
matrix, R:
_
  ___> 1
  Observer eigenvalues:
  _____

  -1.0000
  -0.4551 + 1.0987i
  -0.4551 - 1.0987i

  Gain matrix K that places the poles of A + KC at above
  _____

   -1.9102
   -1.3244
    0.4960

S o l v i n g   f o r   t h e   Q u a d r a t i c   O b s e r v e r

----- Please choose the method for minimization: -----

     1) Identity norm
     2) Norm weighed by a normal distribution

  Select a menu number: 2

----- M A I N   M E N U -----

     1) Help
     2) Enter nonlinear system (plant and output equations)
     3) Select the type of problem to be solved
     4) Simulate and plot results
     5) Quit

Select a menu number: 3

Initial conditions for the plant to be followed by the observer:

_
_ Enter 3x1 vector of initial conditions, x10c:

_
___> [.03 .03 .03]

Simulation will start from initial time t0c = 0.

 Enter final time (*scaled*), tfc: 12

Simulation with external disturbances:

Please define input u1

----- Disturbance input type -----

        1) Zero disturbance on the above input
        2) Impulse (not implemented yet)
        3) Step
        4) Sinusoid
        5) Random noise (not implemented yet)
        6) External data file (not implemented yet)

 Select a menu number: 1

_
_ Enter 3x1 Vector of observer initial conditions, x10o:

_
___> [0 0 0]

To integrate the systems:
1- Please quit MATLAB.
2- Double click on the application "intg_68020" in the folder
3- Restart MATLAB
4- After opening the QC folder, type "qc_continue"
    at which point the program will resume plotting.

Initial Conditions for the plant : [0.03   0.03   0.03]

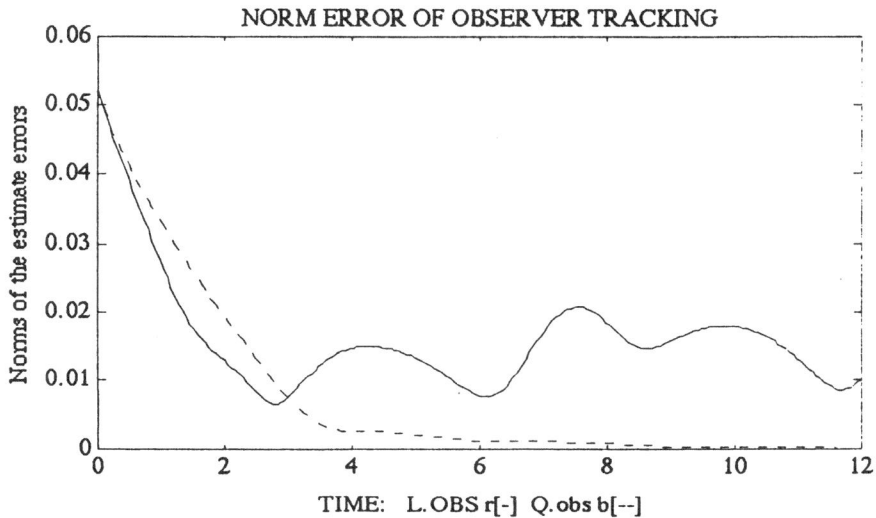

Initial Conditions for the plant : [-0.03 -0.03 -0.03]

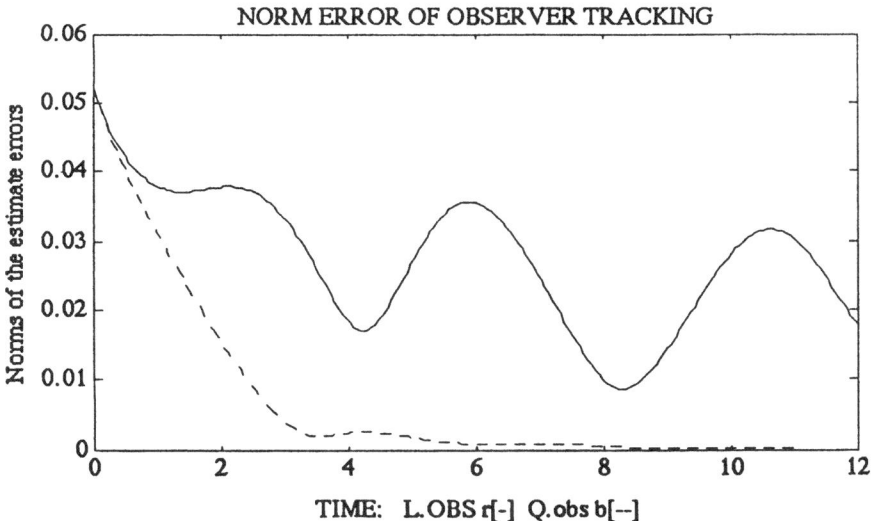

References

[Al]   Al'brekht, E. G., On the optimal stabilization of nonlinear systems. PMM–J. Appl. Math. Mech., 25, 1961, 1254–1266.

[Ar]   Arnold, V. I., Geometrical Methods in the Theory of Ordinary Differential Equations, Springer–Verlag, NY, 1983.

[BZ]   Bestle, D. and M. Zeitz, Canonical form observer design for nonlinear time variable systems, Int. J. Con., 38, 1983, pp. 419–431.

[DM]   Doyle, F. J. III and M. Morari, A conic sector–based methodology for nonlinear controller design, Proceedings of the American Control Conf., 1990, San Diego.

[DS]   Doyle, J. B. and G. Stein, Robustness with observers, IEEE Trans. on Auto. Control, 24, 1979, pp. 607–611.

[Dw]   Dwyer, T. A. W. III, Exact nonlinear control of large angle rotational maneuvers, IEEE Trans. Auto. Control, 29, 1984, pp. 769–774.

[GC]   Garrard, W. L. and L. G. Clark, On the synthesis of suboptimal, inertia wheel control systems, Automatica, 5, 1969, pp. 781–789.

[GJ]   Garrard, W. L. and J. M. Jordan, Design of nonlinear automatic flight control systems, Automatica, 13, 1977, pp. 497–505.

[Gl]   Glad, S. T., Robustness of nonlinear state feedback — a survey, Automatica, 23, 1987, pp. 425–435.

[GH]   Guckenheimer, J. and P. Holmes, Nonlinear Oscillations, Dynamical Systems and Bifurcations of Vector Fields. Springer–Verlag, NY, 1983.

[HS]   Hunt, R. and R. Su, Linear equivalents of nonlinear time varying systems, Proc. Int. Symp. MTNS. Santa Monica, 1981, pp. 119–123.

[JR]   Jakubczyk, B. and W. Respondek, On the linearization of control systems, Bull. Acad. Polon., Sci., Ser. Sci., Math., Astron., Phy., 28, 1980, pp. 517–522.

[Ka]   Karahan, Sinan, Higher degree linear approximations of nonlinear systems, Ph.D. Thesis, University of California, Davis, 1988.

[KI]   Krener, A. J. and A. Isidori, Linearization by output injection and nonlinear observers, Systems and Control Letters, 3, 1983, pp. 47–52.

[KKHF]   Krener, A.J., S. Karahan, M. Hubbard and R. Frezza, Higher order linear approximations to nonlinear control systems, Proc., IEEE CDC, Los Angeles, 1987, pp. 519–523.

[KR]   Krener, A. J. and W. Respondek, Nonlinear observers with linearizable error dynamics, SIAM J. Control Opt., 23, 1985, pp. 197–216.

[Kr] Krener, Arthur J. Nonlinear controller design via approximate normal forms. In Signal Processing, Part II: Control Theory and Its Applications, A. Grunbaum. J. W. Helton, P. Khargonekar, (eds.) Springer Verlag, New York, 1990, pp 139–154.

[Li] Ling, C. K., Quasi–optimum design of an aircraft landing control system, J. Aircraft, 1970, pp. 38–43.

[LWP] Luh, J., B. Walker and R. Paul, Resolved acceleration control of mechanical manipulators, IEEE TAC 25, 1980, pp. 468–474.

[Lu] Lukes, Optimal regulation of nonlinear dynamical systems, SIAM J. Control, 7, 1969, pp. 75–100.

[MC] Meyer, G. and L. Cicolani, A journal structure for advanced flight control systems, NASA TN D–7940, 1975.

[P] Phelps, A., A Simplification of Nonlinear Observer Theory, Ph.D. Dissertation, Department of Mathematics, University of California, Berkeley, 1987.

Program for Symbolic and Rule–Based Analysis and Design of Nonlinear Systems

J. Birk and M. Zeitz

Institut für Systemdynamik und Regelungstechnik, Universität Stuttgart,

Pfaffenwaldring 9, D-W-7000 Stuttgart 80, Germany.

Abstract

With the aid of the symbolic calculating language MACSYMA, a program system has been realized for analyzing and designing the nonlinear system equations $\dot{x} = f(x, u)$, $y = h(x, u)$. The special features of the program are on one hand the examination of controllability and observability of nonlinear systems, and on the other hand the design of nonlinear state feedback systems and state estimators. A rule–based handling of the program helps the user to select and to execute suitable analysis and design procedures. For this purpose, special structures of the nonlinear system equations, e.g. bilinear or canonical forms are recognized by symbolic calculations. Moreover, several heuristic and system theoretic rules are implemented in the MACSYMA program for application of the nonlinear methods. For a numeric evaluation and simulation of the symbolic results, interfaces to the external program systems ACSL and MATLAB are available.

1. Introduction

Recently, numerous analytical or geometrical methods for the analysis and design of non-linear systems have been developed [IS]. The application of these methods requires extensive analytical calculations – even for low order systems – which are not practicable without computer assistance. For this purpose, a symbolic calculating programming language such as MACSYMA or REDUCE can be used. First experiences with the language REDUCE for the controllability analysis of elastic robots and the stabilization of an electrical power system have been presented in [CM1] and [CM2]. In [AB], a MACSYMA program for the nonlinear controller design has been introduced. The computer–aided design of nonlinear observers with the aid of MACSYMA is shown in [BFZ], [PK], [BZ1], and [BZ2].

Methods for the design of nonlinear systems are often limited to certain conditions which are very extensive to examine and for which some experience is required. Hence, a rule-based handling of the program can be taken into consideration. Similar advising or intelligent help functions have also been introduced in [L], [RK], and [TJF]. — In this paper, the symbolic, numeric, and rule–based functionalities of the MACSYMA program MAC-NON (MACsyma program for NONlinear systems) [BZ1], [BZ2] and their realizations are described with emphasis on the nonlinear observation problem.

2. Symbolic Calculating Program System MACNON for Analysis and Design of Nonlinear Systems

The symbolic calculating program system MACNON which has been developed at the 'Institut für Systemdynamik und Regelungstechnik' of the University of Stuttgart enables the analysis and design of nonlinear multiple–input multiple–output systems of the form

$$\dot{x} = f(x, u) \quad t > 0 \; , \quad x(0) = x_0 \; , \tag{1}$$

$$y = h(x, u) \tag{2}$$

where the state x is an n-vector, the input u is a p-vector, and the output y is a q-vector. With the aid of MACNON, nonlinear system equations can be analyzed with respect to controllability, observability, or canonical forms. Moreover, nonlinear state feedback and state estimators can be designed using several methods. In MACNON, linear systems with symbolic parameters are handled as special cases of nonlinear systems.

As represented in figure 1, MACNON consists of calculating functions, input/output functions, user interface functions, a rule base for the selection and execution of appropriate methods for design, and an environment of the symbolic calculating language MACSYMA. The calculating functions include analysis, design, and mathematical basic functions. They are ordered corresponding to the mathematical hierarchy of the solution steps.

The mathematical basic functions are written using the standard functions available in MACSYMA. They build together with other standard MACSYMA functions also the ingredients for the analysis and design functions. Typical examples for basic functions are the calculation of a Jacobian matrix or the realization and application of differential operators or Lie brackets. The required explicit and recursive operations for an analysis and a design of nonlinear systems such as differentiations and matrix inversions can be handled without difficulties by MACSYMA. Opposed to this, implicit calculations, e.g. the solution of nonlinear equations or the determination of the rank of non–constant matrices cause troubles. For such operations, several algorithms with corresponding hints for an application are offered by MACNON so that the user can select an appropriate

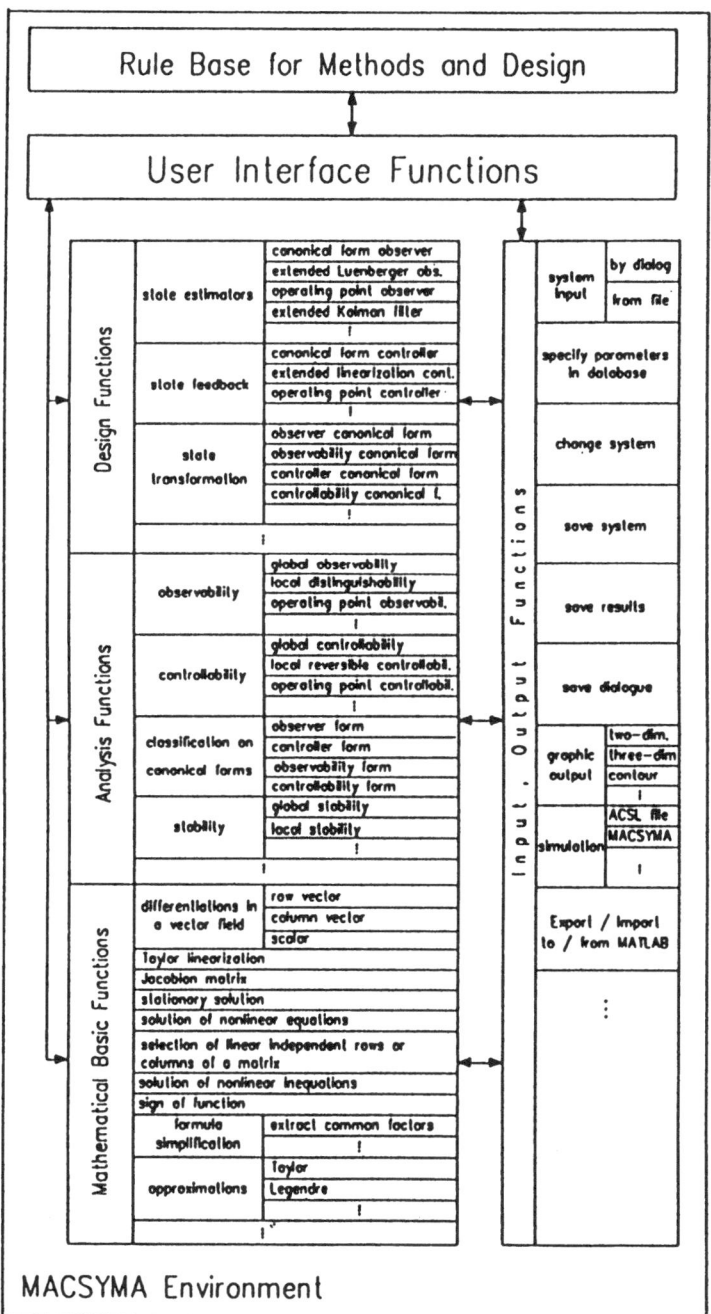

Fig. 1: Program structure of MACNON in the MACSYMA environment with calculating functions, input/output functions, user interface functions, and a rule base for the application of nonlinear methods.

method. In this context, the interactive facilities of the program are very important so that the user can intervene with his knowledge and experience.

In the program, the analysis and design functions are sorted with respect to classes, tasks, and solution methods. For instance, the design functions are separated into the tasks of designing state estimators, state feedback, and transformations into canonical forms. The design methods for state estimators contain, amongst others, procedures for the design of a canonical form observer [KF], an extended Luenberger observer [BZ1], an operating point observer, and an extended Kalman filter. During the investigation of a system, several analysis and design methods can be applied. Within this process, the user has the possibility to simplify or to approximate intermediate and final results. The calculated results are saved automatically in global variables. Therefore, possibly following analysis and design procedures can utilize these intermediate results to avoid repeated time consuming calculations.

The input/output functions which can be called from the user interface as well as from the calculating functions are used to enter, to change, or to save the system equations (1), (2) symbolically. Moreover, the protocol of a session can be drawn up or analytical results can be plotted. Besides, symbolic design results can be evaluated using numerical simulations within the MACSYMA environment or with aid of the simulation language ACSL. In addition, an interface to the control engineering software MATLAB is realized in MACNON.

All parameters belonging to the system, design results, and numerical calculations can be specified in a special database. This database is realized internally by a global list in which the parameter symbols and the corresponding values are sorted in groups. The parameters are hidden so that they are not substituted in MACSYMA expressions. Using special functions, the values in this database can be changed and substituted into symbolic MACSYMA results. Thereby, all analysis and design problems are solved symbolically as far as possible. If needed, the parameters are substituted in order to perform a numerical evaluation, a graphic representation, or a simulation. If desired by the user, the calculations can be continued with symbolic parameters, or the numerical calculations can be repeated with different parameters.

MACNON is handled by mouse–supported menus which are structured hierarchically in analogy to the program structure depicted in figure 1. Moreover, the menu–driven program sequence can be interrupted at every point in order to work interactively with the whole repertory of the MACSYMA functions without losing the results already calculated. On this lowest program level, elementary computations or operations not realized in higher levels can additionally be built up and executed with the aid of available MACSYMA functions. After finishing the interactive MACSYMA calculations, the program sequence can be continued at the interrupt point. In figure 2, the sequence of menus for the design of a nonlinear observer is shown. At every point, the menu structure allows the return into higher level menus or the call of I/O functions and MACSYMA functions. Using the command 'DESCRIBE', short descriptions of the chosen analysis or design functions can be printed out. The descriptions contain informations regarding the method, assumptions, problems, degrees of freedom, and references of the function.

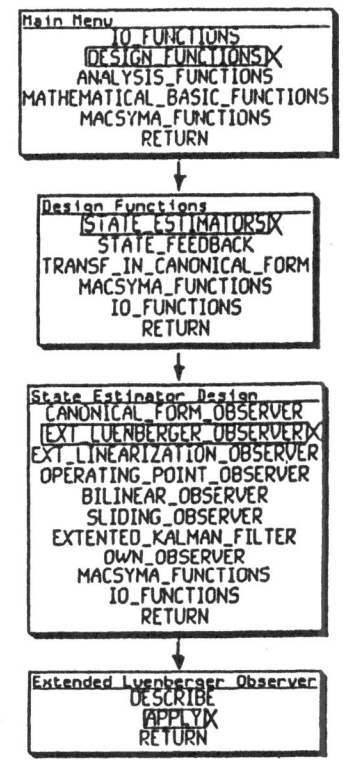

Fig. 2:   Call of an observer method in MACNON via the main menu and the menus for design functions and state estimator functions. The activated menu functions are framed and marked by an X.

In spite of the detailed documentation, the experience made up to now with the computer–aided analysis and design of nonlinear systems show that it is necessary to advise the user in the selection of appropriate methods and to support him/ her in solving analysis and design problems. Such an instruction is necessary since the application of nonlinear methods is characterized by a great deal of assumptions, restrictions, and degrees of freedom which must be proven at several points. An important aspect for the selection of appropriate methods concerns the presence of special system structures whose recognition is shown in the next chapter.

3. Symbolic Recognition and Application of Special Nonlinear System Structures

Nonlinear analysis and design problems can be solved very easily if the nonlinear system equations have a special structure or can be transformed into such a form [Z1], [K]. For the example of solution of the nonlinear observation problem, some particularly appropriate system structures $f^*(x, u)$ and $h^*(x, u)$ are shown in table 1 together with the corresponding characteristic assumptions and properties.

Name	$f^*(x, u)$	$h^*(x, u)$	Properties
Bilinear systems	$Ax + \sum_{i=1}^{p} u_i N_i x + Bu$	$Cx + Du$	bilinear analysis and design methods [DN]
Generalized observer form	$Ax + Bu + a(y, u)$	$h^*(Cx, u), Cx = h^{*-1}(y, u)$ (C, A) observable	input/output injection [Z1]
Sliding observer form	$Ax + Bu + Vw(x, u)$ $\| w(x, u) \| < \rho$	$Cx + Du$ $V^T P = FC$ $P \geq 0, F$ – arbitrary	robust observer design [WZ]
Generalized observability form	$\begin{pmatrix} f_1^*(x_1, \ldots, x_{q+1}, u) \\ f_2^*(x_1, \ldots, x_{q+2}, u) \\ \vdots \\ f_{n-q}^*(x_1, \ldots, x_n, u) \\ \vdots \\ f_n^*(x_1, \ldots, x_n, u) \end{pmatrix}$	$\begin{pmatrix} h_1^*(x_1, u) \\ h_2^*(x_1, x_2, u) \\ \vdots \\ h_q^*(x_1, \ldots, x_q, u) \end{pmatrix}$ $\dfrac{\partial h_j^*}{\partial x_j} \neq 0 \ , \ j = 1(1)q,$ $\dfrac{\partial f_i^*}{\partial x_{q+i}} \neq 0 \ , \ i = 1(1)n - q$	globally observable [Z1]

Table 1: Nonlinear system structures $\dot{x} = f^*(x, u)$, $y = h^*(x, u)$, for which the analysis and design of the observation problem are especially simple.

For the nonlinear control problem, there also exist especially suitable system structures which are structurally controllable and for which a state feedback can be designed very easily [Z1], [K]. — Before applying nonlinear analysis and design methods, it is recommended to examine first if the considered nonlinear system is already given in an appropriate form. Since such special forms – especially in the multiple–input multiple–output case – can not be identified easily by a non–expert, this recognition is carried out automatically by the program MACNON. These methods are also realized symbolically. Using the computer–aided structure recognition, the practically important question can be examined how alternative or additional actuators and sensors effect the system structure and thereby the solution of the analysis and design problems.

Following, the procedures for the symbolic structure recognition and determination are described exemplarily for the system structures of table 1. Corresponding to the structures of $f^*(x,u)$ and $h^*(x,u)$, the analysis of the structure starts with a separation into linear and nonlinear expressions

$$f(x,u) = Ax + Bu + \bar{f}(x,u) \ , \tag{3}$$

$$h(x,u) = Cx + Du + \bar{h}(x,u) \ . \tag{4}$$

For this purpose, the Jacobian matrix $\partial f/\partial x$ is calculated analytically using a mathematical basic function mentioned in chapter 2. The matrix elements which are independent from x and u yield the matrix A. The matrices B, C, D are obtained in an equivalent manner. The nonlinear parts $\bar{f}(x,u)$ and $\bar{h}(x,u)$ are calculated by subtracting the linear expressions from the original nonlinearities $f(x,u)$ and $h(x,u)$.

Canonical forms and their generalizations as well as bilinear systems are characterized by a special dependence on x and u of the new nonlinearities $\bar{f}(x,u)$ and $\bar{h}(x,u)$. Such dependencies can be identified directly from the Jacobian matrices $\partial \bar{f}/\partial x$, $\partial \bar{f}/\partial u$, $\partial \bar{h}/\partial x$, and $\partial \bar{h}/\partial u$. For example, the system (1), (2) is bilinear if the following conditions are satisfied: $\bar{h}(x,u) = 0$, $\partial \bar{f}/\partial x = \partial \bar{f}/\partial x(u) \neq \partial \bar{f}/\partial x(x)$, $\partial \bar{f}/\partial u = \partial \bar{f}/\partial u(x) \neq \partial \bar{f}/\partial u(u)$. The matrices N_i of the bilinearities are obtained by computing the Jacobian matrices $N_i = \frac{\partial}{\partial x} \partial \bar{f}/\partial u_i, \ i = 1(1)p$.

Whereas for bilinear systems, several special analysis and design methods exist, nonlinear normal forms and their generalizations have characteristic properties which are appropriate especially for one analysis or design problem. The first example of table 1 concerns the generalized nonlinear observer form for which an observer design is very easy [Z1]. In this form, the nonlinearities $a(y,u)$ depend only on y and u, hence, they must be measurable. Moreover, the output equation $y = h(Cx, u)$ must be solvable with respect to $Cx = h^{-1}(y,u)$. In this case, a nonlinear observer with an input and output injection $\dot{\hat{x}} = A\hat{x} + Bu + a(y,u) + G[h^{-1}(y,u) - C\hat{x}]$ leads back to the design of the linear error system $\dot{\tilde{x}} = \dot{\hat{x}} - \dot{x} = (A - GC)\tilde{x}$. — For the recognition of the generalized observer form, it is first examined if the output equation (2) is solvable with respect to Cx or to some state coordinates x_i, respectively. Afterwards, these coordinates are substituted in $f(x,u)$ so that a new function $\bar{f}(x,y,u)$ is obtained. Then, the Jacobian matrix $\partial \bar{f}/\partial x$ is examined. The constant elements yield the matrix A, and the remaining elements must be independent from x. If the system is not given in this generalized observer form, it is printed out which state variables x_j are contained in $\bar{f}(x,y,u)$. Thereby, the user is hinted with which additional sensors the system would be in generalized observer form.

If – similarly to the generalized observer form – the function $f(x,u)$ can be separated into a linear kernel $Ax + Bu$ and into a part of not measurable nonlinearities $Vw(x,u)$, a variable structure observer or sliding observer [WZ] of the form $\dot{\hat{x}} = A\hat{x} + Bu + \rho\frac{V^T P(x-\hat{x})}{\|V^T P(x-\hat{x})\|} + G[y - C\hat{x} - Du]$ can be set up with an approximate injection of the nonlinearity. It only has to be supposed that the norm $\| w(x,u) \| < \rho$ of the unknown m-dimensional function $w(x,u)$ which can contain nonlinearities, disturbances, and parameter uncertainties is limited. — The matrices A, B, C in this sliding observer form can again be determined symbolically as described above. Moreover, $\bar{h}(x,u) = 0$ must be fulfilled in (4). The nonlinearity in $\bar{f}(x,u) = Vw(x,u)$ must be splitted into a constant distribution matrix V and a minimal number m of uncertainties or nonlinearities $w(x,u)$. Because of the unknown value of x, the term $V^T P(x - \hat{x})$ contained in the observer can be realized only if it can be written in a form depending only on the output error weighted by a constant matrix F. Taking the output equation into consideration, the so called 'Matching Condition' $V^T P = FC$ follows. This condition means that the distribution matrix V of the nonlinearities and the output matrix C must be connected by an arbitrary (m,q) weighting matrix F and a symmetric and positive definite (n,n)-matrix P. This condition supposes also that the number q of measurements must be greater than or equal to the number m of nonlinearities. For the symbolic evaluation of the 'Matching Condition', the matrix F and the symmetric matrix P are set up generally. The conditions for a positive definite matrix P lead to a system of nonlinear inequalities for whose solution symbolic conversions are also very helpful.

The last canonical form contained in table 1 is the generalized nonlinear observability form which is characterized by a triangular state dependency of the functions $f^*(x,u)$ and $h^*(x,u)$. The observability map for a system in this form can be solved uniquely for the state x. Therefore, a system in this form is structurally globally observable [N], [Z1]. Often, nonlinear systems can be brought into this form by an exchange of indices; this holds especially for multiple-output systems. In order to change the indices for the recognition of this form, a computer assistance is also very helpful. In this process, it is first tried whether the vector of the output equations $h(x,u)$ can be written in the required triangular structure by an exchange of rows and renaming the states $x \to x^*$. It is examined if there exists one row of the output vector $h(x,u)$ which depends uniquely merely on one state variable. In this case, that equation is renamed internally to the first output equation $h_1^*(x_1^*,u)$ and the corresponding state is denoted internally by x_1^*. Afterwards, all other output equations are examined successively if they depend only on one more state variable. Hence, the internal state variables $x_1^*, \ldots, x_q^*$ are fixed. A similar renaming algorithm is used for the recognition of a triangular structure of the system equations $f(x, u)$ and for the determination of the internal state variables $x_{q+1}^*, \ldots, x_n^*$ and the new system vector f^*.

If the considered nonlinear system does not have one of the structures shown in table 1, it might be transformed by a unique nonlinear state transformation $x^* = v(x, u)$, $x = v^{-1}(x^*, u)$ from physical coordinates x into new appropriate coordinates x^* [Z1]. If such a transformation exists it can also be computed symbolically by MACNON. The conditions for the existence and the defining equations of the transformations are often simpler if a system in one canonical form should be transformed into another.

4. Rule Base for the Application of Nonlinear Design Methods

The rule-based advising functions for the application of nonlinear design methods of the program system MACNON shall support the user in the selection of appropriate design methods in dependency of the system equations as well as in the execution of the methods. In figure 3, the rule-based design procedure of nonlinear systems is shown as a flow diagram.

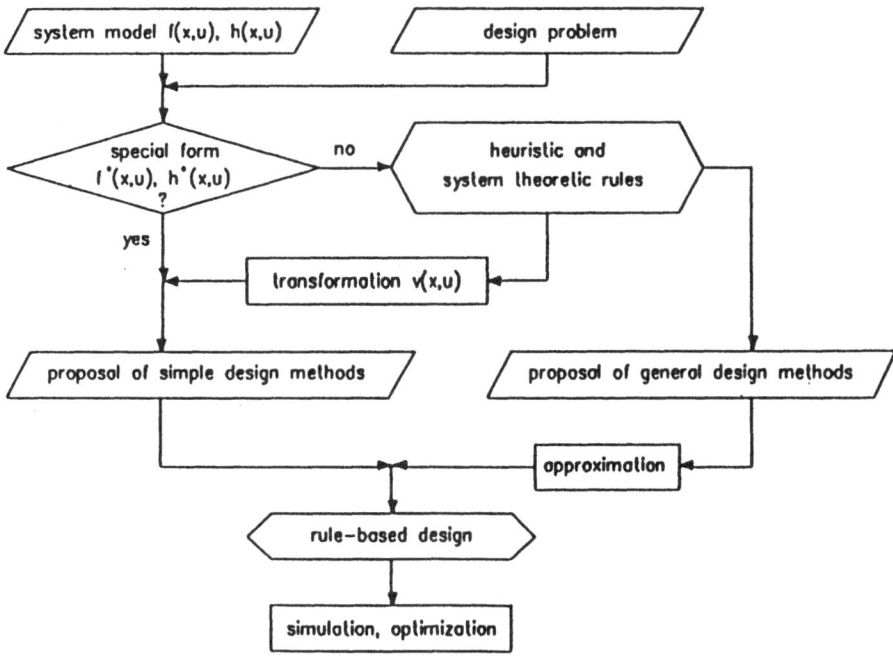

Fig. 3: Rule-based selection and execution of nonlinear design methods.

First, the symbolic examination of the structure of the considered system is carried out as described in chapter 3. If a classification yields an appropriate structure of the system, special design methods are proposed. If the given system does not have a structure suitable for a simple design, several heuristic and system theoretic rules are applied in order to enable a selection of appropriate design methods. The heuristic rules take the number of states, inputs, and outputs, the kind of appearing measurement noise, possible parameter uncertainties, and the expense for realization into account. Moreover, the system equations are examined with respect to possible approximations and linearizations. The quality of the approximations is checked analytically and numerically by estimations of remainders. The possibility for approximations is an important system property since nearly all general nonlinear design methods are based on several approximations. The implemented heuristic rules have been obtained by testing the design methods for selected linear and nonlinear test problems. For this, the design and simulation functions mentioned in chapter 2 have also been proven to be very useful. If the application of heuristic rules does not lead to a contradiction, system theoretic rules are applied, too. These rules include for instance the examination of the transformability into a canonical form. For this, several integrability conditions and compatibility conditions must be proven analytically. If a state transformation into the corresponding nonlinear normal form does exist, the characteristic properties of the normal forms can be utilized for a simple and exact design.

Because of the required analytical calculations for the application of the rules and with regard to a uniform software, the rules are formulated in the symbolic language MAC-SYMA. The rule base of the advising system for design of nonlinear systems consists of a list of all realized methods together with the corresponding rules for their application. Figure 4 shows an extract of this rule base realized in MACSYMA for the selection of appropriate observer design methods for nonlinear systems which do not have a special form of table 1.

The main advantage of this representation lies in the clear description so that the rule base can be extended easily. The assumptions of the methods are contained in sublists following the corresponding name of the method. A simple rule interpreter realized in MACSYMA evaluates the special rule base.

```
[[Canonical_form_observer,  [dimensions_n_q,[2,1],[3,1],[3,2],[4,2]],
                            [measurement_noise, weak, medium],
                            [on_line_not_time_critical],
                            [polynomial_condition_of_observability_form],
                            [integrability_conditions_for_observer_form]],
  [Ext_Luenberger_observer, [measurement_noise, weak, medium],
                            [on_line_not_time_critical],
                            [linearizability_around_reconstructed_trajectory]],
  [Operating_point_observer,[measurement_noise, weak, medium],
                            [linearizability_around_one_stat_solution]],
  [Extended_Kalman_filter,  [measurement_noise, medium, strong],
                            [on_line_not_time_critical],
                            [linearizability_around_reconstructed_trajectory]],
  ...
]
```

Fig. 4: Extract of the rule base for the selection of nonlinear observers as a list of realized methods and corresponding application rules.

By each assumption of a method, a MACSYMA function with the same name is called which must either yield 'TRUE' or a result contained in the corresponding sublist. For instance, for the application of an operating_point_observer, the measurement noise must either be characterized as being weak or medium, and the system must be linearizable around one stationary solution. Those design methods for which all assumptions are fulfilled are proposed to the user in a selection menu. Then, the user has the possibility to execute several designs automatically with aid of the design functions mentioned in chapter 2.

Beside the selection of methods, the application of design methods also requires some empirical knowledge. MACNON contains some heuristic rules for the choice of degrees of freedom. For example, these rules enable proposals for an appropriate utilization of the redundancy of MIMO–systems, the choice of integrating factors for the integration of defining equations of nonlinear state transformations, or dynamic system parameters to be specified. The default values for dynamic parameters are based on an analysis of the time constants of the linearized system model.

At the end of a design process, the results must be checked and optimized using simulation studies. The designed systems can be examined with the aid of the simulation functions mentioned in chapter 2. To compare the design methods, the simulation results can be evaluated with respect to their transient behaviour and their robustness against measurement noise and parameter uncertainties.

5. Experiences with Applications

The program system MACNON is a very important tool for the practical application of nonlinear analysis and design methods. The extensive calculations are done quickly and yield correct results. However, the solutions may contain very large terms. The complexity of the solution depends strongly on the chosen design method as well as on the order and the special structure of the system. For testing the analysis and design methods in MACNON, a catalogue of nonlinear system models of order 2 – 8 from several domains of application is available. In order to give an idea of calculation complexity and required computation time, corresponding results for the analysis and design of the direct current motor model

$$\dot{x} = \begin{pmatrix} -a_1 x_1 + b_2 u_2 \\ -a_{21} x_2 - a_{22} x_1 x_3 + b_1 u_1 \\ a_3 x_1 x_2 \end{pmatrix} \quad , \quad y = \begin{pmatrix} x_1 \\ x_2 \end{pmatrix} \tag{5}$$

are summarized in table 2.

Method	Symbolic calculations	Result	CPU time
Recognition of generalized observability form (table 1)	exchange of indices	globally observable for $a_{22} \neq 0, x_1 \neq 0$	2 s
Recognition of generalized observer form (table 1)	splitting of system into linear and nonlinear part	not in this form, x_3 remains nonlinear	7 s
Recognition of sliding observer form (table 1)	splitting of system into linear and nonlinear part	matching condition can not be satisfied	12 s
Transformation into nonlinear observer canonical form [KF], [PK]	repeated applications of differential operators, determination of rank of non–constant matrix, solution of nonlin. eqns., solution of ode's	consists of 94 MACSYMA–objects	12 s
Canonical form observer [KF], [PK]	transformation into observer canonical form	consists of 183 MACSYMA–objects	120 s
Extended Luenberger observer [BZ1]	repeated applications of differential operators, determination of rank of non–constant matrix	consists of 75 MACSYMA–objects	30 s
Operating point observer	determination of an operating point, Jacobian matrices,	constant observer gain matrix	55 s
Extended Kalman filter	Jacobian matrices matrix operations	consists of 433 MACSYMA–objects	24 s

Table 2: Symbolic calculations required by some important methods for analysis and design of the nonlinear observation problem for example (5). The CPU–times refer to a Symbolics–3620–Workstation which is comparable to a VAX–station 3200.

6. Conclusions

The experiences obtained with the MACSYMA program system MACNON can be summarized as follows:

- Symbolic calculating languages as MACSYMA or REDUCE enable the development of powerful tools for computer–aided application of analytical analysis and design functions for nonlinear systems.

- The interfaces between symbolical and numerical programs such as simulation languages and CACSD–systems are very helpful for the examination of the symbolic design results.

- A rule–based support for the user in selection and execution of the analysis and design procedures is very meaningful since nonlinear methods are characterized by several assumptions and degrees of freedom.

- Symbolic calculating languages are also suitable for the formulation of rules, especially if they are based on the language LISP. This uniform realization has the advantage that the required analytical calculations for the application of the rules are also implemented in the symbolic languages.

Acknowledgement

This research is supported by Deutsche Forschungsgemeinschaft (DFG). The authors would like to acknowledge the engagement of their former students R. Elberskirch and D. Schmelzle who have realized some essential functions of MACNON.

References

[AB]   AKHRIF, O.; BLANKENSHIP, G.L.: Computer algebra for analysis and design of nonlinear control systems. Proceedings American Control Conference, Minneapolis 1987, 547-554.

[BFZ]   BÄR, M., FRITZ, H.; ZEITZ, M.: Rechnergestützter Entwurf nichtlinearer Beobachter mit Hilfe einer symbolverarbeitenden Programmiersprache. Automatisierungstechnik 35 (1987), 177-183.

[BZ1]   BIRK, J.; ZEITZ, M.: Computer-aided design of nonlinear observers. In: *Nonlinear Control Systems Design – Selected Papers from the IFAC-Symposium* Capri/ Italy 1989 (A. Isidori, ed.), Pergamon Press, Oxford, 1990, 1-6.

[BZ2] BIRK, J.; ZEITZ, M.: Anwendung eines symbolverarbeitenden Programmsystems zur Analyse und Synthese von Beobachtern für nichtlineare Systeme. messen-steuern-regeln 12 (1990), 536-543.

[CM1] CESAREO, G.; MARINO, R.: On the controllability properties of elastic robots. In: BALAKRISHNAN, A.V.; THOMA, M. (Eds.): Lecture Notes in Control and Information Sciences 63. Springer-Verlag, Berlin 1984, 352-363.

[CM2] CESAREO, G.; MARINO, R.: The use of symbolic computation for power system stabilization: An example of computer aided design. In: BALAKRISHNAN, A.V.; THOMA, M. (Eds.): Lecture Notes in Control and Information Sciences 63. Springer-Verlag, Berlin 1984, 598-611.

[DN] DERESE, I.; NOLDUS, E.J.: Nonlinear control of bilinear systems. IEE Proceedings Part D 127 (1980), 169-175.

[IS] ISIDORI, A.: Nonlinear Control Systems, 2. Edition. Springer-Verlag Berlin, Heidelberg, New York 1989.

[K] KRENER, A.J.: Normal forms for linear and nonlinear systems. In M. LUKSIC, C. MARTIN, W. SHADWICK (Eds.), Differential Geometry: The Interface between Pure and Applied Mathematics. Contemporary Mathematics 68 (1987), 157-189.

[KF] KELLER, H.,FRITZ, H.: Design of nonlinear observers by a two-step-transformation. In FLIESS, M., HAZEWINKEL (Eds.), Algebraic and Geometric Methods in Nonlinear Control Theory, Reidel, Dordrecht (1986), 89-98.

[L] LUNZE, J.: Wissensbasierte Beratung beim rechnergestützten Entwurf von Automatisierungssystemen. messen-steuern-regeln 32 (1989), 204-209, 258-263.

[N] NIJMEIJER, H.: Observability of a class of nonlinear systems: a geometric approach. Ric. Automatica 12 (1981), 50-68.

[PK] PHELPS, A.R.; KRENER, A.J.: Computation of observer normal form using MACSYMA. In: BYRNES, C.; MARTIN, C.; SEAKS, R. (eds.): Analysis and Control of Nonlinear Systems. North Holland 1988, 475-482.

[RK] RIMVALL, M.; KÜNDIG, M.: Intelligent help for CACE applications.Preprints 11th IFAC–Congress, Tallinn 1990, Vol. 10, 85-90.

[TJF] TAYLOR, J.H.; JAMES, J.R.; FREDERICK, D.K.: Expert–aided control engineering environment for nonlinear systems. Preprints 10th IFAC–Congress, Munich 1987, Vol. 6, 363-368.

[WZ] WALCOTT, B.L.; ZAK, S.H.: State observation of nonlinear uncertain dynamical systems. IEEE Transactions on Automatic Control AC-32 (1987), 166-170.

[Z1] ZEITZ, M.: Canonical forms for nonlinear systems. In: *Nonlinear Control Systems Design – Selected Papers from the IFAC-Symposium* Capri/Italy 1989 (A. Isidori, ed.), Pergamon Press, Oxford, 1990, 33-38.

ALGEBRAIC COMPUTATIONS FOR DESIGN OF NONLINEAR CONTROL SYSTEMS

O. Akhrif
Systems Engineering Department
Case Western Reserve University
Cleveland, Ohio 44106

G. L. Blankenship
Electrical Engineering Department
and Systems Research Center
University of Maryland
College Park, Maryland 20742

December 11, 1990

Abstract

This paper describes computer algebra algorithms for the application of certain differential geometric tools to the analysis and design of nonlinear control systems. Feedback equivalence among nonlinear systems is used to linearize and thereby control certain classes of nonlinear control systems. Left and right invertibility of nonlinear systems is used to solve the output tracking problem. The algorithms include functions that perform basic differential geometric computations, two modules for study and analysis of nonlinear control systems, and two packages for the design of nonlinear controllers for the output tracking problem.

1 Introduction

In this paper, we describe computer algebraic algorithms for the analysis and design of nonlinear control systems based on differential geometric methods. Differential geometry provides a very useful framework for the analysis of nonlinear control systems. It permits the generalization of many constructions for linear control systems to the nonlinear case. However, differential geometric objects (Lie brackets, Lie derivatives, etc.) are not easily manipulated by hand. The algorithms we describe, embodied in a software system called CONDENS, use computer algebra (MACSYMA) as a powerful tool for manipulating differential geometric objects. Our objective is to explain how some of these algorithms are realized as MACSYMA functions.

The programs that form CONDENS fall into three different categories:

1. A set of user-defined functions that perform certain differential geometric computations, including straightforward computations such as *Lie brackets* and *Lie derivatives* and more complex computations such as *Kronecker indices* or *relative order* of nonlinear systems.

2. CONDENS contains more sophisticated modules that use the basic functions to address two theoretical issues in geometric control theory – *feedback equivalence* and *left-invertibility* – for nonlinear systems affine in control, i.e., systems described (in local coordinates) by:

$$\frac{dx}{dt} = f(x) + \sum_{i=1}^{m} u_i(t) g_i(x) \tag{1}$$

3. Finally, CONDENS contains two design packages. Their purpose is to use either feedback linearization or invertibility of nonlinear systems to design a control law that forces the output of a nonlinear system to follow some desired trajectory.

These design packages include the capability to automatically generate Fortran programs for the solution of problems posed in symbolic form or for simulation purposes. Thus, for a new design problem, the time involved in analyzing the analytical structure of the problem and then writing the Fortran code to execute the design algorithm is eliminated. The time required to write, test, and debug the associated Fortran code is eliminated.

The advantages of CONDENS are clear. It can make available to the design engineer design methods that rely on more or less sophisticated mathematical tools. Most importantly, the system allows the engineer to interact with the computer for design at the symbolic manipulation level. In this way, he can modify his analysis or design problem by modifying the symbolic functional form of the model. The Fortran subroutines that he might have to modify by hand to accomplish this in conventional design procedures are written automatically for him.

In the next section, we recall briefly the concepts of state and feedback equivalence among control systems and right and left invertibility of nonlinear systems. We also show how we can use these concepts to design an output tracking controller for the nonlinear system, Eq. 1.

In the third section, we summarize the structure of the basic algorithms that form CONDENS. In the fourth section, we present two case studies from Attitude and Process Control, and show in this way the advantages that such a software system may present.

2 Design Methods Used in CONDENS

2.1 Exact Linearization

Exact linearization is a consequence of the concept of feedback equivalence which generalizes the idea of linear feedback group in linear system theory (Wonham (Ref. 18)) leading, among other things, to the Brunovsky canonical form and the definition of controllability indices. Krener (Ref. 12), Brockett (Ref. 4), Jacubczyck and Respondek (Ref. 9), Hunt et al (Ref. 7) and Su (Ref. 15) introduced the concept of feedback linearization or feedback equivalence for nonlinear systems and gave necessary and sufficient conditions determining those nonlinear systems which are feedback equivalent to linear controllable ones.

The nonlinear system affine in control:

$$\dot{x} = f(x) + \sum_{i=1}^{m} g_i(x)u_i \qquad x \in R^n \tag{2}$$

is said to be *exactly linearizable* in a neighborhood U of the origin ($f(0) = 0$) if it can be transformed via change of coordinates in the state space, change of coordinates in the control space and feedback to a controllable linear system in Brunovsky canonical form:

$$\dot{z} = Az + Bv \qquad z \in R^n \tag{3}$$

with *Kronecker* indices $k_1, \ldots, k_m$.

Hunt, Su and Meyer (Ref. 7) showed that a necessary and sufficient condition for the existence of such transformations is the existence of a mapping T satisfying:

$$\begin{cases} < dT, ad_f^i(g) >= 0, & i = 0, 1, \ldots, n-2 \\ < dT, ad_f^{n-1}(g) \neq 0 \end{cases} \tag{4}$$

where $ad_f^0(g) = g$, $ad_f^1(g) = [f, g]$ and $ad_f^k(g) = [f, ad_f^{k-1}(g)]$, $k > 1$.

Using the well-known theorem of Frobenius (Ref. 1) they concluded that the latter condition is equivalent to

$$\begin{cases} g, ad_f^1(g), \ldots, ad_f^{n-1}(g) \text{ are linearly independent} \\ \text{The set of vector fields } \{g, ad_f^1(g), \ldots, ad_f^{n-2}(g)\} \\ \text{is involutive} \end{cases} \tag{5}$$

Once a transformation T satisfying Eq. 4 is found, the new set of coordinates $(z_1, \ldots, z_n)^T$ is constructed in a straightforward fashion:

$$\begin{aligned} z_1 &= T(x) \\ z_2 &= < dz_1, f > \\ &\vdots \\ z_n &= < dz_{n-1}, f > \\ v &= < dz_n, f > + < dz_n, g > u \end{aligned} \tag{6}$$

In the multi-input case, we obtain m subsystems of dimensions $k_1, k_2, \ldots, k_m$, respectively. The indices $k_1, \ldots, k_m$, being invariant under the transformations considered as shown by Jacubczyck and Respondeck in Ref. 9, are the same as the Kronecker indices of the linearized system, Eq. 3. We then obtain m sets of partial differential equations of the form, Eq. 4. Hunt and Su (Ref. 7) also gave a procedure for computing such a transformation. This procedure

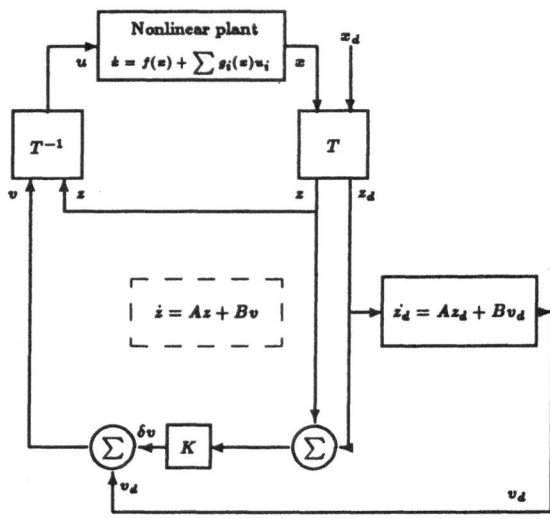

Figure 1: Design Scheme Used by FENOLS.

involves constructing from the set of partial differential equations, Eq. 4, n sets of n ordinary differential equations. This is not always possible or easy to do. However, in some cases–the single input case Brockett (Ref. 4) or systems in block triangular form Meyer (Ref. 14) – the construction of the transformation is simple.

The main application of the exact linearization concept is in the output tracking problem where the goal is to force the output of a given system to follow some desired trajectory. The design proceeds in three steps. First, the given nonlinear system is transformed into a decoupled, controllable linear representation. Second, standard linear design techniques, such as pole placement, are used to design an output tracking control for the equivalent linear system, forcing its output to follow a desired trajectory. Third, the resulting control law is transformed back out into the original coordinates to obtain the control law in terms of the original controls.

Meyer, at NASA Ames Research Center, proposed this scheme in the design of exact model following automatic pilots for vertical and short take-off aircraft and has applied it to several aircraft of increasing complexity (Meyer (Ref. 14)).

This design procedure is performed by FENOLS, one of the design packages contained in CONDENS (see Fig. 1) along with the module TRANSFORM. A description of FENOLS and examples of its functions are given in the third section.

2.2   Inverse Dynamics Method

In this subsection, we briefly review some of the results on left and right invertibility for nonlinear systems that are also implemented in CONDENS.

We consider systems of the following form:

$$
\begin{aligned}
\dot{x}(t) &= f(x(t)) + \sum_{i=1}^{m} g_i(x(t))u_i(t) \qquad x(0) = x_0 \in M \\
y(t) &= h(x(t)) \qquad u \in R^m, y \in R^m
\end{aligned}
\tag{7}
$$

where the state space M is a connected real analytic manifold, f, g are analytic vector fields on M, $h_i, i = 1, \ldots, m$ is a real analytic mapping, and $u \in U$, the class of real analytic functions from $[0, \infty)$ into R^m.

The left-invertibility problem is the problem of injectivity of the input-output map of system, Eq. 7. If the input-output map is invertible from the left, then it is possible to reconstruct uniquely the input acting on the system from knowledge of the corresponding output. The main application of this is in the output tracking problem where we try to control a system so that its output follows some desired path. The inverse system is used to generate the required control given the desired output function.

The problem of right-invertibility is the problem of determining functions $f(t)$ which can be realized as output of the nonlinear system, Eq. 7, driven by a suitable input function. While the left-invertibility is related to the *injectivity* of Input/Output map of the nonlinear system, Eq. 7, the right-invertibility is related to the *surjectivity* of the Input/Output map.

Hirschorn (Ref. 6) gave necessary and sufficient conditions of left and right invertibility of system, Eq. 7, as well as a procedure to construct the left-inverse system. The notion of *relative degree* has a very important role in this construction. For the multivariable case, refer to Akhrif and Blankenship (Ref. 1).

It is easy to see how the notion of left and right invertibility can be applied to the output tracking problem. The application of the inversion algorithm to the nonlinear system (see (Ref. 15) for the algorithm in the multivariable case) gives rise to a left-inverse system which, when driven by appropriate derivatives of the output y, produces $u(\cdot)$ as its output. The question of trajectory following by the output is related to right-invertibility of the nonlinear input-output map, and the ability of the nonlinear system to reproduce the reference path as its output. To obtain robustness in the control system under perturbations, design of a servocompensator around the inner loop using servomechanism theory is suggested.

TRACKS, the second design package contained in CONDENS, uses this procedure along with the module INVERT to design an output tracking controller (see Fig. 2).

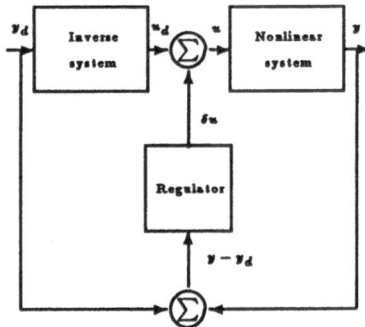

Figure 2: Design Scheme Using TRACKS.

3   Structure of CONDENS

In this section we present the structure of CONDENS as a collection of computer algebra functions. Employing the analytical methods presented above, CONDENS can, given a nonlinear system affine in control, answer questions such as: Is this system feedback-equivalent to a controllable linear one? If so, can we construct the diffeomorphism that makes this equivalence explicit? Is the nonlinear system invertible? What is the relative order of the system? In case the system is invertible, can we construct a left-inverse for it? Given a real analytic function $y(t)$, can $y(t)$ appear as output of the nonlinear system? If so, what is the required control?

CONDENS tries to answer these questions and uses this knowledge to solve a design problem: the output tracking problem for the given nonlinear control system. If we are interested in simulation results, the system can automatically generate Fortran programs for that purpose.

Before describing CONDENS in more detail, we shall make some general remarks relevant to all worked examples in each section.

1. Input lines always begin with a "(C_n)," which is the prompt character of MACSYMA, indicating that the system is waiting for a command.

2. Results of commands terminated by a ";" are printed.

3. Results of commands terminated by a "$" are not printed.

4. If we are working with a single-input, single-output system:

$$\dot{x} = f(x) + g(x)u$$
$$y = h(x)$$

then the vector fields f and g are entered in the form of lists:

$$f \quad : \quad [f_1(x), f_2(x), \ldots, f_n(x)];$$
$$g \quad : \quad [g_1(x), g_2(x), \ldots, g_n(x)];$$
$$h \quad : \quad h(x);$$

5. If we are working with a multivariable control system:

$$\dot{x} \quad = \quad f(x) + \sum_{i=1}^{m} g_i(x) u_i$$
$$y \quad = \quad h(x)$$

$x \in R^n$, $u \in R^m$, $y \in R^m$, then the vector field f is entered in the form:

$$f : [f_1(x), \ldots, f_n(x)];$$

but the m vector fields $g_1, g_2, \ldots, g_m$ are entered in the compact form:

$$g : [g_1, \ldots, g_m]; \quad \text{where each } g_i \text{ is} \quad g_i : [g_{i1}, \ldots, g_{in}];$$

3.1 Package Facilities

CONDENS is formed by a set of large functions which, in turn, are built up of smaller subfunctions. In order to save time, we would like, before using a certain function, to be able to automatically load into MACSYMA only the subfunctions that this specific function uses. This way, we will not have to load the entire package into MACSYMA each time we want to use it.

For this, the **setup-autoload** MACSYMA command is used. All the functions in CONDENS are collected in a special file (``initfile.mac'') with file addresses for all the functions. Once this file is loaded into MACSYMA, a CONDENS function, not previously defined, will be automatically loaded into MACSYMA when it is called. So the first thing the user has to do once in MACSYMA, is to load the login file (load ``initfile.mac'';). He can then start the package using the command ``condens();'' which will give him some hints on how to use CONDENS.

CONDENS also contains some help facilities. The standard help functions in MACSYMA are DISPLAY(``fct-name'') and APROPOS(``fct-name''). We added two *help* functions: MENU and HELP. MENU will display the list of all the functions the package contains. HELP(``fct-name'') returns an information text describing the function, its syntax, how to enter its arguments, and an example.

Here is how to start a session in CONDENS:

```
(c1) load("initfile.mac");
Batching the file initfile.mac
(d1)                          initfile.mac
(c2) condens();
```

```
**********************************************************************
*                                                                    *
*  Hello ! WELCOME to CONDENS :  CONtrol DEsign of Nonlinear Systems  *
*                                                                    *
*  a MACSYMA package for the design of controllers for the output    *
*  tracking problem using differential geometric concepts.           *
*                                                                    *
*      by:           OUASSIMA AKHRIF                                  *
*                                                                    *
*      under the supervision of  Prof. GILMER BLANKENSHIP            *
*                                                                    *
**********************************************************************
```

f,g and h,whenever mentioned, always designate the nonlinear
dynamics of the nonlinear system affine in control:

$$dx/dt = f(x) + g(x) u$$
$$y = h(x)$$

Type MENU(), to get a list of all functions available in CONDENS.

(d2)   done

(c3) menu();

HELP(fun-name): generates an information text describing the function.
JACOB(f):computes the Jacobian matrix of f.
LIE(f,g): computes the Lie brackets of the vector fields f and g.
ADJ(f,g,k): computes the k-th adjoint of f and g.
LIDEV(f,h): computes the Lie derivative of the real valued function h
            along the direction defined by the vector field f.
NLIDEV(f,h,k): computes the k-th Lie derivative of h along f.
BTRIANG(f,g): checks if the nonlinear system is in block triangular form
RELORD(f,g,h): computes the relative order of the scalar control system
            dx/dt=f(x)+g(x)u, y=h(x).
TRANSFORM(f,g): treats the feedback linearization problem,that is checks
            if the system is linearizable and solves for the nonlinear
            transformation.
FENOLS(): module that treats the output tracking problem using feedback
            linearization.
INVERT(f,g,h): finds the left inverse to the original nonlinear system.
TRACKS(): module that treats the output tracking problem using the left-
            inverse system.

(d3)   done

(c4) help(relord);

relord(f,g,h) : computes the relative order of the scalar nonlinear
            system: dx/dt = f(x) + g(x) u
                        y = h(x)
            f is entered in the form   f:[f1(x),....,fn(x)],
            example:(if n=2)   f:[x1,x2**2],
            g is entered in the form   g:[g1(x),.....,gn(x)],
            example:(if n=2)   g:[x2,x1+x2],
            h is entered in the form   h:h(x),
            example: h:x1+x2,

(d4)   done

All functions not explicitly described in this section are already available in the basic MACSYMA system. We recommend that the reader consult the MACSYMA Reference Manual (Ref. 16) for details.

```
jacob(arg1):=block([p,temp1,temp2,temp3],
            if arg1=help then return((
print(" jacob(f) : computes the Jacobian matrix of f. "),
print("         f is entered in the form  f:[f1(x),...,fn(x)],      "),
print("         example:(if n=2)  f:[x1**2,x2],      "),done)),
            p:length(arg1),temp1:[],
            for i1:1 thru p do
            (temp2:makelist(diff(arg1[i1],concat('x,j1)),j1,1,p),
            temp1:endcons(temp2,temp1)),
            temp3:apply(matrix,temp1),
            'df/dx= temp3)$
```

Figure 3: MACSYMA Code for Computation of the Jacobian.

3.2 Differential Geometric Tools in CONDENS

In this subsection, we describe the basic functions available in CONDENS. Since we are using concepts from differential geometry, we have to manipulate vector fields and differential forms. Manipulations with vector fields and forms are systematic and well suited for symbolic computations. The functions described in this section perform some straightforward differential geometric computations such as *Lie brackets* and *Lie derivatives*, and more complex computations such as *Kronecker indices* or *relative degree* of nonlinear systems.

- JACOB(f): computes the *Jacobian* of f, that is, JACOB(f) returns the matrix:

$$\begin{pmatrix} \frac{\partial f_1}{\partial x_1} & \cdots & \frac{\partial f_1}{\partial x_n} \\ \vdots & \cdots & \vdots \\ \frac{\partial f_n}{\partial x_1} & \cdots & \frac{\partial f_n}{\partial x_n} \end{pmatrix}$$

Example: For the following nonlinear system:

$$\begin{pmatrix} \dot{x}_1 \\ \dot{x}_2 \end{pmatrix} = \begin{pmatrix} x_1^2 \\ x_1 x_2 \end{pmatrix} + \begin{pmatrix} 2x_1 \\ 1 \end{pmatrix} \tag{8}$$

$$y = x_1^2 + x_2^2$$

```
(c1) f:[x1**2,x1*x2];
                                2
(d1)                        [x1 , x1 x2]
(c2) jacob(f);
                            df  [ 2 x1  0 ]
(d2)                        -- = [         ]
                            dx  [ x2    x1 ]
```

The code for this function, which illustrates some of the basic features of MACSYMA, is shown in Fig. 3.

- LIE(f,g): computes the *Lie brackets* of the vector fields f and g:

$$[f, g] = \frac{\partial g}{\partial x} f - \frac{\partial f}{\partial x} g$$

- ADJ(f,g,k): computes the kth adjoint of f and g:

$$ad_f^k g = [f, ad_f^{k-1} g]$$
$$ad_f^0 g = g$$

Example: For system, Eq. 8:

```
(c3) g:[2*x1,1];
(d3)                    [2 x1, 1]
(c4) lie(f,g);
                            2
(d4)              [f, g] = [- 2 x1 , - 2 x1 x2 - x1]
(c5) adj(f,g,2);
                            2
(d5)        ad (f, g) = [0, x1  (- 2 x2 - 1) - x1 (- 2 x1 x2 - x1)]
            2
```

- LIDEV(f,h): computes the *Lie derivative* of the real valued function h along the direction defined by the vector field f: $L_f h$.

- NLIDEV(f,h,k): computes the kth *Lie derivative* of h along f:

$$L_f^k h = L_f \left(L_f^{k-1} h \right)$$
$$L_f^0 h = h$$

Example: For system, Eq. 8:

```
(c6) lidev(f,h);
                              2      3
(d6)              lf(h) = 2 x1 x2  + 2 x1
(c7) nlidev(f,h,3);
              2        2      2          2       3        3  2
(d7) lf (h) = x1  (2 x1 (2 x2  + 6 x1 ) + 8 x1 x2  + 12 x1 ) + 12 x1  x2
     3
(c8) ratsimp(%);
                              3  2       5
(d8)              lf (h) = 24 x1  x2 + 24 x1
     3
```

Next, we have three more complicated functions. They use the "sub-functions" LIE, ADJ, etc., to compute quantities such as *Kronecker* indices or *relative degree* of nonlinear systems.

- KRONECK(f,g): used for multivariable systems, where g represents the set of vector fields $g_1, \ldots, g_m$

This function is useful when the nonlinear system is transformable to a linear controllable system in Brunovsky canonical form. It computes the *Kronecker indices* of the linear system equivalent to the original nonlinear system. It returns a set of numbers:

$$k_1 \geq k_2 \geq \cdots \geq k_m$$

$$k_1 + k_2 + \cdots + k_m = n$$

- BTRIANG(f,g): This function checks if the nonlinear system $\dot{x} = f(x) + \sum_{i=1}^{m} g_i(x)u_i$ is in *block triangular form* (see (Meyer (Ref. 14)) for a definition). The argument g of the function represents the m vector fields $g_1, \ldots, g_m$. This function is perhaps the most "complicated" program included in CONDENS. The algorithm is sketched in Fig. 4; the code is in Fig. 5.

- RELORD(f,g,h): computes the *relative order* (see Ref. 6 for definition) of the single-input, single-output nonlinear system:

$$\dot{x} = f(x) + g(x)u$$
$$y = h(x)$$

Example: For system, Eq. 8:

```
(c7) h:x1**2+x2**2;

                          2    2
(d7)                    x2  + x1
(c8) relord(f,g,h);
relative order of system = 1

(d8)                     done
```

3.3 Analysis Tools in CONDENS

Using the above functions, CONDENS contains two independent modules for analysis of nonlinear control systems, TRANSFORM and INVERT. They treat the two problems: feedback linearization and invertibility of nonlinear control systems, respectively.

TRANSFORM(f,g): TRANSFORM treats the feedback linearization problem presented in the subsection on Exact Linearization.

Given the nonlinear control system:

$$\dot{x} = f(x) + \sum_{i=1}^{m} g_i(x)u_i \tag{9}$$

TRANSFORM proceeds in the following manner:

- Using LIE, ADJ and KRONECK, it computes the *Kronecker* indices and checks the three necessary and sufficient conditions (see Hunt, Su, Meyer (Ref. 8)) for feedback linearizability.

- If the system is feedback linearizable, TRANSFORM investigates properties of the nonlinear system which simplify computation of the transformation. For example, TRANSFORM checks to whether the system is in block triangular form (see Meyer (Ref. 14)) (using the function BTRIANG), or if it is scalar and satisfies the more restrictive conditions presented in (Brockett (Ref. 4)).

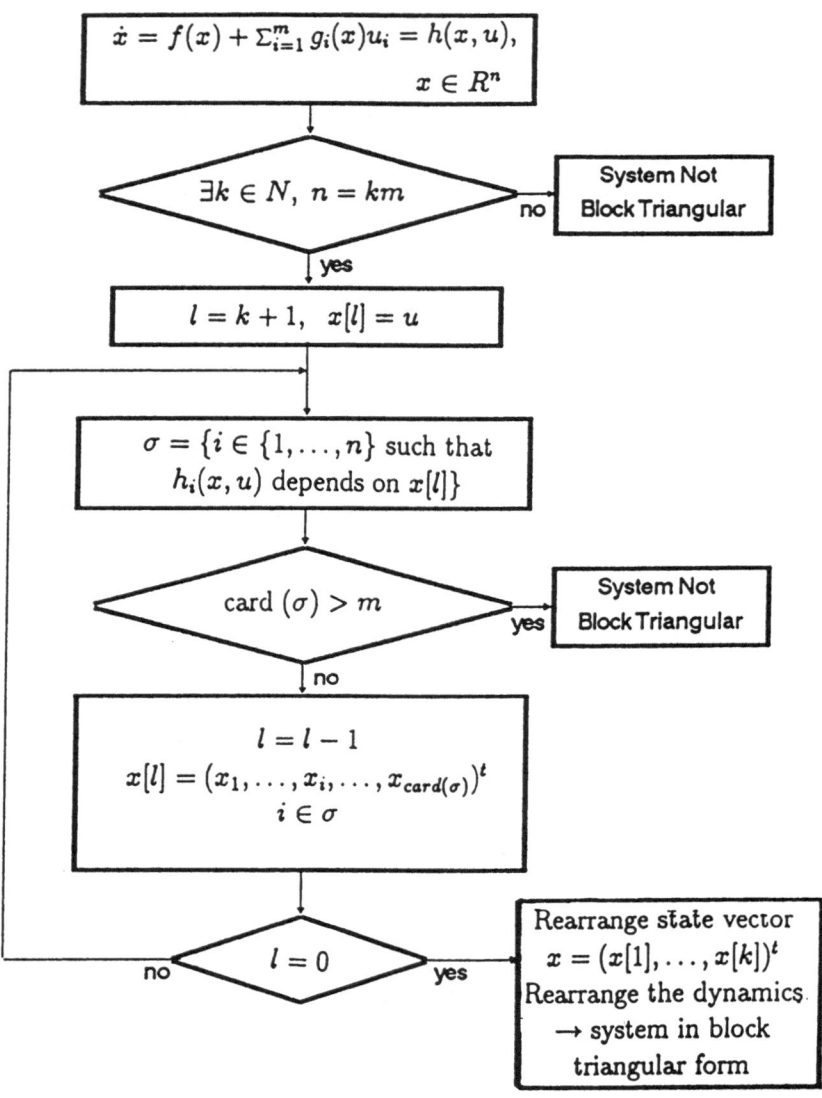

Figure 4: Flow Diagram for btriang(f,g).

```
/*****************************************************************************/
/*     This function checks if the system is in block triangular form        */
/*****************************************************************************/

btriang(arg1,[arg2]):=block([x,xl,u,ind],
                    if arg1=help then return((
print("Btriang(f,g) : checks if the multivarible system dx/dt=f(x)+g(x)u is "),
print("                in block triangular form. "),
print("                f and g are entered in the form f:[f1(x),...,fn(x)],   "),
print("                                                g:[[g1(x)],..[gm(x)]], "),
print("                Example:(if n=2,m=2)            f:[x1,x2**2],           "),
print("                                                g:[[x1,1],[x2,x1]],  ")
,done)),
                    f:arg1,g:part(arg2,1),n:length(f),m:length(g),
                    x:makelist(concat('x,i),i,1,n),
                    ind:makelist(concat('ind,i),i,1,n),
                    ind[1]:0,d:1,p1:1,it:n,
                    u:makelist(concat('u,i),i,1,m),
                    xl:makelist([],i,1,n),xl[p1]:u,
                    for i:1 thru n do
                    equ[i]:f[i]+sum(g[j][i]*u[j],j,1,m),
                    for p:1 thru n/m do(l:0,los:0,
                        for i:n thru 1 step -1 do (
                    for id :1 thru d do
                    if i=ind[id] then (if i>1 then i:i-1
                    else (los:1,id:d)),
                    if los #1 then for r:1 thru length(xl[p]) do
                        if equ[i]#subst(0,xl[p][r],equ[i]) then (
                        l:l+1,cx[it]:x[i],it:it-1,
                        xl[p+1]:endcons(x[i],xl[p+1]),
                        ind[d]:i,if d=n then (i:1,p:n/m),d:d+1,
                        r:length(xl[p]))),
                    if l#m then
                    (print("system is not in block triangular form"),
                    resul:0,p:n/m)
                    else resul:1 ),if resul=1 then
                    print("system in block triangular form"),done)$
```

Figure 5: MACSYMA Code for Checking Block Triangular Form.

- In the more general case, TRANSFORM proceeds to solve for the transformation (change of coordinates and feedback), say T and its inverse T^{-1} by using the MACSYMA function "desolve". This function tries to solve the set of n ordinary differential equations defining the transformation (see Hunt, Su, Meyer (Ref. 8)).[1]

A flow chart setting out the implementation of transform(f,g) is shown in Fig. 6. Referring to the figure, there are five major operations that take place in transform.

BLOCK 1: The function btriang(f,g) is called; if successful, it rearranges the system into the ordered form:

$$
\begin{aligned}
\dot{x}_1 &= f_1(x_1, x_2) \\
\dot{x}_2 &= f_2(x_1, x_2, x_3) \\
&\vdots \\
\dot{x}_k &= f_k(x_1, \dots, x_k) + \sum_{i=1}^{m} g_{ki}(x)u_i
\end{aligned}
\tag{10}
$$

where $k = n/m$, and $x_1, \dots, x_k$ are m-dimensional sub-vectors.

BLOCK 2: In case the system is in block triangular form, the function feed(f,g) is called to construct the transformation T and the feedback pair α, β. feed(f,g) constructs the following transformation:

$$
\begin{aligned}
z_1 &= x_1 \\
z_2 &= f_1(x_1, x_2) \\
z_3 &= \dot{z}_2 = \frac{\partial f_1}{\partial x_1} \cdot f_1(x_1, x_2) \frac{\partial f_1}{\partial x_2} \cdot f_2(x_1, x_2, x_3) \\
&\vdots \\
z_{k-1} &= \dot{z}_{k-2} = F_{k-1}(x_1, x_2, \dots, x_{k-1}) \\
z_k &= \dot{z}_{k-1} = F_k(x_1, x_2, \dots, x_k) \\
v &= \dot{z}_k = \sum_{i=1}^{k} \frac{\partial F_k}{\partial x_i} \dot{x}_i \\
&= \sum_{i=1}^{k} \frac{\partial F_k}{\partial x_i} \cdot f_i(x_1, \dots, x_{i+1}) + \frac{\partial F_k}{\partial x_k} \cdot \sum_{i=1}^{m} g_{ki}(x)u_i \\
v &= \alpha_1(x) + \beta_1(x)u
\end{aligned}
\tag{11}
$$

Finally, feed(f,g) arranges the vector $z = [z_1, \dots, z_k]^T$ so that the system

$$
\dot{z} = Az + Bv
\tag{12}
$$

is in canonical form.

---

[1] If the MACSYMA function "desolve" fails, then TRANSFORM fails. It would be possible to augment this function to solve the ODE's numerically; however, we have not implemented these enhancements.

143

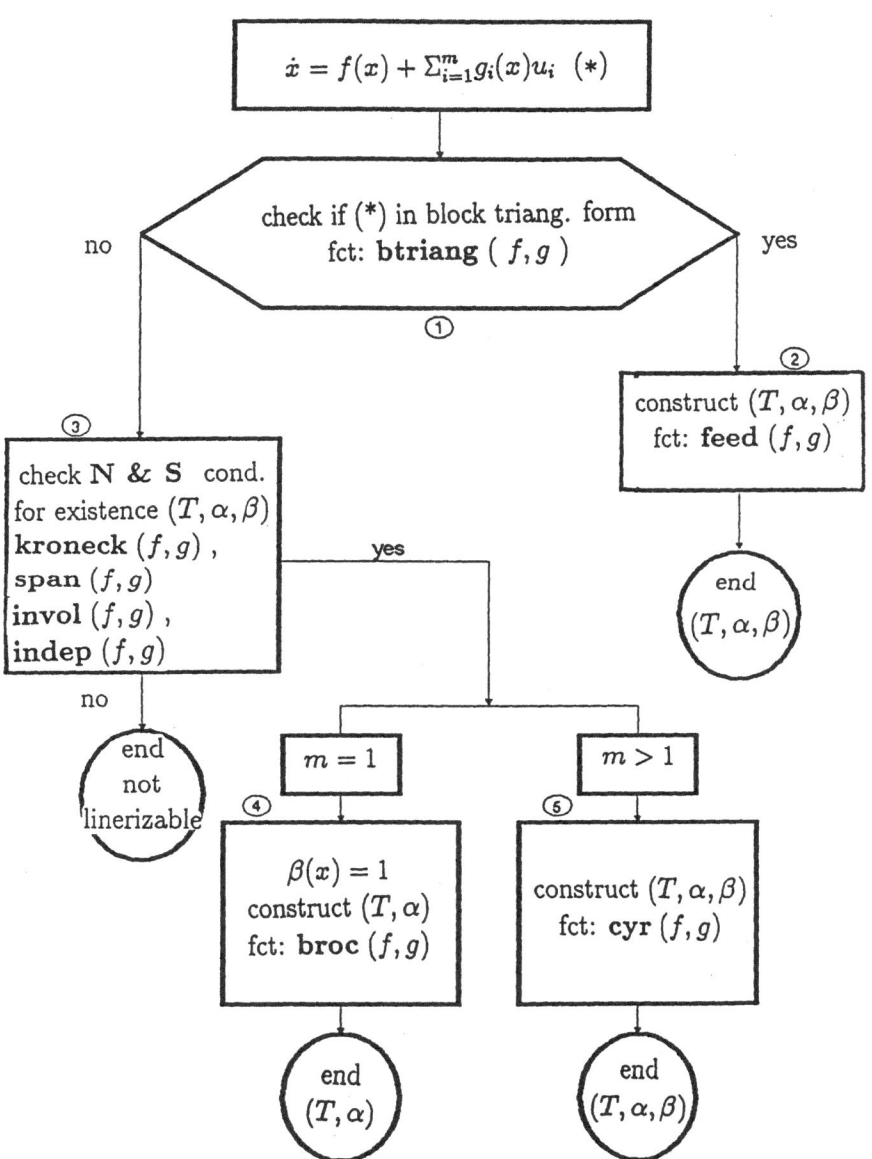

Figure 6: Flow Chart of the Function `transform(f,g)`.

BLOCK 3: This block contains four functions: kroneck(f,g), indep(f,g), invol(f,g), and span(f,g). In the event that the original nonlinear system is not in block triangular form, we use the last three functions to check the necessary and sufficient conditions of tranformability. To apply these functions, it is first necessary to use the function kroneck(f,g) to compute the Kronecker indices of the nonlinear system. This involves a sequence of rank computations:

$$\begin{aligned}
\alpha_0 &= \text{rank}\{g_1, \ldots, g_m\} \\
\alpha_1 &= \text{rank}\{g_1, \ldots, g_m, [f, g_1], \ldots, [f, g_m]\} \\
&\vdots \\
\alpha_{n-1} &= \text{rank}\{g_1, \ldots, g_m, [f, g_1], \ldots, ad_f^{n-1} g\}
\end{aligned} \tag{13}$$

$$r_0 = \alpha_0, r_1 = \alpha_1 - \alpha_0, \ldots, r_{n-1} = \alpha_{n-1} - \alpha_{n-2}$$

$$k_i = \text{ the number of } r_j \geq i$$

The function indep(f,g) forms the set of vectors

$$C = \{g_1, [f, g_1], \ldots, ad_f^{k_1-1} g_1, g_2, \ldots, ad_f^{k_2-1} g_2, \ldots, g_m, \ldots, ad_f^{k_m-1} g_m\} \tag{14}$$

and checks if they are linearly independent.

The function invol(f,g) forms the sets

$$C_j = \{g_1, \ldots, ad_f^{k_j-2} g_1, \ldots, g_m, ad_f^{k_j-2} g_m\} \quad j = 1, \ldots, m \tag{15}$$

and checks to see if each is involutive. This is done by checking the linear dependence of

$$C_j \cup \{[X, Y]\} \quad \forall X, Y \in C_j \tag{16}$$

The function span(f,g) forms the sets $C, C_j, j = 1, \ldots, m$ as above and checks that

$$\text{span } (C_j \cap C) = \text{span } C_j \quad \forall j = 1, \ldots, m. \tag{17}$$

Since $(C_j \cap C) \subset C$, we only have to check

$$\text{rank } (C_j \cap C) = \text{rank } C \quad \forall j = 1, \ldots, m. \tag{18}$$

In each of the functions, kroneck, indep, invol, span, we essentially check either the linear dependence or the linear independence of a set of vectors. The associated code makes heavy use of the MACSYMA function RANK(A) which computes the rank of a matrix A. Since the vectors we are dealing with are actually vector fields, the ranks we compute in this way will not be constant over the manifold M, the state space of the original nonlinear system. However, by smoothness, they will be constant over a dense submanifold $M' \subset M$. Thus, M' is the effective state space of the system. The code for invol is shown in Fig. 7. The other functions are similar. Note the use of the internal MACSYMA function RANK.

```
invol(f,g,w):=block([dep1,dep2,dep3,dep4,l,m1,r1,r2],dep3:[] ,
                    for d:1 thru m do
                    (dep1:[] ,for i5:1 thru m do(
                    dep2:g[i5],dep1:endcons(dep2,dep1),
                    if k[d]>w then for j5:1 thru k[d]-w do
                    (dep2:plie(f,dep2),dep1:endcons(dep2,dep1))),
                    l:apply(matrix,dep1),if (m*(k[d]-1)+1)>n
                    then dep4:[concat(c,d),"involutive"]
                    else for j6:1 thru m*(k[d]-1)-1 do
                    for j7:j6+1 thru m*(k[d]-1) do
                    (r1:makelist(l[j6,t],t,1,n),
                     r2:makelist(l[j7,t],t,1,n),
                    m1:addrow(l,plie(r1,r2)),
                    if rank(m1)<(m*(k[d]-1)+1)
                    then dep4:[concat(c,d),"is involutive"]
                    else(dep4:[concat(c,d),"not invol"],resul:1)),
                    dep3:endcons(dep4,dep3)),dep3)$
```

Figure 7: MACSYMA Code to Test the Involutivity of a Set of Vector Fields.

BLOCK 4: This block constructs the feedback linearization in the single-input case using a function broc(f,g)[2]. This function works as follows: Since the transformation $T : R^n \to R^n$ satisfies

$$
\begin{aligned}
< dT_1, g > &= 0 \\
< dT_1, [f,g] > &= 0 \\
&\vdots \\
< dT_1, ad_f^{m-2}g > &= 0 \\
< dT_1, ad_f^{m-1}g > &= \beta \neq 0
\end{aligned}
\tag{19}
$$

$$
v = < dT_n, f(x) + g(x)u > = \alpha(x) + \beta(x)u.
\tag{20}
$$

If we restrict attention to the case $\beta(x) \equiv 1$, then the PDE's can be rewritten as

$$
\begin{aligned}
\frac{\partial T_1}{\partial x_1}g_1(x) + \ldots \frac{\partial T_1}{\partial x_n}g_n(x) &= 0 \\
&\vdots \\
\frac{\partial T_1}{\partial x_1}ad_f^{m-2}g_1(x) + \ldots \frac{\partial T_1}{\partial x_n}ad_f^{m-2}g_n(x) &= 0 \\
\frac{\partial T_1}{\partial x_1}ad_f^{m-1}g_1(x) + \ldots \frac{\partial T_1}{\partial x_n}ad_f^{m-1}g_n(x) &= 1
\end{aligned}
\tag{21}
$$

and so, we have the vector equation:

$$
\left(\frac{\partial T_1}{\partial x}\right)^T = \left[g, ad_f g, \ldots, ad_f^{m-1}g\right]^{-1} \cdot (0, 0 \ldots, 0, 1)^T
\tag{22}
$$

---

[2]Named for Roger Brockett.

```
broc(f,g):=block([x,v,s,a,b,T,u,eqa],x:makelist(concat('x,i),i,1,n),
                u:[concat('u,1)],
                zr:makelist(concat('z,i),i,1,n),dep:g[1],
                for j:1 thru n do s[1,j]:dep[j],
                if n>=2 then for i:2 thru n do ( dep:plie(f,dep),for j:1
                thru n do
                s[i,j]:dep[j]),a:genmatrix(s,n,n),
                b:a^^-1,for i:1 thru n do T[i]:b[i,n],
                zr[1]:integrate(T[1],x[1],0,x[1]),for i:2 thru n do (
                    for j:1 thru i-1 do T[i]:subst(0,x[j],T[i]),
                    zr[1]:zr[1]+integrate(T[i],x[i],0,x[i])),
                for i:2 thru n do zr[i]:makelist(diff(zr[i-1],x[j]),j,1,n).
                                    transpose(f+g[1]*u[1]),
                vr[1]:makelist(diff(zr[1][n],x[j]),j,1,n).transpose
                (f+g[1]*u[1]),
                print("The new state variables are:"),
                for i:1 thru n do print ("z[",i,"]=",zr[i]),
                print("The new control variables are:"),
                for i:1 thru m do print ("v[",i,"]=",vr[i]),
                eqa:concat('v,1)=vr[1],
                dd2[1]:rhs(solve(eqa,u[1])[1]))$
```

Figure 8: MACSYMA Code for the Function broc.

This expression is "integrated" to find T_1, and the remaining elements $T_2, \ldots, T_n$ are found from this. The code for the function broc is given in Fig. 8. Note the use of the MACSYMA functions INTEGRATE and SOLVE. If these internal functions fail in a specific case, then the design cannot be completed.

BLOCK 5: This block, called cyr(f,g) in our code, constructs the feedback linearizing transform (T, α, β) in the general case by implementing the algorithm of Hunt, Su, and Meyer converting the set of PDE's for the transformation into n-sets of ODE's. This is a complex block, and we shall omit its construction here. (See Ref. 7 for details.) A key element of its implementation is the use of the MACSYMA function DESOLVE to solve the set of ODE's generated in the Hunt–Su–Meyer algorithm. DESOLVE implements several methods for first order ODE's which are tested in the following order: linear, separable, exact (perhaps requiring an integrating factor), homogeneous, Bernoulli's equation, and a generalized homogeneous method. If DESOLVE cannot obtain a solution it returns the value "false." One way to extend the capabilities of the function transform would be to add methods to DESOLVE applicable to the cases encountered in control system design. Alternately, one could easily add a facility which would integrate the ODE's of the Hunt–Su–Meyer algorithm numerically, by "automatically" generating a Fortran program for this purpose, similar to the simulation codes generated by the design packages FENOLS and TRACKS. The function cyr(f,g) also uses the MACSYMA function SOLVE to invert the Hunt–Su–Meyer ODE's to produce the transformation T from the solutions of the ODE's. If the ODE's were solved numerically, this function would have to be implemented in a recursive algorithm.

Example

Check if the following nonlinear system is feedback-linearizable to a controllable linear system and if so, find the F-transformation.

$$\begin{pmatrix} \dot{x}_1 \\ \dot{x}_2 \\ \dot{x}_3 \\ \dot{x}_4 \\ \dot{x}_5 \end{pmatrix} = \begin{pmatrix} \sin(x_2) \\ \sin(x_3) \\ x_4^3 \\ x_5 + x_4^3 - x_1^{10} \\ 0 \end{pmatrix} + \begin{pmatrix} 0 & 0 \\ 0 & 0 \\ 1 & 0 \\ 0 & 0 \\ 0 & 1 \end{pmatrix} \begin{pmatrix} u_1 \\ u_2 \end{pmatrix}$$

```
(c2) load("initfile.mac");
(d2)                      initfile.mac
(c3) f:[sin(x2),sin(x3),x4**3,x5+x4**3-x1**10,0];
                    3       3    10
(d3)        [sin(x2), sin(x3), x4 , x5 + x4  - x1 , 0]
(c4) g:[[0,0,1,0,0],[0,0,0,0,1]];
(d4)        [[0, 0, 1, 0, 0], [0, 0, 0, 0, 1]]
(c5) transform(f,g);

Hello,TRANSFORM tries to solve the problem:
        Given the non linear system:
            dx
            -- = f(x) + u  g (x) + u  g (x)
            dt             2  2      1  1
find a non singular transformation that takes this system to a controllable
linear system:
                dz
                -- = b . v + a . z
                dt
checking if the system is in block triangular form....
system not in block triangular form ==> trying the general method...
the system is multi-input ==> computing first the Kronecker indices of the equf
ivalent controllable linear system
Kronecker indices are:
k[ 1 ]= 3
k[ 2 ]= 2
checking the first condition of transformability.....
checking the second condition of transformability...
checking the third condition of transformability....
span of c1 =span of c1 /C
span of c2 =span of c2 /C
All the conditions are satisfied
trying to construct (if possible) the transformation....
The new state variables are :
z[ 1 ]= x1
z[ 2 ]= sin(x2)
z[ 3 ]= cos(x2) sin(x3)
z[ 4 ]= - x4
                3    10
z[ 5 ]= - x5 - x4  + x1
The new control variables are :
                  3                      2
v[ 1 ]= cos(x2) cos(x3) (x4  + u1) - sin(x2) sin (x3)
            2       3    10          9
v[ 2 ]= - 3 x4  (x5 + x4  - x1  ) + 10 x1  sin(x2) - u2

(d5)                          done
```

INVERT (f,g,h): Given the nonlinear control system:

$$\dot{x} = f(x) + \sum_{i=1}^{m} g_i(x)u_i \qquad (23)$$

$$y = h(x)$$

INVERT tries to check if this system is strongly invertible by investigating the relative order of the system. In case it is invertible (that is, in case the relative order is finite), INVERT returns this relative order; furthermore, it computes the left-inverse to system, Eq. 23, and returns: $A_{inv}(x), B_{inv}(x), C_{inv}(x), D_{inv}(x)$ where:

$$\dot{x} = A_{inv}(x) + B_{inv}(x)\hat{y} \tag{24}$$
$$u = C_{inv}(x) + D_{inv}(x)\hat{y}$$

3.4 Control Design Packages in CONDENS

FENOLS and TRACKS are two design packages contained in CONDENS. Their purpose is essentially to use, respectively, the two modules TRANSFORM and INVERT to design a control law that forces the output of a nonlinear system to follow some desired trajectory. They are both "user-friendly," asking progressively for all the data they need. They are capable of automatically generating, upon request, Fortran codes for simulation purposes.

FENOLS (): This package follows exactly the design scheme of Fig. 1, that is:

- It takes as input the nonlinear dynamics $f, g_1, \ldots, g_m$, the desired path $x_d(t)$ and the desired eigenvalues for the linear regulator.

- It uses the package TRANSFORM to transform the system into a controllable linear one, then FENOLS uses the desired eigenvalues (entered as inputs) to design a linear feedback control (by pole placement). This control solves the output tracking problem for the equivalent linear system.

- If the user is interested in simulation results, all he has to do is answer "yes" to a question posed by FENOLS at the end: "Are you interested in simulation results?".

TRACKS (): This package follows the design scheme in Fig. 2. It uses the module INVERT to generate the output tracking controller. It presents the same interaction facilities as FENOLS; that is, takes as input the nonlinear dynamics $f, g_1, \ldots, g_m, h_1, \ldots, h_m$ and the desired trajectory $y_d(t)$ and generates, upon request, a Fortran program for simulation purposes.

4   Example: Control of Spacecraft Slewing Maneuvers

In this section, we consider the problem of nonlinear control of spacecraft slewing maneuvers with internal momentum transfer as studied by Dwyer (Ref. 12).

Control laws for sequential single-axis spacecraft rotational maneuvers can be designed effectively with linear techniques. The problem becomes harder when the nonlinearity of the equations of rotational motion is taken into account, to satisfy the increasingly stringent requirements of agile spacecraft undergoing fast multiaxial slews of large amplitude. Such maneuvers are usually formulated as nonlinear optimal control problems. Suboptimal torque commands are then either generated by continuation techniques from open-loop laws arising from Pontryagin's maximum principle, or obtained by polynomial truncation of analytic feedback laws arising from Bellman's optimality principle. Such techniques have been found to be sufficiently accurate for sufficiently slow maneuvers. However, being not exact techniques, they cannot guarantee zero terminal error with bounded controls in the absence of disturbances.

Dwyer (Ref. 5) showed that it is possible to apply the exact linearization method to this problem and obtain exact command generation and tracking by standard linear methods through the prior construction of nonlinear coordinate transformations together with nonlinear feedback. The nonlinear equations of motion of a spacecraft controlled by internal reaction wheels are thereby transformed into three decoupled double integrators, driven by acceleration commands in attitude parameter space. Here, we come to the same result without any computations and in a much faster way: by using CONDENS. We also apply the second method (the inverse dynamics) to obtain the left inverse to the spacecraft model and obtain in this way a second controller for path following.

The spacecraft model Dwyer used is the one considered by Vadali and Junkins in Ref. 13. The model consists of a rigid main body equipped with three reaction wheels mounted coaxially with the yaw, pitch and roll axes originating from the spacecraft's center of mass.

The reduced equation of motion can then be written as:

$$\begin{aligned} \dot{\gamma} &= \Gamma(\gamma)\omega \\ \dot{\omega} &= J^{-1}\{h(\gamma) \times \omega - \tau\} \end{aligned} \tag{25}$$

where:

$$\begin{aligned} J &= I^0 - I^A \\ h(\gamma) &= [2\gamma\gamma^T + (\gamma_0^2 - \gamma^T\gamma)I - 2\gamma_0\gamma\times]h^I \\ \Gamma(\gamma) &= 1/2(\gamma_0 I + \gamma\times) \\ \gamma_0 &= \sqrt{1 - \gamma^T\gamma} \end{aligned} \tag{26}$$

$\gamma \in R^3$ attitude coordinate
$\omega \in R^3$ angular velocity
$h(\gamma) \in R^3$ total spacecraft angular momentum
I^0 system inertia matrix
I^A diagonal matrix of axial wheel moments alone
$\tau \in R^3$ control
h^I constant vector, total angular momentum in inertial coordinates
$\times$ cross product

To enter the data of the model, Eq. 25, in CONDENS, we have to use the standard (x, u) notation:

$$\dot{x} = f(x) + \sum_{i=1}^{i=3} g_i(x)u$$

where:

$$
\begin{aligned}
x &= [\gamma, \omega]^T = [x_1, x_2, x_3, x_4, x_5, x_6]^T \\
u &= \tau = [u_1, u_2, u_3]^T \\
f(x) &= [\frac{x_2 x_6}{2} - \frac{x_3 x_5}{2} + \frac{\sqrt{-x_3^2 - x_2^2 - x_1^2 + 1}\, x_4}{2}, -\frac{x_1 x_6}{2} + \frac{\sqrt{-x_3^2 - x_2^2 - x_1^2 + 1}\, x_5}{2} + \frac{x_3 x_4}{2}, \\
&\quad \frac{\sqrt{-x_3^2 - x_2^2 - x_1^2 + 1}\, x_6}{2} + \frac{x_1 x_5}{2} - \frac{x_2 x_4}{2}, 2x_1 x_2 x_6 - 2x_1 x_3 x_5 - 2\sqrt{-x_3^2 - x_2^2 - x_1^2 + 1} \\
&\quad (x_3 x_6 + x_2 x_5), (2x_3^2 + 2x_2^2 - 1)x_6 + (2x_2\sqrt{-x_3^2 - x_2^2 - x_1^2 + 1} + 2x_1 x_3)x_4, \\
&\quad (-2x_3^2 - 2x_2^2 + 1)x_5 + (2x_3\sqrt{-x_3^2 - x_2^2 - x_1^2 + 1} - 2x_1 x_2)x_4] \\
g_1(x) &= [0, 0, 0, 1, 0, 0] \\
g_2(x) &= [0, 0, 0, 0, 1, 0] \\
g_3(x) &= [0, 0, 0, 0, 0, 1]
\end{aligned}
$$

To show that this model is equivalent to three double integrators driven by acceleration commands in attitude parameter space, we run the CONDENS function TRANSFORM to obtain the session:

```
(c32) transform(f,g);

Hello,TRANSFORM tries to solve the problem:


checking if the system is in block triangular form....

system in block triangular form

the transformation is easy to construct

                  *******************************
                  The new state variables are:
                  *******************************
x[ 1 ]= x1

x[ 2 ]= x2

x[ 3 ]= x3

        x2 x6 - x3 x5 + sqrt(- x3  - x2  - x1  + 1) x4
x[ 4 ]= ---------------------------------------------
                              2


                 2     2     2
        - x1 x6 + sqrt(- x3  - x2  - x1  + 1) x5 + x3 x4
x[ 5 ]= ---------------------------------------------
                              2
```

$$z[\ 6\ ]= \frac{\text{sqrt}(-\ x3^2\ -\ x2^2\ -\ x1^2\ +\ 1)\ x6\ +\ x1\ x5\ -\ x2\ x4}{2}$$

```
            ********************************
                The new control variables are:
            ********************************
```

v[1]= (- x1 x6² + sqrt(- x3² - x2² - x1² + 1)

  (4 x1 x2 x6 - 4 x1 x3 x5 + 2 u1) + (4 x1² - 2) x3 x6² - x1 x5² +

  (4 x1² - 2) x2 x5 - x1 x4² + (- 4 x1 x3 - 4 x1 x2²) x4 - 2 u2 x3 + 2 u3 x2)/4

v[2]= - (sqrt(- x3² - x2² - x1² + 1) ....
                  .
                  .
                  .

enter the desired trajectory in the form [xd(1),...xd(n)]
[exp(t),t**2,t,exp(t),t,t];

enter the desired eigenvalues of the linear controller in the form
[delta(1),...,delta(n)] followed by a semi colon
[-2,-2,-2,-2,-2,-2];

```
            *******************************
                The tracking controller is :
            *******************************
```

u(x,xd)[1]= - (sqrt(- x3² - x2² - x1² + 1)

  + (8 x2² + 8 x1 x2 + 8 x1² - 8) x3) x4 + (16 x2 + 16 x1 - 20 %e^t - 16 t²) x3²

  + (((8 t - 8 t²) %e^t + 4) x1 - 16 x1 x2) x3 + 16 x2³ + . . .
                  .
                  .
                  .

Are you interested in simulation results ? (answer y or no)
y;
enter filename of fortran code
file1;
enter initial time you would like the simulation to start from
0.0;
enter final time tf
5.;
enter step size
0.05;
enter initial condition in the form[xo[1],...xo[n]]
[0,0,0,0,0,0];

```
      dimension x( 6 ),dx( 6 ),datad(1000, 6 ),data(1000,
    1 6 ),u( 3 ),y( 6 )
c     set no of equations
      n= 6
      m= 3
c     set initial conditions
      x(1) = 0
      x(2) = 0
            .    ||
            .    ||       Fortran code generated
```

```
        .    \/


    return
    end
```

If we are interested in following some desired attitude trajectory $\gamma = \gamma^*$, we can either apply standard linear methods for tracking to the linear model obtained from TRANSFORM, or apply the second tracking nonlinear method offered by CONDENS, namely, the inverse dynamics method. By running the module TRACKS, the tracking controller obtained is given in this session:

```
(c32) tracks();

Hello, TRACKS tries to solve the problem:
            .   ||
            .   ||      Entering Input data
            .   \/

            *********************************
            relative order of system = 2
            *********************************


    ***************************************************************
    The inverse system to our non linear control system is:
    ***************************************************************

            dx
            -- = binv(x) . yi(t) + ainv(x)
            dt
            u(t) = Cinv(x) + Dinv(x) . yi(t)

                            2    2
                    dy3  d y1  d y2
    where     yi(t) = [---, ----, ----]
                    dt     2    2
                          dt   dt
                                        2      2     2
                    x2 x6 - x3 x5 + sqrt(- x3  - x2  - x1  + 1) x4
    Ainv(x) = matrix([------------------------------------------------],
                                            2
              2       2     2
    - x1 x6 + sqrt(- x3  - x2  - x1  + 1) x5 + x3 x4
    [-----------------------------------------------],
                        2
          2     2     2
    sqrt(- x3  - x2  - x1  + 1) x6 + x1 x5 - x2 x4
    [----------------------------------------------],
                        2
            2      2     2                2     2      2
    [- (sqrt(- x3  - x2  - x1  + 1) (x1 (x6  + x5  + x4 )
        2          2              3          2
    + x3 (4 x2  x6 + 4 x1  x6 - 4 x6) + 4 x3  x6 + x2 (4 x1  x5 - 4 x5)
           2          3               2        2      2
    + 4 x2 x3  x5 + 4 x2  x5) + x2 x3 (- x6  - x5  - x4 )
          2     2     2    3/2              2      2
    + (- x3  - x2  - x1  + 1)   (4 x3 x6 + 4 x2 x5))/(2 x3  + 2 x1  - 2)], [0],
          2              2     2      2                 2      2     2
    [(x1 x2 (- x6  - x5  - x4 ) + sqrt(- x3  - x2  - x1  + 1)
          2     2      2             2          2             3
    (x3 (- x6  - x5  - x4  + 4 x2 x4 + 4 x1  x4 - 4 x4) + 4 x3  x4)
```

$$+ 4 x3^2 (- x3^2 - x2^2 - x1^2 + 1)^{3/2} \quad x4)/(2 x3^2 + 2 x1^2 - 2)])$$

$$Dinv(x) = \begin{bmatrix} -\dfrac{x3\ A + x1\ x2}{B} & -\dfrac{2\ A}{B} & \dfrac{2\ x2}{B} \\[2mm] 1 & 0 & 0 \\[2mm] \dfrac{x1\ A - x2\ x3}{B} & -\dfrac{2\ x2}{B} & -\dfrac{2\ A}{B} \end{bmatrix}$$

where $A = \text{sqrt}(- x3^2 - x2^2 - x1^2 + 1)$

$B = x3^2 + x1^2 - 1$

Binv(x) = matrix([. . .

.
.
.

enter the desired trajectory you want the output of your system to track
yd[1](t)=
t**3;

yd[2](t)=
t**3;

yd[3](t)=
t**3;

Checking if yd(t) is trackable by our system (right-invertibility)

yd(t) can be tracked by the output y(t)

```
        ****************************************************
        The control ud(t) that makes y(t) track yd(t) is:
        ****************************************************
```

$$ud(t)[1] = - (\text{sqrt}(- x3^2 - x2^2 - x1^2 + 1) (x1 (x6^2 + x5^2 + x4^2)$$
$$+ x3 (4 x2^2 x6 + 6 t^2) + 4 x2^2 x5 + 24 t) + x3 (x2 (- x6^2 - x5^2 - x4^2)$$
$$+ 4 x1 x2 x5) + (- x3^2 - x2^2 - x1^2 + 1)^{3/2} (4 x3 x6 + 4 x2 x5)$$
$$+ (- x3^2 - x2^2 - x1^2 + 1) (4 x1 x3 x5 - 4 x1 x2 x6) - 4 x1 x2^2 x6$$
$$+ (6 t^2 x1 - 24 t) x2)/(2 x3^2 + 2 x1^2 - 2)$$

$$ud(t)[2] = - 2 x3^2 x6 - 2 x2^2 x6 + x6$$
$$- 2 x2 \text{sqrt}(- x3^2 - x2^2 - x1^2 + 1) x4 - 2 x1 x3 x4 + 3 t^2$$

$$ud(t)[3] = (\text{sqrt}(- x3^2 - x2^2 - x1^2 + 1)$$
$$(x3 (- x6^2 - x5^2 - x4^2 + 4 x2 x4) + 6 t^2 x1 - 24 t)$$
$$+ x2 (x1 (- x6^2 - x5^2 - x4^2) - 24 t) + (- x3^2 - x2^2 - x1^2 + 1)$$
$$(- 4 x3^2 x5 - 4 x2^2 x5 + 2 x5 - 4 x1 x2 x4) - 4 x2 x3^3 x5 - 4 x2 x5^2$$

```
 + 2 x2  x5 + 4 x3 (- x3  - x2  - x1  + 1)    x4 - 4 x1 x2  x4 - 6 t  x2 x3)
        2        2
/(2 x3  + 2 x1  - 2)
```

Are you interested in simulation results ? (answer y or n)
n;

5 Conclusions

Differential geometric tools have proven to be very useful for analysis of nonlinear systems. Manipulations of vector fields and forms are essentially straightforward; however, doing the computations by hand, in particular on higher-dimensional spaces, is hard since the amount of labor grows rapidly and becomes rather tedious. Moreover, there is every chance that errors will be introduced. By using a computer algebra system like CONDENS, much valuable time can be saved and design methods that rely on rather sophisticated mathematical tools can be made available to a wider class of users.

If a reader is interested in using CONDENS, he or she should contact one of the authors.

Acknowledgments: This research was supported in part by NSF Research Centers Grant CDR-88-003102. The authors would like to thank J. P. Quadrat for introducing them to the potential of MACSYMA.

References

1. Akhrif, O. and Blankenship, G. L., "Using Computer Algebra for Design of Nonlinear Control Systems," Proceedings of the American Control Conference, Minneapolis, MN, 1987, pp. 547-554.

2. Akhrif, O. and Blankenship, G. L., "Computer Algebra Algorithms for Nonlinear Control Systems," Advanced Computing Concepts and Techniques in Control Engineering, Denham, M. and Laub, A. (eds.), NATO ASI Series, Vol. 47, Springer-Verlag, New York, 1988, pp. 53-80.

3. Boothby, W. M., An Introduction to Differentiable Manifolds and Riemannian Geometry, Academic Press, New York, 1975.

4. Brockett, R. W., "Feedback Invariants for Nonlinear Systems," Proceedings of the VIIth IFAC World Congress, Helsinki, 1978, pp. 1115-1120.

5. Dwyer, T. A. W. III, "Exact Nonlinear Control of Spacecraft Slewing Maneuvers with Internal Momentum Transfer," AIAA Journal of Guidance, Control and Dynamics, Vol. 9, No. 2, 1986, pp. 917-953.

6. Hirschorn, R. M., "Invertibility of Nonlinear Control Systems," SIAM Journal on Control, Vol. 17, 1979, pp. 289-297.

7. Hunt, R. L., Su, R. and Meyer, G., "Design for Multi-Input Systems," Differential Geometric Control Theory, Brockett, R., Millman R. and Sussman H. J., (eds.), Birkhauser, Boston, Vol. 27, 1983, pp. 268-298.

8. Hunt, R. L., Su, R. and Meyer, G., "Global Transformations of Nonlinear Systems," IEEE Transactions on Automatic Control, AC-28, No. 1, 1983, pp. 24-31.

9. Jakubczyk, B. and Respondek, W., "On linearization of control systems," Bulletin de L' Academie Polonaise des Sciences, Serie des Sciences Mathematiques, Vol. 28, 1980, pp. 517-522.

10. Jutan, A. and Uppal, A., "Combined Feedforward-Feedback Servo Control Scheme for an Exothermic Batch Reactor," Industrial Engineering and Chemical Processes, Vol. 23, pp. 597-610.

11. Kravaris, C. and Chung, C. B., "Nonlinear State Feedback Synthesis by Global Input/Output Linearization," AICE Journal, Vol 33, No. 4, 1987, pp. 592-603.

12. Krener, A. J., "On the Equivalence of Control Systems and the Linearization of Nonlinear Systems," SIAM Journal on Control, Vol. 11, 1973, pp. 670-676.

13. Meyer, G. and Cicolani, L., "Application of Nonlinear System Inverses to Automatic Flight Control Design-System Concepts and Flight Evaluations," AGARDograph 251, Theory and Applications of Optimal Control in Aerospace Systems, P. Kent (ed.), reprinted by NATO, 1980.

14. Meyer, G., "The Design of Exact Nonlinear Model Followers," Proceedings of the Joint Automatic Control Conference, Charlottsville, VA, 1981, pp. 110-114.

15. Su, R., "On the Linear Equivalents of Nonlinear Systems," Systems and Control Letters, Vol. 2, No. 1, 1982, pp. 48-52.

16. The Matlab Group Laboratory for Computer Science, "MACSYMA Reference Manual," Massachussetts Institute of Technology, Cambridge, MA, 1983.

17. Vadali, S. R. and Junkins, J. L., "Spacecraft Large Angle Rotational Maneuvers with Optimal Momentum Transfer," AIAA Paper 82-1469, 1982.

18. Wonham, W. M., Linear Multivariable Control: A Geometric Approach, Springer-Verlag, New York, 1979.

Joint Use of Maple and Basile:
Using Computer Algebra
in a CACSD Environment

G. Le Vey - J. Masse

Simulog

1, Rue James Joule 78182 St Quentin en Yvelines

Abstract

Basile is a *CACSD* software of the Matlab family, covering the field of automatic control, signal processing and optimization, allowing calculus with polynomial and rational matrices. For real-life, industrial processes, the size and complexity of underlying mathematical models necessitates intensive use of computer algebra. This paper presents two realizations involving Maple (the well-kown symbolic package developped at University of Waterloo, Canada) and Basile. The first application concerns optimal control with *ICSE*, the optimal control toolbox of Basile. In the second, we show an integrated use of Maple and Basile, implemented as an interface between these two packages, allowing data communication, automatic procedure translation and Fortran code generation for simulation and control purposes.

1   Introduction

At the present time, two classes of computer packages exist, which are devoted: the first to numerical purposes such as numerical simulation, control systems design, parameter identification etc.. *Matlab* or *Basile* are examples of this class for high-level computations software in control systems design and linear algebra computations. The second class is constituted of computer algebra packages such as *Macsyma*, *Reduce*, *Maple* e.g. To our knowledge, no real tool exists that combines the powerfulness of these two classes in an integrated environment, although there is a great need for such a tool in control theory. We shall present two examples that illustrate how access can be given to powerful numerical and symbolic capabilities, via *Basile* (for numerical computations) and *Maple* (for symbolic computations). The first example deals with optimal control with Basile, while the second shows a more general view of what has been developed jointly by Simulog and INRIA as a computer interface between those two packages.

2 Using Maple for optimal control problems with Basile

ICSE (*Identification et Contrôle des Systèmes Evolutifs*) is the optimal control tool-box of Basile. It can solve optimal control problems for semi-explicit systems of *DAE* (Differential Algebraic Equations):

$$\begin{cases} 0 = f_i(t, u, y) & \text{for } i \le n_{ea} \\ \frac{dy_j}{dt} = f_i(t, u, y) & \text{for } n_{ea} < j \le n_y \\ y(t_0) = y_0 \end{cases} \qquad (1)$$

taking into account fairly general nonlinear criteria of the following type:

$$J = \phi(y(tf), tf) + \int_{t0}^{tf} g[y(t), u(t), t] dt \qquad (2)$$

where ϕ and g are real functions. In the preceding expressions, u is the control, y the state (dimension n_y), f is a differentiable nonlinear function (components f_i), n_{ea} the number of algebraic equations. When solving practically an optimal control problem with ICSE, a user must supply some jacobian matrices as we briefly recall now.

2.1 The adjoint state method

To solve the optimal control problem, i.e. to find the control u that minimizes the criterion $J(u)$, the usual method (the adjoint state method) consists in computing $J'(u)$ by integration of the adjoint state equation; it is a well-known fact that there exists a vecteur $\lambda(t)$ (adjoint vector) solution of the following system (adjoint system):

$$-\frac{d\lambda}{dt} = (\frac{\partial f}{\partial y})^T . \lambda + (\frac{\partial g}{\partial y})^T \qquad (3)$$
$$\lambda(t_1) = 0$$

we then have for the derivative of J with respect to the control u:

$$J'(u) = \lambda^T \frac{\partial f}{\partial u} + \frac{\partial g}{\partial u} \qquad (4)$$

Knowing $J'(u)$, we can find the optimal control u with an optimization algorithm.

2.2 Solving it with ICSE

By inspection of the system in 4, we see that the necessary derivatives are the following:

$$\begin{cases} f(y(t), u(t), t) \\ \frac{\partial f(y(t), u(t), t)}{\partial y} \\ \frac{\partial f(y(t), u(t), t)}{\partial u} \end{cases} \qquad \begin{cases} g(y(t), u(t), t) \\ \frac{\partial g(y(t), u(t), t)}{\partial y} \\ \frac{\partial g(y(t), u(t), t)}{\partial u} \end{cases}$$

We can easily see that for systems even with few equations and variables (state and control), the formal derivation of these four jacobian matrices will be long and subject to a good lot of errors and verifications, followed by the writing of a simulation code for numerical integration of the adjoint state equation (4). To circumvent this difficulties, we have developped utility functions under *Maple* that derive rapidly and correctly those jacobian matrices and automatically generate Fortran code for the corresponding simulator. This way, when you have to solve an optimal control problem with *Basile*, you can concentrate on the criterion to chose properly, saving debugging time and code writing time. The two following examples show the results obtained with *ICSE* for two classical problems from which we know the exact solution: The dead-beat control (minimum time) of the double integrator and an aerodynamic problem known as first posed by *I.Newton*. The point for these two examples is that they have been treated using Maple for the formal description and Fortran code generation; so, several criteria were tried, with variations on state constraints, taken into account by a penalty method.

2.3   Examples

Minimum time control of the double integrator   This problem may be simply described as follows: the goal is to bring the system from an initial state $x(t_0) = x_0$ to the final state $x(t_1) = 0$ in minimum time. The control u is such that $|u| \leq 1$. The system is described by the following equations:

$$\begin{cases} \ddot{x} = u \\ J = \int_{t0}^{T} dt \\ |u| \leq 1 \\ x(t0) = d, \dot{x}(t0) = v0, x(T) = 0, \dot{x}(T) = 0 \\ T \text{ free} \end{cases} \tag{5}$$

The solution for u is known as the deadbeat control, switching from 1 to -1 depending on initial conditions. The switching curves in the phase space are two parabolas along which the system goes to zero when crossing. Figure 1 gives the time evolution of the control and figure 2 shows the phase plot of this simple system with the following numerical values:

$$\begin{aligned} \dot{x}_0 &= 1m/s \\ x_0 &= 10m \end{aligned}$$

the theoretical and numerical solutions are superposed on the same plot to allow comparison.

Newton's aerodynamic problem   : This problem was first posed and solved by *I.Newton* in his *Principes de la philosophie naturelle* in 1687. He considered the movement of a truncated cone in a rarified medium, which he supposed to be constituted of small fixed balls with mass m and found the best shape giving the minimum resistance in the medium. This problem is exposed in [1] and is still relevant in today aerodynamics. It can be described by the following equations:

$$\begin{cases} \dot{x} = u \\ J = \int_0^T \frac{t}{1+u^2} dt \\ x(0) = 0, x(T) = d, \dot{x}(t0) = 0 \\ u \in R_+ \end{cases} \tag{6}$$

where d is the distance to be covered by the cone and u is the control, that is the speed. The last condition on x constraints the shape to be monotone.

The cost function to be minimized is the resistance of the medium to the penetration by the cone, that is a force; it is nonnegative. The solution (given in [1]) is given by the following system:

$$\begin{cases} x^*(t, p0) = \frac{-p0}{2}(ln(\frac{1}{u}) + u^2 + \frac{3}{4}u^4) + \frac{7}{8}p0 \\ t = \frac{-p0}{2}(\frac{1}{u} + 2u + u^3) \\ p0 < 0 \end{cases} \tag{7}$$

which is a parametric equation for the shape. In this equations, p_0 is the Lagrange multiplier for x. It can be shown ([1]) that the time τ where $x(\tau) = 0$ and $u(\tau) = 1$ is given by $\tau = -2p_0$. We can see from (7) that the curve $x(., p_0)$ may be obtained from $x(., -1)$ by a dilatation with center $(0,0)$ and ratio $|p_0|$. This problem has been treated with *Maple* for the description and code generation and then u has been computed with *ICSE*. The necessary numerical values are:

$$\begin{aligned} d &= 1m \\ t_f &= 1s \end{aligned}$$

$$\tag{8}$$

The results appear in figures 3 (the cost) 4 (the optimal shape of the cone) and 5 (the control or speed), where numerical solutions only have been plotted. The optimal angle for the cone shape gives 135 degrees, the value found by Newton.

3   Interfacing Maple and Basile

In the preceding section, we saw how we could save time when using *Maple* for symbolic computations, before numerical treatment with *Basile* for optimal control purposes. We shall see now that this particuliar use of computer algebra can be generelized to a wide variety of situations appearing in automatic control as well as general mathematical treatments: the idea is to give access simultaneously to symbolic and numerical facilities, namely to *Maple* and *Basile*, in the present situation. Basile already allows some symbolic calculus, but for polynomials and rationals only. Although it covers a fairly large fiels in control theory, it would be pleasant to access to more powerfull functions, as indefinite integration, differential algebra, *Lie* groups and so on, asked for by recent methods in control. A first answer to that problem is given by the recent development, jointly by *INRIA* and *Simulog*([2]), of an interface between *Maple* and *Basile*. At the present time, a good lot of functions is available with that interface: data communication, instructions sending, automatic procedure translation (a procedure in Maple may be translated into a macrofunction of Basile). Furthermore, the availability of *Macrofort*, a *Maple* library, developed at Inria, for Fortran code generation, gives high-level facilities for automatic Fortran code generation. The situation may then be described by the following diagram:

3.1   Practicalities of the interface

The interface allows data communications between *Maple* and *Basile* transparently for the user; it has four main functions: under *Basile*:

tomaple to send data to *Maple*

frmaple to get data from *Maple*

under *Maple*

tobasile to send data to *Basile*

frbasile to get data from *Maple*

those functions can take as arguments simple variables as well as instructions or complicated procedures.

3.2 Examples

To illustrate what can be done with this interface, we present some simple sessions:

Computation of gradient : this is the most common situation. The expression to differentiate is:

$$f(x) = sin(x_1 + x_2) - 3(x_1 x_2 + e^{x_3})/(x_2 + x_3)$$

the following program is first loaded into *Maple*:

```
with(linalg);

f:=sin(x[1]+x[2])-3*(x[1]*x[2]+exp(x[3]))/(x[2]+x[3]);

gf:=grad(f,[x[1],x[2],x[3]]);

tobasile(f,'f',[x]);
tobasile(gf,'grad',[x]);
```

Then we just have to execute under *Basile*:

```
<>frmaple()

<>disp(f)

<out>=f(x)
out=sin(x(1)+x(2))-3*(x(1)*x(2)+exp(x(3)))/(x(2)+x(3)),

<>disp(grad)

<out>=grad(x)
out=<cos(x(1)+x(2))-3*x(2)/(x(2)+x(3)),cos(x(1)+x(2))-3*x(1)/(x(2)+x(3))+3*(x(1\
)
*x(2)+exp(x(3)))/(x(2)+x(3))**2,-3*exp(x(3))/(x(2)+x(3))+3*(x(1)*x(2)+exp(x(3))\
)
/(x(2)+x(3))**2>,
```

Help to curve plotting : If we want to plot a curve implicitly defined by $f(y, t = 0$, we cannot do it with *Basile*. We can then use the implicit functiuns theorem giving the derivative of y with respect to t. This will be done with *Maple*. the differential equation will then be solved with *Basile* and the result will be plotted. In the example, we took the simple function $f(x, y) = y - t^2$. It works as follows:

First, the following program is loaded into *Maple*:

```
f:=y-t^2;
d:=-diff(f,t)/diff(f,y);
tobasile(d,kdot,[t,y]);
```

Then we simply execute the following under *Basile*:

```
<>frmaple()

<>disp(kdot)

<out>=kdot(t,y)
out=2*t,

<>tt=<0,1,2,3,4,5,6,7,8,9,10>;

<>yt=ode(0,0,tt,kdot);

<>plot(tt,yt)
```

4   Conclusion

We have shown in this paper a practical realization of friendly joint use of a computer algebra package, *Maple* and a *CACSD* package, *Basile*. The new functionalities offered by the resulting interface will allow more complex treatments than those given by each of the packages solely. The main advantage of this new environment is time saving in formal description of control system applications and in code generation for numerical computations (simulation and control). The second advantage is a greater safety in application development, connected to time saving in an evident way (less debugging e.g.). Typical applications of such an interface could be for example control of mechanical systems, where complicated nonlinear equations must be differentiated. We believe that the field of application is large, due to the fact that the usual way in systems analysis and control is first formal (model design) and then numerical (simulation and/or control).

References

[1] V. Alexeev, V. Tikhomirov, S. Fomine, *Commande Optimale*, Mir, Moscou, 1982.

[2] C. Gomez, G. Le Vey, C. Rougerie, *Interface Basile-Maple, Version 0.0*, unpublished report, Feb. 1991.

162

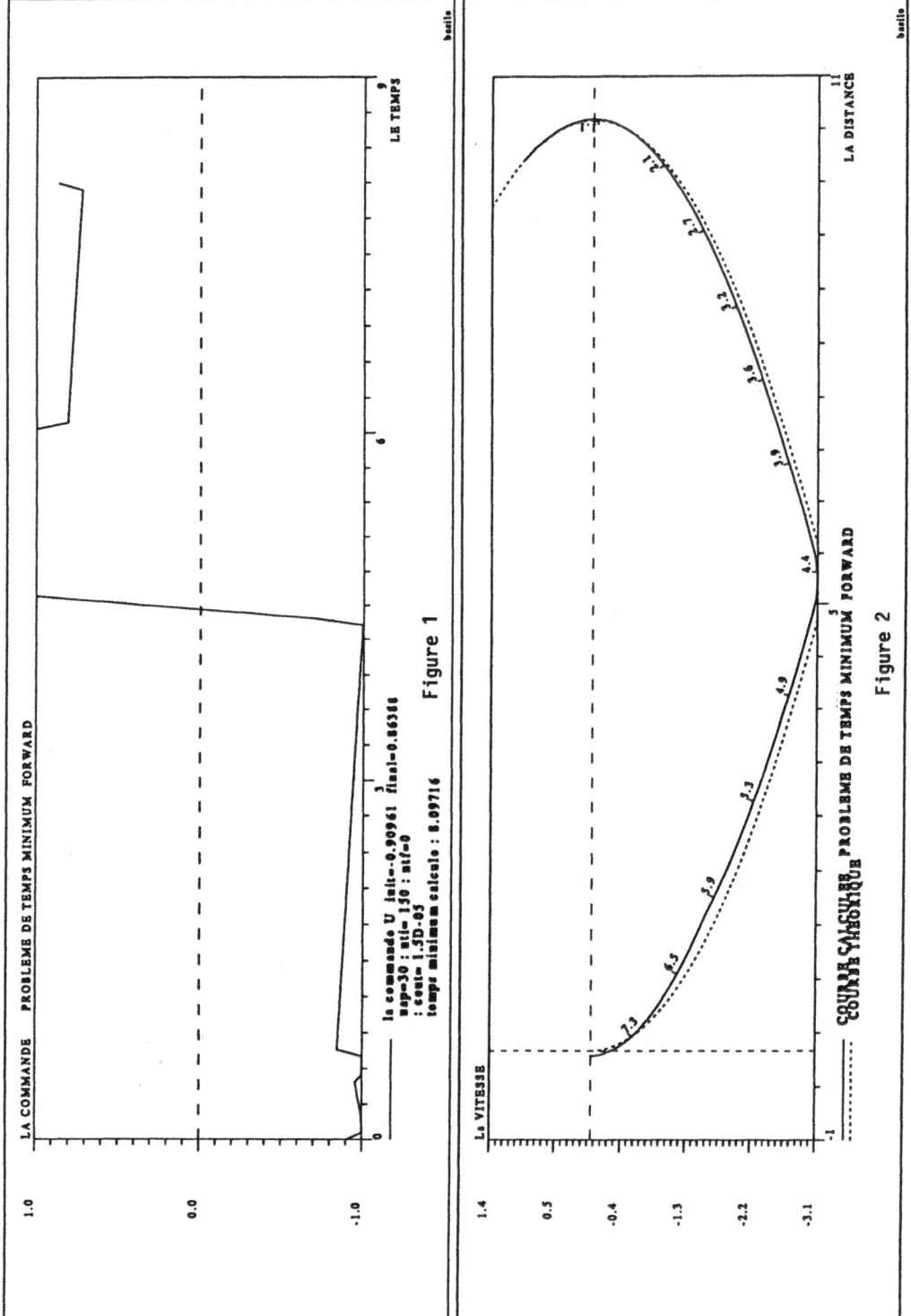

Figure 1

Figure 2

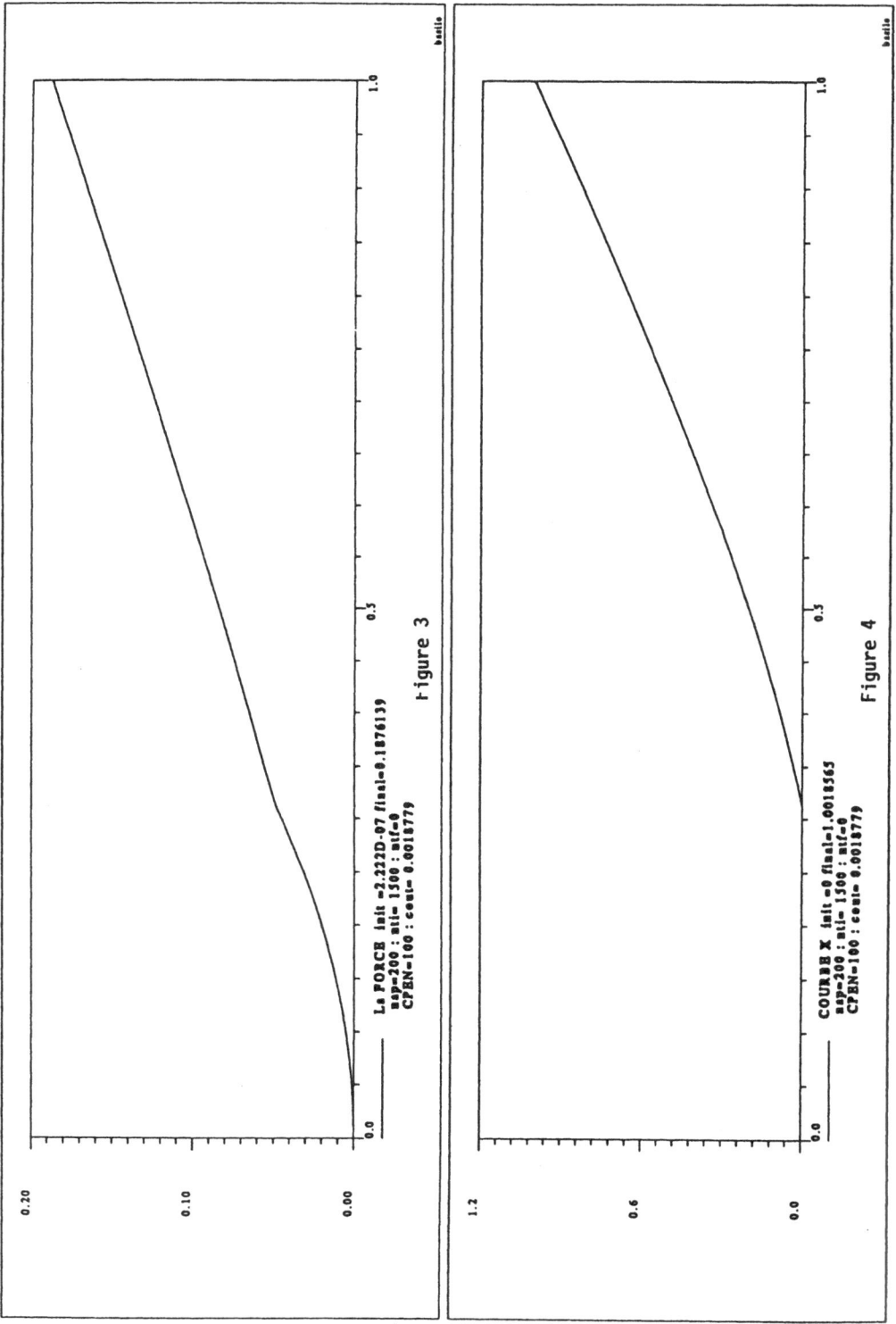

La FORCE init =2.222D-07 final=0.1876139
sap=200 : nti= 1500 : ntf=0
CPEN=100 : cest= 0.0018779

Figure 3

COURBE X init =0 final=1.0018565
sap=200 : nti= 1500 : ntf=0
CPEN=100 : cest= 0.0018779

Figure 4

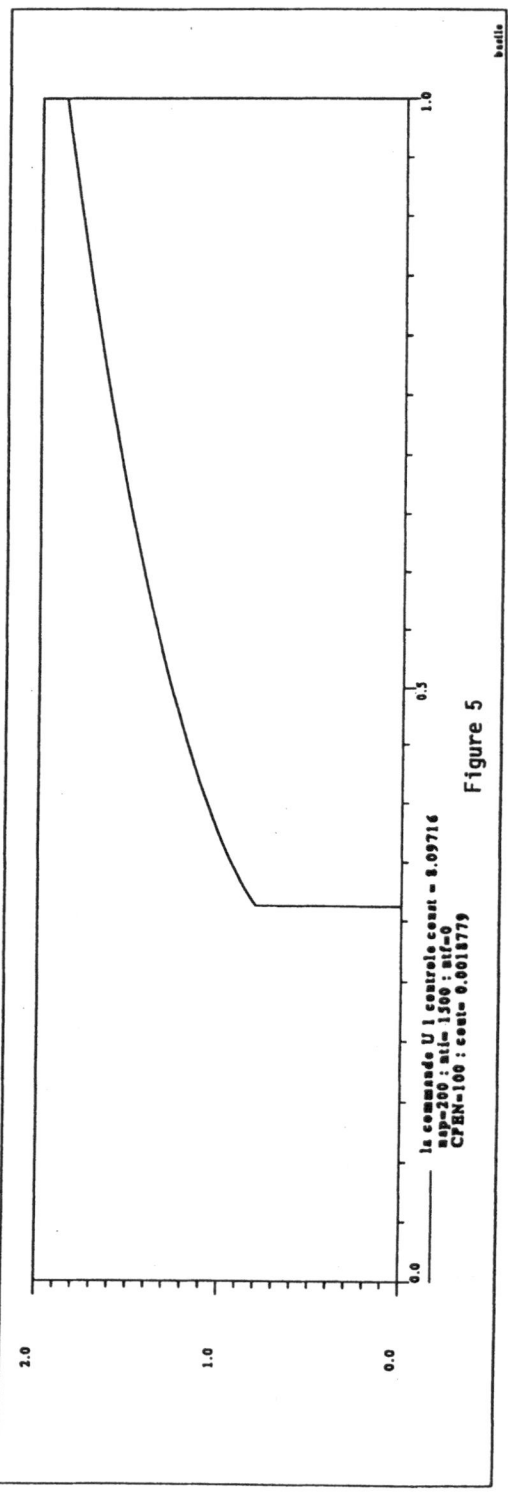

la commande U 1 controle cout = 8.09716
map=200 ; nit= 1500 ; mtf=0
CPBN=100 ; cout= 0.0018779

Figure 5

COMPUTER ALGEBRA FOR NONLINEAR OUTPUT FEEDBACK STABILIZATION

RICCARDO MARINO, PATRIZIO TOMEI

Dipartimento di Ingegneria Elettronica, Seconda Università di Roma
"Tor Vergata", Via O. Raimondo, 00173 Roma, ITALY.

Abstract. We determine via geometric conditions a class of single-input, single- output nonlinear systems which are globally stabilizable by a dynamic observer-based output feedback control. The only restriction on the nonlinearities is their dependence, in suitable coordinates, on the output only. The differential geometric conditions which identify this class of systems can be automatically verified by means of symbolic languages such as MACSYMA. The construction of the stabilizing output feedback control involves an iterative procedure which can also be implemented on computers.

1. INTRODUCTION

We study single-input, single-output nonlinear systems and obtain sufficient conditions for global, output-feedback stabilization, that is for the existence of a dynamic observer-based control such that the origin of the closed loop system is globally asymptotically stable. The sufficient geometric conditions include: the conditions which guarantee the existence of a global observer with linear error dynamics (obtained locally in [1] and globalized according to the results in [2,3]); additional conditions which guarantee that the zero dynamics (see [4], ch. 4) be asymptotically stable and linear in suitable coordinates. The only restriction imposed on the nonlinearities is their dependence, in suitable state coordinates, on the output only: no structural condition, boundedness or growth restriction are required.

The only result so far available for global output feedback stabilization is provided in [5], where necessary and sufficient conditions for local output-feedback linearization, which can be globalized according to [2,3], are obtained (output-feedback linearization means transformation into a linear, observable and controllable system by output feedback and state space change of coordinates). When the state is available for measuraments, the class of systems we are considering can be globally stabilized by state feedback following the results presented in [6].

The differential geometric conditions (i), (ii) and (iii) given in Theorem 2.1 which identify the class of nonlinear systems considered can be checked automatically (see [7,8]) by means of symbolic languages such as MACSYMA. The conditions involve Lie bracket computations which may be difficult already for low dimensional systems. The computation of those conditions is preliminary and needed to determine whether the control scheme proposed in this paper applies to a given system. When the conditions are satisfied, linear partial differential equations are to be solved symbolically in order to compute the state space change of coordinates in which the system is expressed in a simple form. Finally in Section 4 we provide an iterative procedure, which can be deduced from the induction proof of Theorem 4.1, to compute the dynamic observer-based control which can also be implemented on computers by means of symbolic languages.

2. Basic Results and Definitions

We consider nonlinear single-input single-ouput systems

$$\dot{x} = f(x) + g(x)u, \quad x \in R^n, u \in R$$
$$y = h(x), \quad y \in R \tag{Σ}$$

where x is the state vector, u is the control, $h : C^\infty(R^n, R)$, $h(0) = 0$, is the output function, f, g, are smooth vector fields on R^n, with $f(0) = 0$, $g(x) \neq 0$, $\forall x \in R^n$.

Theorem 2.1 [9] The system (Σ) is transformable by a global state space diffeomorphism

$$\zeta = T(x), \quad T(0) = 0 \tag{2.1}$$

into

$$\dot{\zeta} = A_c\zeta + b\sigma(y)u + \psi(y)$$
$$y = C_c\zeta \tag{S}$$

with (A_c, b, C_c) in observer canonical form

$$A_c = \begin{bmatrix} 0 & 1 & 0 & \cdots & 0 \\ 0 & 0 & 1 & \cdots & 0 \\ \vdots & \vdots & \vdots & \ddots & \vdots \\ 0 & 0 & 0 & \cdots & 1 \\ 0 & 0 & 0 & \cdots & 0 \end{bmatrix}, \quad b = \begin{bmatrix} b_1 \\ b_2 \\ \vdots \\ b_{n-1} \\ b_n \end{bmatrix}$$
$$C_c = \begin{bmatrix} 1 & 0 & 0 & \cdots & 0 \end{bmatrix}$$

if, and only if,

(i) $\qquad\qquad [ad_f^i r, ad_f^j r] = 0, \qquad 0 \leq i, j \leq n - 1$

(ii) $\qquad\qquad [g, ad_f^k r] = 0, \qquad 0 \leq k \leq n - 2$

(iii) $\qquad\qquad g = \sigma(y) \sum_{i=1}^{n} b_i(-1)^{n-i} ad_f^{n-i} r$

(iv) $\qquad\qquad$ the vector fields f and r are complete,

where r is the vector field satisfying

$$< \begin{bmatrix} dh \\ \vdots \\ d(L_f^{n-1}h) \end{bmatrix}, r > = \begin{bmatrix} 0 \\ \vdots \\ 1 \end{bmatrix}.$$

Proof. Conditions (i) are shown in [1] to be necessary and sufficient for the existence of a local state space change of coordinates

$$\zeta_i = t_i(x), \quad 1 \le i \le n \tag{2.2}$$

in which $ad_f^i r = (-1)^i \frac{\partial}{\partial \zeta_{n-i}}$, $0 \le i \le n-1$, and the system

$$\begin{aligned} \dot{x} &= f(x) \\ y &= h(x) \end{aligned} \tag{2.3}$$

takes the form

$$\begin{aligned} \dot{\zeta} &= A_c \zeta + \psi(y) \\ y &= C_c \zeta \end{aligned} \tag{2.4}$$

Conditions (i) and (ii) are shown in [10] to be necessary and sufficient for the existence of a local parameter independent state space diffeomorphism (2.2) transforming system (Σ) into

$$\begin{aligned} \dot{\zeta} &= A_c \zeta + \beta(y)u + \psi(y) \\ y &= C_c \zeta \end{aligned} \tag{2.5}$$

Condition (iii) implies that the vector field $\beta(y)$ in (2.5) becomes

$$\beta(y) = \sigma(y) \begin{bmatrix} b_1 \\ b_2 \\ \vdots \\ b_{n-1} \\ b_n \end{bmatrix} \tag{2.6}$$

The proof of the above statement is an easy consequence of the main result in [1]: in the coordinates $\zeta = T(x)$ in which system (2.3) takes the form (2.4) we have

$$ad_f^i r = (-1)^i \frac{\partial}{\partial \zeta_{n-i}},$$

and (iii) becomes in ζ-coordinates

$$g = \sigma(y) \sum_{i=1}^{n} b_i \frac{\partial}{\partial \zeta_i},$$

i.e. (2.6). On the other hand, the necessity of conditions (i), (ii) and (iii), which are coordinate free, is readily verified on system (S). Condition (iv) is necessary and sufficient, according to [2], for the above change of coordinates to be global. When (iv) is violated the change of coordinates is a local one. $\qquad\square$

Let us recall ([4,11]) the definition of relative degree.

Definition 2.1 Let ρ, the relative degree of the output y with respect to the control u for system (Σ) be defined as

$$L_g L_f^i h(x) = 0, \quad 0 \le i \le \rho - 2, \forall x \in R^n$$
$$L_g L_f^{\rho-1} h(x) \ne 0, \quad \text{for some } x \in R^n.$$

$\square$

Definition 2.2 A vector $b = [b_1, \dots, b_n]^T$ is said to be Hurwitz of degree ρ if the associated polynomial

$$b_1 s^{n-1} + b_2 s^{n-2} + \cdots + b_n.$$

is of degree $n - \rho$ ($b_1 = \cdots = b_{\rho-1} = 0, b_\rho \ne 0$) and Hurwitz, i.e. all its zeroes have real part less than zero. $\square$

In what follows we shall make use of the following assumption

(v)   the vector b in (iii) is an Hurwitz vector.

The class of systems (Σ) satisfying conditions (i)-(iv) of Theorem 2.1 and, in addition, assumption (v) have the remarkable property that the zero dynamics (see [4], ch. 4, for a definition) can be expressed, in suitable global coordinates, by a linear asymptotically stable system.

Theorem 2.2 The zero dynamics of system (Σ) of relative degree ρ satisfying conditions (i)-(v), can be expressed in suitable global coordinates as

$$\dot{z} = \begin{bmatrix} -\frac{b_{\rho+1}}{b_\rho} & 1 & 0 & \cdots & 0 \\ -\frac{b_{\rho+2}}{b_\rho} & 0 & 1 & \cdots & 0 \\ \vdots & \vdots & \vdots & \ddots & \vdots \\ -\frac{b_{n-1}}{b_\rho} & 0 & 0 & \cdots & 1 \\ -\frac{b_n}{b_\rho} & 0 & 0 & \cdots & 0 \end{bmatrix} z \tag{2.7}$$

Proof. Conditions (i)-(iv) guarantee by Theorem 2.1 the existence of a global change of coordinates (2.1) in which system (Σ) is expressed as (S). Since the relative degree is ρ, we have $b_1 = \cdots = b_{\rho-1} = 0, b_\rho \ne 0$ and, by assumption (v) the polynomial

$$b_\rho s^\rho + \cdots + b_{n-1} s + b_n \tag{2.8}$$

is Hurwitz. Choosing the the control law

$$u = \frac{1}{b_\rho \sigma(y)}(-\zeta_{\rho+1}) \tag{2.9}$$

the manifold $\{\zeta \in R^n : \zeta_1 = \cdots = \zeta_\rho = 0\}$ is made invariant and the dynamics on it (the zero dynamics according to [4], ch. 4) is given by (2.7) and is asymptotically stable since (2.8) is Hurwitz. $\square$

Definition 2.3 A global output feedback stabilizing controller for system (Σ) is a finite dimensional system

$$\dot{w} = \alpha(w, y(t)), \quad \alpha(0,0) = 0, \ w(0) = w_0, \ w \in R^r,$$
$$u = \beta(w, y(t)), \quad u \in R,$$

such that ($x = 0, w = 0$) is a globally asymptotically stable equilibrium point for the closed loop system. □

3. GLOBAL OUTPUT FEEDBACK STABILIZATION ($\rho = 1$)

If the assumptions of Theorem 2.1 and assumption (v) are satisfied and $\rho = 1$, then b in (iii) is an Hurwitz vector of degree $\rho = 1$. The following Theorem shows how to design a global stabilizing output feedback control for the relative-degree-one case.

Theorem 3.1 Let system (Σ) be of relative degree $\rho = 1$. If conditions (i)-(v) are satisfied, then there exists a global output feedback stabilizing controller.

Proof. Since conditions (i)-(iv) are satisfied, Theorem 2.1 applies and guarantees that system (Σ) can be transformed into (S) with b an Hurwitz vector of degree one according to assumption (v). Since $f(x)$ is assumed to be smooth and $f(0) = 0$ also $\psi(y)$ in (S) is smooth with $\psi(0) = 0$ and, therefore, we can write

$$\psi(y) = \left. \frac{d\psi(y)}{dy} \right|_{y=0} y + R(y) \tag{3.1}$$

where $R(y)$ is such that

$$\lim_{y \to 0} \left\| \frac{R(y)}{y} \right\| = 0$$

Hence, $\psi(y)$ can be written as

$$\psi(y) = y\phi(y) \tag{3.2}$$

where $\phi(y) = \psi(y)/y$. Using (3.2), system (S) becomes

$$\dot{\zeta} = A_c \zeta + b\sigma(y)u + y\phi(y)$$
$$y = C_c \zeta \tag{3.3}$$

Let u be defined as

$$u = -\frac{1}{\sigma(y)} [K_c \hat{\zeta} + ky\phi^T(y)\phi(y)] \tag{3.4}$$

where $\hat{\zeta}$ is provided by the observer

$$\dot{\hat{\zeta}} = A_c \zeta + b\sigma(y)u + \psi(y) + K_o(y - C_c \hat{\zeta}) \tag{3.5}$$

in which K_o is chosen so that all the eigenvalues of the matrix $A_c - K_o C_c$ have negative real part (the pair (A_c, C_c) is observable). The extended dynamics is given by ($\tilde{\zeta} = \zeta - \hat{\zeta}$)

$$\dot{\zeta} = (A_c - bK_c)\zeta + y\phi(y) + bK_c \tilde{\zeta} - bky\phi^T(y)\phi(y)$$
$$\dot{\tilde{\zeta}} = (A_c - K_o C_c)\tilde{\zeta} \tag{3.6}$$

Let the matrix K_c be chosen so that (λ_c is a positive real)

$$\det[sI - (A_c - bK_c)] = (s + \lambda_c)(b_1 s^{n-1} + \cdots + b_n), \tag{3.7}$$

i.e. $-\lambda_c$ and the zeroes of the polynomial $b_1 s^{n-1} + \cdots + b_n$, which is Hurwitz by assumption (v), are the eigenvalues of the matrix $(A_c - bK_c)$. Such a K_c exists since the pair (A_c, b) is controllable (b is Hurwitz and therefore $b_n \neq 0$). Consider the candidate Lyapunov function

$$V(\zeta, \tilde{\zeta}) = \zeta^T P_c \zeta + \eta \tilde{\zeta}^T P_o \tilde{\zeta}$$

where η is a suitable positive real and P_c is the symmetric positive definite solution of

$$(A_c - bK_c)^T P_c + P_c(A_c - bK_c) = -\epsilon I - qq^T$$
$$P_c b = C_c^T \tag{3.8}$$

with ϵ a positive real, which exists since by virtue of (3.7) the triple $(A_c - bK_c, b, C_c)$ satisfies the Meyer-Kalman-Yacubovich Lemma (see [12], p. 66). The matrix P_o is the symmetric positive definite solution of the Lyapunov equation

$$(A_c - K_o C_c)^T P_o + P_o(A_c - K_o C_c) = -I \tag{3.9}$$

The time derivative of V is

$$\dot{V} = -\zeta^T(\epsilon I + qq^T)\zeta + 2\zeta^T P_c \phi(y)y + 2\zeta^T P_c b K_c \tilde{\zeta} - 2ky^2 \phi^T(y)\phi(y) - \eta \tilde{\zeta}^T \tilde{\zeta}$$

and satisfies the inequality

$$\dot{V} \leq - \begin{bmatrix} \|\zeta\| \\ \|\tilde{\zeta}\| \end{bmatrix}^T \begin{bmatrix} \epsilon/2 & \|C_c^T K_c\| \\ \|C_c^T K_c\| & \eta \end{bmatrix} \begin{bmatrix} \|\zeta\| \\ \|\tilde{\zeta}\| \end{bmatrix}$$
$$- \begin{bmatrix} \|\zeta\| \\ |y| \end{bmatrix}^T \begin{bmatrix} \epsilon/2 & \|P_c\|\|\phi(y)\| \\ \|P_c\|\|\phi(y)\| & 2k\|\phi(y)\|^2 \end{bmatrix} \begin{bmatrix} \|\zeta\| \\ |y| \end{bmatrix} \tag{3.10}$$

Hence, if we choose k so that

$$k \geq \frac{\|P_c\|^2}{\epsilon} \tag{3.11}$$

and η sufficiently large, from (3.10) we have

$$\dot{V} \leq -c_L \left\| \begin{bmatrix} \zeta \\ \tilde{\zeta} \end{bmatrix} \right\|^2$$

where c_L is a positive constant. Since V is radially unbounded we have thus proved the global asymptotic stability of the equilibrium point $\zeta = 0, \tilde{\zeta} = 0$. $\qquad \square$

Example 3.1 Consider the system

$$\begin{aligned}
\dot{x}_1 &= x_2 + (e^y - 1) \\
\dot{x}_2 &= x_3 \\
\dot{x}_3 &= u \\
y &= x_1 + 2x_2 + x_3
\end{aligned} \tag{3.12}$$

This system is globally stabilizable by an output feedback control since it satisfies conditions (i)-(v), it has relative degree $\rho = 1$ and therefore Theorem 3.1 applies. Indeed, the linear transformation

$$\zeta_1 = x_1 + 2x_2 + x_3$$
$$\zeta_2 = x_2 + 2x_3$$
$$\zeta_3 = x_3$$

takes system (3.12) into

$$\dot{\zeta}_1 = \zeta_2 + u + (e^y - 1)$$
$$\dot{\zeta}_2 = \zeta_3 + 2u$$
$$\dot{\zeta}_3 = u$$
$$y = \zeta_1$$

which is in the form (S), with b being an Hurwitz vector of degree one. The stabilizing control is given by

$$u = -K_c \hat{\zeta} - k \frac{(e^y - 1)^2}{y}$$

with $\hat{\zeta}$ given by the observer (3.5). □

4. GLOBAL OUTPUT FEEDBACK STABILIZATION ($\rho > 1$)

In this section we generalizes Theorem 3.1 to the case of any relative degree ρ.

Theorem 4.1 Let system (Σ) of relative degree ρ, $2 \leq \rho \leq n$. If assumptions (i)-(v) are satisfied, then there exists a global output feedback stabilizing controller.

Proof. Conditions (i)-(iv) are satisfied and therefore Theorem 2.1 allows us to transform system (Σ) into (S), in which, by virtue of assumption (v), b is an Hurwitz vector of degree ρ. Let us introduce the the filters ($\lambda_i > 0$, $1 \leq i \leq \rho - 1$)

$$
\begin{bmatrix} \dot{\xi}_1 \\ \dot{\xi}_2 \\ \vdots \\ \dot{\xi}_{\rho-1} \end{bmatrix} = \begin{bmatrix} -\lambda_1 & 1 & 0 & \cdots & 0 \\ 0 & -\lambda_2 & 1 & \cdots & 0 \\ \vdots & \vdots & \vdots & \vdots & \vdots \\ 0 & 0 & 0 & \cdots & -\lambda_{\rho-1} \end{bmatrix} \begin{bmatrix} \xi_1 \\ \xi_2 \\ \vdots \\ \xi_{\rho-1} \end{bmatrix} + \begin{bmatrix} 0 \\ 0 \\ \vdots \\ 1 \end{bmatrix} \sigma(y)u
$$

$$\overset{\triangle}{=} \Lambda \xi + b_c \sigma(y)u, \qquad \xi(0) = \xi_0, \tag{4.1}$$

Define the vectors $d_i \in R^n$, $1 \leq i \leq \rho$,

$$d_\rho = b \tag{4.2}$$
$$d_{i-1} = A_c d_i + \lambda_{i-1} d_i, \qquad \rho \geq i \geq 2 \tag{4.3}$$

Note that (4.3) can be written in matrix form as

$$D_1 = A_c D - D\Lambda \tag{4.4}$$

where $D = [d_2, \ldots, d_\rho]^T$ and $D_1 = [d_1, 0, \ldots, 0]^T$. Consider the linear transformation

$$\begin{bmatrix} z \\ \xi \end{bmatrix} = \begin{bmatrix} I & -D \\ 0 & I \end{bmatrix} \begin{bmatrix} \zeta \\ \xi \end{bmatrix} \overset{\triangle}{=} T \begin{bmatrix} \zeta \\ \xi \end{bmatrix} \tag{4.5}$$

The extended system (S), (4.1) can be written in compact form as

$$\begin{bmatrix} \dot{\zeta} \\ \dot{\xi} \end{bmatrix} = \begin{bmatrix} A_c & 0 \\ 0 & \Lambda \end{bmatrix} \begin{bmatrix} \zeta \\ \xi \end{bmatrix} + \begin{bmatrix} b \\ b_c \end{bmatrix} \sigma(y)u + \begin{bmatrix} \psi(y) \\ 0 \end{bmatrix} \tag{4.6}$$

Applying (4.5) to (4.6) and taking into account that by virtue of (4.2) and (4.4),

$$T \begin{bmatrix} A_c & 0 \\ 0 & \Lambda \end{bmatrix} T^{-1} = \begin{bmatrix} A_c & A_cD - D\Lambda \\ 0 & \Lambda \end{bmatrix} = \begin{bmatrix} A_c & D_1 \\ 0 & \Lambda \end{bmatrix}$$

$$T \begin{bmatrix} b \\ b_c \end{bmatrix} = \begin{bmatrix} b - d_\rho \\ b_c \end{bmatrix} = \begin{bmatrix} 0 \\ b_c \end{bmatrix}$$

$$T \begin{bmatrix} \psi(y) \\ 0 \end{bmatrix} = \begin{bmatrix} \psi(y) \\ 0 \end{bmatrix}$$

$$[C_c \quad 0] T^{-1} = [C_c \quad 0]$$

we obtain in new coordinates

$$\begin{aligned} \dot{z} &= A_c z + d_1 \xi_1 + \psi(y) \\ y &= C_c z \\ \dot{\xi} &= \Lambda \xi + b_c \sigma(y) u \end{aligned} \tag{4.7}$$

in which d_1 is now an Hurwitz vector of degree one by construction. Using (3.2), we rewrite the first of (4.7) as

$$\dot{z} = A_c z + d_1 \xi_1 + y\phi(y) \tag{4.8}$$

Let us introduce the observer

$$\dot{\hat{z}} = A_c \hat{z} + d_1 \xi_1 + \psi(y) + K_o(y - C_c \hat{z}) \tag{4.9}$$

with error equation $(\tilde{z} = z - \hat{z})$

$$\dot{\tilde{z}} = (A_c - K_o C_c)\tilde{z} \tag{4.10}$$

Consider the extended system (S_1) consisting of (4.8), (4.10) and

$$\dot{\xi}_1 = -\lambda_1 \xi_1 + \xi_2 \tag{4.11}$$

Perform the change of coordinates

$$\tilde{\xi}_1 = \xi_1 - \xi_1^* \tag{4.12}$$

with

$$\xi_1^* = -K_c \hat{z} - k y \phi^T(y)\phi(y) = \xi_1^*(\hat{z}, y) \tag{4.13}$$

System (4.8), (4.11) becomes

$$\begin{aligned} \dot{z} &= (A_c - d_1 K_c)z - d_1 k y \phi^T(y)\phi(y) + d_1 K_c \tilde{z} + y\phi(y) + d_1 \tilde{\xi}_1 \\ \dot{\tilde{\xi}}_1 &= -\lambda_1 \tilde{\xi}_1 - \dot{\xi}_1^* - \lambda_1 \xi_1^* + \xi_2 \end{aligned} \tag{4.14}$$

Consider the function

$$V_1 = z^T P_c z + \eta_1 \tilde{z}^T P_o \tilde{z} + \tilde{\xi}_1^2 \tag{4.15}$$

with P_c and P_o solutions of (3.8),(3.9) and η_1 a suitable positive real. Recalling the proof of Theorem 3.1, by choosing η_1 sufficiently large and k satisfying (3.11) the time derivative of V_1 satisfies the inequality

$$\dot{V}_1 \le -c_0 \left\| \begin{bmatrix} z \\ \tilde{z} \end{bmatrix} \right\|^2 - \lambda_1 \tilde{\xi}_1^2 + 2\tilde{\xi}_1(y - \dot{\xi}_1^* - \lambda_1 \xi_1^* + \xi_2) \tag{4.16}$$

where c_0 is a positive real. From (4.13), we have

$$\dot{\xi}_1^* = \frac{\partial \xi_1^*}{\partial \hat{z}} \dot{\hat{z}} + \frac{\partial \xi_1^*}{\partial y} \dot{y} \tag{4.17}$$

Noting that (see (4.8))

$$\dot{y} = C_c \dot{z} = C_c A_c z + C_c d_1 \xi_1 + C_c y \phi(y) \tag{4.18}$$

we define

$$\xi_2^* = -y + \lambda_1 \xi_1^* + \frac{\partial \xi_1^*}{\partial \hat{z}} \dot{\hat{z}} + \frac{\partial \xi_1^*}{\partial y} [C_c A_c \hat{z} + C_c d_1 \xi_1 + C_c y \phi(y) - \alpha_1(t)] \tag{4.19}$$

so that, from (4.16), (4.17), (4.19) and setting $\xi_2 = \xi_2^*$, we obtain

$$\dot{V}_1 \le -c_0 \left\| \begin{bmatrix} z \\ \tilde{z} \end{bmatrix} \right\|^2 - \lambda_1 \tilde{\xi}_1^2 - 2\tilde{\xi}_1 \frac{\partial \xi_1^*}{\partial y} (C_c A_c \tilde{z} + \alpha_1(t)) \tag{4.20}$$

Choosing

$$\alpha_1(t) = \tilde{\xi}_1 \frac{\partial \xi_1^*}{\partial y} \tag{4.21}$$

and η_1 in (4.15) sufficiently large, we obtain the inequality

$$\dot{V}_1 \le -c_1 \left\| \begin{bmatrix} z \\ \tilde{z} \\ \tilde{\xi}_1 \end{bmatrix} \right\|^2 \tag{4.22}$$

with c_1 a positive real. If $\rho = 2$ we compute the control variable u by means of $\xi_2^* = \xi_2^*(\hat{z}, y, \xi_1)$ as

$$u = \frac{1}{\sigma(y)} \xi_2^*(\hat{z}, y, \xi_1) \tag{4.23}$$

If $\rho > 2$ the proof proceeds by induction.
Claim. Assume that for a given index $i < \rho$, for the extended system (S_i), consisting of (4.8), (4.10) and

$$\dot{\xi}_1 = -\lambda_1 \xi_1 + \xi_2$$

$$\vdots \tag{4.24}$$

$$\dot{\xi}_i = -\lambda_i \xi_i + \xi_{i+1}$$

there exist a function

$$\xi_{i+1}^* = \xi_{i+1}^*(\hat{z}, y, \xi_1, \ldots, \xi_i) \tag{4.25}$$

and a change of coordinates

$$(z, \tilde{z}, \xi_1, \ldots, \xi_i) \to (z, \tilde{z}, \tilde{\xi}_1, \ldots, \tilde{\xi}_i) \tag{4.26}$$

such that the function

$$V_i = z^T P_c z + \eta_i \tilde{z}^T P_o \tilde{z} + \sum_{j=1}^{i} \tilde{\xi}_j^2 \tag{4.27}$$

with $\xi_{i+1} = \xi_{i+1}^*$ and η_i sufficiently large, has time derivative satisfying the inequality

$$\dot{V}_i \leq -c_i \left\| \begin{bmatrix} z \\ \tilde{z} \\ \tilde{\xi}_1 \\ \vdots \\ \tilde{\xi}_i \end{bmatrix} \right\|^2 \tag{4.28}$$

Then, for the extended system (S_{i+1}) consisting of (4.8), (4.10) and

$$\dot{\xi}_1 = -\lambda_1 \xi_1 + \xi_2$$

$$\vdots$$

$$\dot{\xi}_{i+1} = -\lambda_{i+1} \xi_{i+1} + \xi_{i+2}$$

there exist a function

$$\xi_{i+2}^* = \xi_{i+2}^*(\hat{z}, y, \xi_1, \ldots, \xi_{i+1})$$

and a change of coordinates

$$(z, \tilde{z}, \xi_1, \ldots, \xi_{i+1}) \to (z, \tilde{z}, \tilde{\xi}_1, \ldots, \tilde{\xi}_{i+1})$$

such that the function

$$V_{i+1} = z^T P_c z + \eta_{i+1} \tilde{z}^T P_o \tilde{z} + \sum_{j=1}^{i+1} \tilde{\xi}_j^2 \tag{4.29}$$

with $\xi_{i+2} = \xi_{i+2}^*$ and η_{i+1} sufficiently large, has time derivative satisfying the inequality

$$\dot{V}_{i+1} \leq -c_{i+1} \left\| \begin{bmatrix} z \\ \tilde{z} \\ \tilde{\xi}_1 \\ \vdots \\ \tilde{\xi}_{i+1} \end{bmatrix} \right\|^2 \tag{4.30}$$

where c_{i+1} is a positive real.

Proof of the claim. Consider the extended system (S_{i+1}). Perform the change of coordinates $(z, \tilde{z}, \xi_1, \ldots, \xi_{i+1}) \rightarrow (z, \tilde{z}, \tilde{\xi}_1, \ldots, \tilde{\xi}_{i+1})$ given by (4.26) and

$$\tilde{\xi}_{i+1} = \xi_{i+1} - \xi^*_{i+1}(\hat{z}, y, \xi_1, \ldots, \xi_i) \tag{4.31}$$

By virtue of (4.28), the function V_{i+1} given by (4.29) has time derivative such that

$$\dot{V}_{i+1} \leq -c_i \left\| \begin{bmatrix} z \\ \tilde{z} \\ \tilde{\xi}_1 \\ \vdots \\ \tilde{\xi}_i \end{bmatrix} \right\|^2 + 2\tilde{\xi}_i \tilde{\xi}_{i+1} + 2\tilde{\xi}_{i+1}\dot{\tilde{\xi}}_{i+1} \tag{4.32}$$

Since

$$\dot{\tilde{\xi}}_{i+1} = -\lambda_{i+1}\tilde{\xi}_{i+1} - \lambda_{i+1}\xi^*_{i+1} - \frac{\partial \xi^*_{i+1}}{\partial \hat{z}}\dot{\hat{z}} - \sum_{j=1}^{i} \frac{\partial \xi^*_{i+1}}{\partial \xi_j}\dot{\xi}_j - \frac{\partial \xi^*_{i+1}}{\partial y}\dot{y} + \xi_{i+2} \tag{4.33}$$

and $\dot{y}$ is given by (4.18), we define

$$\xi^*_{i+2} = -\tilde{\xi}_i + \lambda_{i+1}\xi^*_{i+1} + \frac{\partial \xi^*_{i+1}}{\partial \hat{z}}\dot{\hat{z}} + \sum_{j=1}^{i} \frac{\partial \xi^*_{i+1}}{\partial \xi_j}\dot{\xi}_j$$

$$+ \frac{\partial \xi^*_{i+1}}{\partial y}[C_c A_c \hat{z} + C_c d_1 \xi_1 + C_c y \phi(y) - \alpha_{i+1}(t)] \tag{4.34}$$

Choosing

$$\alpha_{i+1}(t) = \tilde{\xi}_{i+1}\frac{\partial \xi^*_{i+1}}{\partial y} \tag{4.35}$$

and η_{i+1} in (4.29) sufficiently large, we obtain with $\xi_{i+2} = \xi^*_{i+2}$ the inequality (4.30), so that the claim is proved. □

We have shown that the assumptions of the claim are satisfied for system (S_1). Applying $(\rho - 2)$-times the claim we can construct a function $\xi^*_\rho(\hat{z}, y, \xi_1, \ldots, \xi_{\rho-1})$ which allows us to define the control

$$u = \frac{1}{\sigma(y)}\xi^*_\rho(\hat{z}, y, \xi_1, \ldots, \xi_{\rho-1})$$

and a change of coordinates $(z, \tilde{z}, \xi_1, \ldots, \xi_{\rho-1}) \rightarrow (z, \tilde{z}, \tilde{\xi}_1, \ldots, \tilde{\xi}_{\rho-1})$, in which $\tilde{\xi}_i = \xi_i - \xi^*_i(\hat{z}, y, \xi_1, \ldots, \xi_{i-1})$, such that the radially unbounded function

$$V_{\rho-1} = z^T P_c z + \eta_{\rho-1}\tilde{z}^T P_o \tilde{z} + \sum_{j=1}^{\rho-1} \tilde{\xi}_j^2$$

is for a sufficiently large $\eta_{\rho-1}$ a Lyapunov function for the closed loop system with

$$\dot{V}_{\rho-1} \leq -c_{\rho-1} \left\| \left[\begin{array}{c} z \\ \tilde{z} \\ \tilde{\xi}_1 \\ \vdots \\ \tilde{\xi}_{\rho-1} \end{array} \right] \right\|^2$$

This proves that $(z = 0, \tilde{z} = 0, \xi = 0)$ is a globally asymptotically stable equilibrium point.   □

Example 4.1 Consider the system

$$\dot{x}_1 = x_2 + x_1^3$$
$$\dot{x}_2 = x_3$$
$$\dot{x}_3 = u$$
$$y = x_1$$

According to Theorem 4.1 this example can be globally stabilized by output feedback: in fact, the example belongs to the class of systems (S) with $b = [0, 0, 1]^T$ an Hurwitz vector of degree $\rho = 3$.   □

ACKNOWLEDGEMENT

This work was supported in part by Ministero dell'Università e della Ricerca Scientifica e Tecnologica.

REFERENCES

1. A.J. Krener and A. Isidori. Linearization by output injection and nonlinear observers. *Systems & Control Letters*, 3:47-52, 1983.
2. W. Respondek. Global aspects of linearization, equivalence to polynomial forms and decomposition of nonlinear control systems. In *Algebraic and Geometric Methods in Nonlinear Control Theory*. M. Fliess and M. Hazewinkel eds., D. Reidel Publishing Company, 1986.
3. W. P. Dayawansa, W.M. Boothby, D. Elliott. Global state and feedback equivalence of nonlinear systems, *Systems & Control Letters*, Vol. 6, pp. 229-234, 1985.
4. A. Isidori. *Nonlinear control systems*. Springer- Verlag, Berlin, 1989.
5. W. Respondek. Linearization, feedback and Lie brackets. Proc. Int. Conf. *Geometric Theory of Nonlinear Control Systems*. Wroclaw Technical University Press, Wroclaw, pp. 131-166, 1985.
6. C.I. Byrnes and A. Isidori. Asymptotic stabilization of minimum phase nonlinear systems, preprint, 1990.
7. G. Cesareo and R. Marino. Nonlinear control theory and symbolic algebraic manipulation, Lecture Notes in Information Sciences, Springer-Verlag, Berlin, Vol. 58, pp. 725-740, 1984.

8. G. Cesareo and R. Marino. On the application of symbolic computation to nonlinear control theory. Lecture Notes in Computer Sciences, Springer-Verlag, Berlin, Vol. 174, pp. 35-46, 1984.

9. R. Marino and P. Tomei. Global adaptive observers for nonlinear systems via filtered transformations. Report R-90.06, University of Rome "Tor Vergata", September 1990. Submitted to *IEEE Trans. Automatic Control.*

10. R. Marino. Adaptive observers for single output nonlinear systems. *IEEE Trans. Automatic Control*, Vol. 35, pp. 1054-1058, 1990.

11. H. Nijmeijer and A. van der Schaft. *Nonlinear dynamical control systems.* Springer-Verlag, Berlin, 1990.

12. K.S. Narendra and A.M. Annaswamy. *Stable Adaptive Systems.* Prentice Hall, Englewood Cliffs, NJ, 1989.

Découplage des systèmes dynamiques non linéaires

V. Hoang Ngoc Minh

Université Lille I, 59655 Villeneuve d'Ascq, France.

Abstract : In this paper, we present a concise implementation (MACSYMA) to study the problem of decoupling for nonlinear control systems (linear in the control) via analytic state feedback. This computation uses the Claude's technic, based on generating power series (as introduced by Fliess). We develop a little more general and unify case of decoupling some subset of outputs with respect to some subset of inputs. We obtain as particular cases the Disturbance Decoupling Problem, the Triangular Decoupling Problem, the Diagonal Decoupling Problem and the Noninteracting Control Problem.

0. Introduction

Dans ce texte, en nous basant sur la technique introduite par D. Claude ([2]), nous proposons une programmation très concise de la construction des lois de bouclage *génériques* assurant le *découplage* des systèmes dynamiques de la forme :

$$(NL) \quad \begin{cases} \dot{q}(t) = \sum_{x \in X} a^x(t) A_x(q) \\ y(t) = h(q(t)), \end{cases}$$

où : - X est un alphabet fini de lettres $X = \{x_0, x_1, \ldots, x_m\}$,
- q est un élément d'une variété analytique réelle Q de dimension finie N,
- pour toute lettre $x \in X$, A_x est un champ de vecteurs analytique sur Q de la forme $\sum_{k=1}^{N} A_x^k(q) \dfrac{\partial}{\partial q_k}$, les A_x^k sont des fonctions analytiques réelles de Q dans $\mathbb{R}$,
- pour toute lettre $x \in X$, a^x est une application réelle continue sur $\mathbb{R}_+$, en particulier, a^{x_0} est l'application constante et égale à 1. On note a le vecteur $(a^{x_0}, a^{x_1}, \ldots, a^{x_m})$,
- l'observation h est une application analytique de Q dans $\mathbb{R}^p$.

Le problème *du rejet de perturbations* a été étudié *syntaxiquement* par D. Claude ([2]) en utilisant les *l'algèbre des champs de vecteurs* et les *séries génératrices* (séries formelles en variables non commutatives sur X, [3]) :

$$\sigma h_{|q(0)} = \sum_{w \in X^*} (A_w \circ h)_{|q(0)} w \in \mathbb{R} \ll X \gg .$$

Rappellons que *la formule fondamentale* de M. Fliess permet d'obtenir $y(t)$ de la manière suivante ([3]):

$$y(t) = \sum_{w \in X^*} (A_w \circ h)_{|q(0)} \mathcal{E}_a(w)(t),$$

où A_w représente l'opérateur différentiel $A_x \circ A_v$ $(w = xv)$ et l'identité si $w = \epsilon$, et $\mathcal{E}_a(w)$ représente l'Evaluation de w pour l'entrée a ([4]).

Comme l'a montré D. Claude, l'obtention d'une *caractérisation algébrique* du rejet de perturbations nécessite l'étude des systèmes *non initialisés*, et conduit à coder les systèmes dynamiques par leurs séries génératrices à coefficients dans $C^\omega(Q)$, l'anneau des fonctions analytiques sur Q. Ainsi, en utilisant donc systématiquement ces séries à coefficients analytiques, nous allons caractériser *le découplage d'un sous-ensemble $\{y_s\}_{s \in \pi}$ de l'ensemble des sorties par rapport à un sous-ensemble $\{a^x\}_{x \in X_1}$ de l'ensemble des entrées.* L'étude de cette nouvelle notion nous permet ici de proposer un seul traitement pour les problèmes distincts, mais voisins, suivants :

- *Le rejet de perturbations : toutes les sorties du système $\{y_s\}_{1 \le s \le p}$ sont indépendantes des entrées $\{a^x\}_{x \in X_1}$.*
- *Le découplage k-bloc triangulaire : chaque bloc de sorties $\{y_s\}_{s \in \pi_l}$, $1 \le l \le k$, ne dépend que des blocs d'entrées $\{a^x\}_{x \in X_j}$, $1 \le j \le l$.*
- *Le découplage k-bloc diagonal : chaque bloc de sorties $\{y_s\}_{s \in \pi_l}$ ne dépend que d'un seul bloc d'entrées $\{a^x\}_{x \in X_l}$, $1 \le l \le k$.*
- *La commande non interactive : chaque sortie ne dépend que d'une seule entrée (c'est une application du cas précédent avec $k = m = p$).*

Ceci nous conduit à réécrire les constructions de D. Claude dans un cadre unifié. Ce cadre nous conduira à une approche par sous-programme pour implanter les quatre problèmes cités par simple instanciation, et il nous permettra d'étendre le théorème de caractérisation du découplage de D. Claude. Enfin, nous construisons les lois de bouclage génériques permettant de découpler chaque fois que c'est possible. Nous montrons ainsi l'avantage de cette méthode par rapport aux autres techniques ([5], [6]) par le fait qu'elle donne *effectivement* des lois de bouclage. Pour cette construction, conduisant à une implantation concise en MACYMA, nous utilisons les *modules* (et les *algèbres*) *de découplage*, et des calculs d'*annulateurs* dans l'*algèbre de Lie libre*.

Les algorithmes que nous proposons diffèrent ceux de D. Claude ([1], [2]) sur les points suivants :

. Nous proposons un algorithme calculant directement *les nombres caractéristiques* avec un critère d'arrêt car lorsqu'ils sont définis, ils sont strictement inférieurs à la dimension de la variété Q.

. Dans le cas d'une *matrice de commande*, Ω, est *singulière*, le calcul des *conditions de compatibilité* ([2]) se traduisent ici simplement par l'existence d'un *inverse à droite* pour Ω.

. Dans l'étude du découplage k-bloc diagonal, en utilisant l'algorithme proposé par C. Claude et P. Dufresne ([1]), nous serions amenés à résoudre simultanément k systèmes linéaires, et donc à assurer la compatibilité des solutions. Ici, le même problème est résolu par un seul système linéaire, et donc sans condition de compatibilité (même remarque pour le découplage k-bloc triangulaire).

Il serait bien sûr agréable de proposer un algorithme permettant de calculer toutes les lois de bouclage possibles pour résoudre chacun des problèmes cités. C'est un problème techniquement difficile, ce travail peut être considéré comme un début pour envisager une implantation des méthodes de *découplage avec contraintes* en lien avec les questions de *stabilité* et de *robustesse*.

1. Polysystème

1.1. Annulateurs ([4])

Soit $f \in C^\omega(Q)$. On défini *l'action à gauche* "$*$" de $I\!R < X >$ sur $C^\omega(Q)$ comme suit :

$$\forall x \in X, \quad z * f = A_x \circ f \quad \text{(dérivée de Lie de } f \text{ par rapport à } A_x),$$

$$\forall w \in X^*, \quad w * f = A_w \circ f,$$

$$\forall P \in I\!R < X >, \quad P * f = \sum_{w \in \text{supp}(P)} < P | w > w * f \quad \text{(somme finie)}.$$

Une implantation de l'action à gauche de $I\!R < X >$ sur $C^\omega(Q)$ en MACSYMA est donnée en Annexe.

Soit $f \in C^\omega(Q)$. On appelle *annulateur* de f, l'ensemble

$$Ann f = \{P \in I\!R < X > | \quad P * f \equiv 0\}.$$

Soit F une partie de $C^\omega(Q)$. On note $Ann F = \bigcap_{f \in F} Ann f$.

1.2. Matrice de commande

Définition 1.2.1. : *Soit $s \in [1..p]$, le nombre caractéristique Ψ_s du système (S) est le plus petit entier ν, lorsqu'il est défini, tel qu'il existe $x \in X_0$, $x x_0^\nu * h_s \not\equiv 0$. Si pour tout entier ν, et pour toute lettre $x \in X_0$, on a $x x_0^\nu * h_s \equiv 0$, alors Ψ_s vaut $+\infty$:*

$$\Psi_s = \begin{cases} \inf\{\nu \in I\!N | \quad \exists x \in X_0, \quad x x_0^\nu * h_s \not\equiv 0\} \\ +\infty \quad si \quad \forall \nu \geq 0, \quad \forall x \in X_0, \quad x x_0^\nu * h_s \equiv 0. \end{cases}$$

On montre que (voir par exemple [4]) lorsque les nombres caractéristiques sont définis, ils sont strictement inférieurs à la dimension de la variété Q. D'où un critère simple d'arrêt pour calculer les nombres caractéristiques (voir Annexe).

Définition 1.2.2. : *Soit $s \in [1..p]$, tel que Ψ_s soit défini. Soit $x \in X_0$, x est une* **lettre significative** *pour h_s si $x x_0^{\Psi_s} * h_s \not\equiv 0$. Nous notons $\text{sgn}(h_s)$ l'ensemble de* **toutes les lettres significatives pour h_s :**

$$\text{sgn}(h_s) = \{x \in X_0 | \quad x x_0^{\Psi_s} * h_s \not\equiv 0\}.$$

Soit $s \in [1..p]$, par convention, $\text{sgn}(h_s)$ est vide si et seulement si $\Psi_s = +\infty$.

Avec ces notations, on obtient les équivalences suivantes :

$$\Psi_s = +\infty \quad \Longleftrightarrow \quad \text{sgn}(h_s) = \emptyset \quad \Longleftrightarrow \quad \forall \nu \in [1..N[, \forall x \in X_0, x x_0^\nu * h_s \equiv 0.$$

Définition 1.2.3. : *Les nombres caractéristiques* $\{\Psi_s\}_{s\in[1..p]}$ *sont supposés défi-nis. La matrice de commande de* (S) *est* $\Omega = \begin{pmatrix} x_1 * \Delta \\ x_2 * \Delta \\ \vdots \\ x_m * \Delta \end{pmatrix}$. *Nous appelons aussi* **matrice de commande étendue** *de* (S) *la matrice* $\Gamma = \begin{pmatrix} x_0 * \Delta \\ \Omega \end{pmatrix}$, *où* Δ *est la ma-trice* $\left(x_0^{\Psi_1} * h_1 \quad x_0^{\Psi_2} * h_2 \quad \ldots \quad x_0^{\Psi_p} * h_p \right)$, *et pour toute lettre* $x \in X$, $x * \Delta$ *représente la matrice* $\left(xx_0^{\Psi_1} * h_1 \quad xx_0^{\Psi_2} * h_2 \quad \ldots \quad xx_0^{\Psi_p} * h_p \right)$.

Nous donnons en Annexe les fonctions permettant de construire l'ensemble des lettres significatives, ainsi que les matrices Ω et Γ.

2. Formulation du problème

On suppose que les nombres caractéristiques $\{\Psi_s\}_{s\in[1..p]}$ de (S) sont tous définis, et $X_1 \subset X_0$ est supposée fixé. On pose :

$$H = \{h_1, \ldots, h_p\}, \quad G = \{x_0^{\Psi_1} * h_1, \ldots, x_0^{\Psi_p} * h_p\}, \quad F = \bigcup_{s=1}^{p} \{x_0^n * h_s, 0 \le n \le \Psi_s\}.$$

Etant donné une partie de π de $[1..p]$, on pose :

$$H_1 = \{h_s\}_{s\in\pi}, \quad G_1 = \{x_0^{\Psi_s} * h_s\}_{s\in\pi}, \quad F_1 = \bigcup_{s\in\pi} \{x_0^n * h_s, 0 \le n \le \Psi_s\},$$

alors nous avons $G_1 \subset F_1$ et $H_1 \subset F_1 \subset (X \setminus X_1)^* * H_1$.

Définition 2.1. : *Les sorties* $\{y_s\}_{s\in\pi}$ *de* (S) *sont* **découplée par rapport aux entrées** $\{a^x\}_{x\in X_1}$ *si et seulement si pour tout* $s \in \pi$, *la sortie* y_s *n'est pas affectée par les entrées* $\{a^x\}_{x\in X_1}$, *quelque soit l'état initial* $q(0) \in Q$.

Pour tout état initial $q(0) \in Q$ de (S), pour $s \in \pi$, si la sortie y_s n'est pas affectée par les entrées $\{a^x\}_{x\in X_1}$, alors toute variation de a^x, $x \in X_1$, n'affecte pas l'observation h_s. La proposition suivante (voir par exemple [4]) montre que si les entrées a^x, $x \in X_1$, n'affectent pas l'observation h_s, $s \in \pi$, alors les lettres de X_1 n'apparaissent pas dans le support de la série génératrice σh_s, $s \in \pi$.

Proposition 2.1. : *Les propriétés suivantes sont équivalentes :*
(a) $\forall f \in H_1$, $\text{supp}(\sigma f) \subset (X \setminus X_1)^*$.
(b) $X_1(X \setminus X_1)^* \subset Ann H_1$.
(c) $\forall x \in X_1$, $x(X \setminus \{x\})^* \subset Ann H_1$.
(d) $\{y_s\}_{s\in\pi}$ *est découplé par rapport à* $\{a^x\}_{x\in X_1}$.
(e) $\forall f \in H_1$, $\sigma f = \displaystyle\sum_{w\in(X\setminus X_1)^*} (w * f)w$.

Désormais on dira tout simplement que H_1 est découplé par rapport à X_1 pour signifier que les sorties $\{y_s\}_{s\in\pi}$ sont découplées par rapport à $\{a^x\}_{x\in X_1}$.

3. Modules et algèbres de découplage

Définition 3.1. : *Soit D_1 un sous-module (resp. une sous-algèbre de Lie) de Lie $< X >$. D_1 est un* **module** *(resp. une* **algèbre***) de découplage de H_1 par rapport à X_1 (ou plus brièvement un module (resp. une algèbre) de découplage), si et seulement si D_1 vérifie les propriétés suivantes :*

$$(a)\ X_1 \subset D_1, \qquad (b)\ D_1 \subset AnnH_1, \qquad (c)\ [X \setminus X_1, D_1] \subset D_1.$$

Le **plus petit sous-module de** Lie $< X >$ qui vérifie les conditions (a) et (c) que l'on note I_1, est engendré par les éléments $(ad_{x_{i_1}} \circ ad_{x_{i_2}} \circ \ldots \circ ad_{x_{i_k}})(x)$, pour tout $x_{i_1} x_{i_2} \ldots x_{i_k} \in (X \setminus X_1)^*$ et pour tout $x \in X_1$. I_1 est inclus dans tout module de découplage. C'est un module de découplage si et seulement si $I_1 \subset AnnH_1$. D'après le théorème de l'élimination de M. Lazard ([7]), **la plus petite sous-algèbre de Lie de** Lie $< X >$ qui vérifie (a) et (c) est l'algèbre de Lie $\Im_1$ engendrée par I_1. $\Im_1$ est inclus dans toute algèbre de découplage. C'est une algèbre de découplage si et seulement si $\Im_1 \subset AnnH_1$.

On peut vérifier aisément que $AnnH_1$ est une sous-algèbre de Lie de Lie $< X >$. Toute algèbre de découplage est incluse dans $AnnH_1$. Si $AnnH_1$ vérifie (a) et (c), alors $AnnH_1$ sera **la plus grande algèbre de découplage**, par conséquent c'est le **plus grand module de découplage**. Il est montré (voir [4]) que si $AnnF_1$ est un module de découplage, $AnnF_1$ sera le plus grand module de découplage.

Le théorème de caractérisation du découplage suivant est dû à D. Claude :

Théorème 3.1. : ([2], voir aussi [4]) *Les propriétés suivantes sont équivalentes :*
(a) H_1 est découplé par rapport à X_1.
(b) $X_1 \subset Ann((X \setminus X_1)^* * H_1)$.
(c) $\Im_1 \subset AnnH_1$.
(d) $I_1 \subset AnnH_1$.
(e) Il existe un module de découplage.
(f) $I_1(X \setminus X_1)^* \subset AnnH_1$.

Ce théorème peut être complété comme suit (voir [4]) :

Proposition 3.1. : *$AnnF_1$ est un module de découplage si et seulement si les deux conditions suivantes sont satisfaites :*

$$(a)\ \forall h_s \in H_1, \mathrm{sgn}(h_s) \subset (X_0 \setminus X_1), \qquad (b)\ AnnF_1 \subset Ann((X \setminus X_1) * G_1).$$

Lorsque H_1 est découplé par rapport à X_1, d'après le théorème 3.1., nous avons $X_1(X \setminus X_1)^* \in AnnH_1$. En particulier, pour tout indice $s \in \pi$ et pour toute lettre $x \in X_1$, $x x_0^{\psi *} * h_s \equiv 0$, la matrice extraite $\left[\Omega_x^s\right]_{x \in X_1}^{s \in \pi}$ est nulle (la réciproque est fausse en général). Inversement, si $\left[\Omega_x^s\right]_{x \in X_1}^{s \in \pi}$ est identiquement nulle, alors pour tout indice $s \in \pi$, $\mathrm{sgn}(h_s) \subset X_0 \setminus X_1$. En plus, si nous avons $AnnF_1 \subset Ann((X \setminus X_1) * G_1)$ alors d'après la proposition 3.1., $AnnF_1$ sera le plus grand module de découplage. Par conséquent, d'après le théorème 3.1., H_1 est découplé par rapport à X_1.

Proposition 4.1. : *Si la matrice de commande Ω d'un système dynamique non linéaire est une matrice constante ($\Omega \in \mathcal{M}_{m,p}(\mathbb{R})$) alors les sorties $\{y_s\}_{s \in \pi}$ sont découplées par rapport aux entrées $\{a^x\}_{x \in X_1}$ si et seulement si la matrice extraite $[\Omega_x^s]_{x \in X_1}^{s \in \pi}$ est nulle.*

4. Etudes particulières

4.1. Rejet de Perturbation

Etant donné $\mathbf{P} \subset X_0$, on pose $\mathbf{A} = X \setminus \mathbf{P}$. Les éléments de $\mathbf{P}$ représentent des **perturbations** et ceux de $\mathbf{A}$ représentent les **commandes admissibles** de H. Pour toute observation $h_s \in H$, le support de σh_s est un sous-langage de $\mathbf{A}^*$.

Définition 4.1.1. : *Le système (S) est **découplé par rapport aux perturbations** si et seulement si H est découplé par rapport à $\mathbf{P}$:*

$$\mathbf{PA}^* \subset AnnH.$$

Par conséquent, pour tout $h_s \in H$, $\mathrm{sgn}(h_s) \subset \mathbf{A} \setminus \{x_0\}$. La matrice de commande Ω de tel système est de la forme :

$$\left(\begin{matrix} \left[x x_0^{\Psi_s} * h_s \right]_{x \in \mathbf{A} \setminus \{x_0\}}^{s \in [1..p]} \\ 0 \end{matrix} \right) = \left(\begin{matrix} \Omega_{a^\cdot} \\ 0 \end{matrix} \right)$$

et la sous-matrice Ω_a est appelée **la matrice de commande admissible**.

$AnnF$ est le plus grand module de découplage de H si et seulement si :

$$\forall s \in [1..m], \mathrm{sgn}(h_s) \subset \mathbf{A} \setminus \{x_0\} \quad \text{et} \quad AnnF \subset Ann(\mathbf{A} * G).$$

4.2. Découplage par k-bloc triangulaire

Pour $k \leq \inf(m,p)$, on se donne une k-partition de X_0 et une k-partition de $[1..p]$:

$$X_0 = X_1 \oplus X_2 \oplus \ldots \oplus X_k$$
$$[1..p] = \pi_1 \oplus \pi_2 \oplus \ldots \oplus \pi_k.$$

Nous posons $\mathbf{P}_1 = \emptyset, \mathbf{P}_j = \mathbf{P}_{j-1} \cup X_{j-1}, j \in [2..k]$. Pour tout $j \in [1..k]$, nous posons aussi :

$$\mathbf{A}_j = X \setminus \mathbf{P}_j, \ \mathbf{H}_j = \{h_s\}_{s \in \pi_j}, \ \mathbf{G}_j = \{x_0^{\Psi_s} * h_s\}_{s \in \pi_j}, \ \mathbf{F}_j = \{x_0^n * h_s \mid 0 \leq n \leq \Psi_s\}_{s \in \pi_j}.$$

Les éléments de $\mathbf{A}_j$ sont considérés comme les commandes admissibles et les éléments de $\mathbf{P}_j$ sont des perturbations de $\mathbf{H}_j$. Pour toute partition π_j, pour toute observation h_s de $\mathbf{H}_j$, le support de σh_s est un sous-langage de $\mathbf{A}_j^*$.

Définition 4.2.1. : *Le système* (S) *est* **découplé** k-**block triangulaire** *si et seulement si pour tout entier* j *de* $[2..k]$, $\mathbf{H}_j$ *est découplé par rapport à* $\mathbf{P}_j$:

$$\mathbf{P}_j \mathbf{A}_j^* \subset Ann\mathbf{H}_j.$$

Par conséquent, pour tout $h_s \in \mathbf{H}_j$, $\mathrm{sgn}(h_s) \subset \mathbf{A}_j \setminus \{x_0\}$. La matrice de commande Ω de tel système est de la forme :

$$\begin{pmatrix} \left[xx_0^{\Psi_s} * h_s\right]_{x \in \mathbf{A}_1 \setminus \{x_0\}}^{s \in \pi_1} & 0 & \cdots & 0 \\ \left[xx_0^{\Psi_s} * h_s\right]_{x \in \mathbf{A}_2 \setminus \{x_0\}}^{s \in \pi_1} & \left[xx_0^{\Psi_s} * h_s\right]_{x \in \mathbf{A}_2 \setminus \{x_0\}}^{s \in \pi_2} & \cdots & 0 \\ \vdots & \vdots & \ddots & \vdots \\ \left[xx_0^{\Psi_s} * h_s\right]_{x \in \mathbf{A}_k \setminus \{x_0\}}^{s \in \pi_1} & \left[xx_0^{\Psi_s} * h_s\right]_{x \in \mathbf{A}_k \setminus \{x_0\}}^{s \in \pi_2} & \cdots & \left[xx_0^{\Psi_s} * h_s\right]_{x \in \mathbf{A}_k \setminus \{x_0\}}^{s \in \pi_k} \end{pmatrix}$$

Pour tout entier $j \in [2..k]$, $Ann\mathbf{F}_j$ est le plus grand module de découplage de $\mathbf{H}_j$ si et seulement si :

$$\forall s \in \pi_j, \mathrm{sgn}(h_s) \subset \mathbf{A}_j \setminus \{x_0\} \quad \text{et} \quad Ann\mathbf{F}_j \subset Ann(\mathbf{A}_j * \mathbf{G}_j).$$

4.3. Découplage par k-bloc diagonal

On se donne de nouveau une k-partition de X_0 et une k-partition de $[1..p]$:

$$X_0 = X_1 \oplus X_2 \oplus \ldots \oplus X_k$$
$$[1..p] = \pi_1 \oplus \pi_2 \oplus \ldots \oplus \pi_k.$$

Pour tout $j \in [1..k]$, nous posons $\mathbf{P}_j = X_0 \setminus X_j$, $\mathbf{A}_j = X \setminus \mathbf{P}_j$,

$$\mathbf{H}_j = \{h_s\}_{s \in \pi_j}, \quad \mathbf{G}_j = \{x_0^{\Psi_s} * h_s\}_{s \in \pi_j}, \quad \mathbf{F}_j = \{x_0^n * h_s | \, 0 \leq n \leq \Psi_s\}_{s \in \pi_j}.$$

Les éléments de $\mathbf{A}_j$ sont considérés comme les commandes admissibles et ceux de $\mathbf{P}_j$ sont des perturbations de $\mathbf{H}_j$. Pour toute partition π_j, pour toute observation h_s de $\mathbf{H}_j$, le support de σh_s est un sous-langage de $\mathbf{A}_j^*$.

Définition 4.3.1. : *Le système* (S) *est* k-**block diagonal découplé** *si et seulement si pour tout entier* j *de* $[1..k]$, $\mathbf{H}_j$ *est découplé par rapport à* $\mathbf{P}_j$:

$$\mathbf{P}_j \mathbf{A}_j^* \subset Ann\mathbf{H}_j.$$

Par conséquent, pour tout $h_s \in \mathbf{H}_j$, $\mathrm{sgn}(h_s) \subset \mathbf{A}_j \setminus \{x_0\}$. La matrice de commande Ω de tel système est k-block diagonal :

$$\begin{pmatrix} \left[xx_0^{\Psi_s} * h_s\right]_{x \in \mathbf{A}_1 \setminus \{x_0\}}^{s \in \pi_1} & 0 & \cdots & 0 \\ 0 & \left[xx_0^{\Psi_s} * h_s\right]_{x \in \mathbf{A}_2 \setminus \{x_0\}}^{s \in \pi_2} & \cdots & 0 \\ \vdots & \vdots & \ddots & \vdots \\ 0 & 0 & \cdots & \left[xx_0^{\Psi_s} * h_s\right]_{x \in \mathbf{A}_k \setminus \{x_0\}}^{s \in \pi_k} \end{pmatrix}$$

Pour tout entier j de $[1..k]$, $Ann\mathbf{F}_j$ est le plus grand module de découplage de $\mathbf{H}_j$ si et seulement si :

$$\forall s \in \pi_j, \mathrm{sgn}(h_s) \subset \mathbf{A}_j \setminus \{x_0\} \quad \text{et} \quad Ann\mathbf{F}_j \subset Ann(\mathbf{A}_j * \mathbf{G}_j).$$

4.4. Système non interactif

On applique le cas précédent avec $k = p = m$. Pour toute observation $h_s \in H$, le support de σh_s est un sous-langage de $\{x_0, x_s\}^*$ (et non pas seulement de x_s^* !).

Définition 4.4.1. : *Le système* (S) *est un* **système non interactif** *si et seulement si pour tout* $s \in [1..p]$, h_s *est découplé par rapport à* $X_0 \setminus \{x_s\}$:

$$(X_0 \setminus \{x_s\})\{x_0, x_s\}^* \subset Annh_s.$$

Dans ce cas, d'après la définition des nombres caractéristiques (supposés définis), nous avons pour tout $s \in [1..p]$, $x_s x_0^{\Psi_s} * h_s \not\equiv 0$. Par conséquent $\mathrm{sgn}(h_s) = \{x_s\}, s \in [1..p]$. Et la matrice de commande Ω de tel système est diagonal :

$$\begin{pmatrix} x_1 x_0^{\Psi_1} * h_1 & 0 & \ldots & 0 \\ 0 & x_2 x_0^{\Psi_2} * h_2 & \ldots & 0 \\ \vdots & \vdots & \ddots & \vdots \\ 0 & 0 & \ldots & x_m x_0^{\Psi_m} * h_m \end{pmatrix}$$

Les termes diagonaux de la matrice de commande Ω ne s'annulent pas, par conséquent, Ω est de rang maximal : $\mathrm{rank}\,\Omega = m$.

Pour tout $s \in [1..p]$, $Ann\{x_0^n * h_s | 0 \le n \le \Psi_s\}$ est le plus grand module de découplage de h_s si et seulement si :

$$\mathrm{sgn}(h_s) = \{x_s\} \quad \text{et} \quad Ann\{x_0^n * h_s | \ 0 \le n \le \Psi_s\} \subset Ann\{x_0 x_0^{\Psi_s} * h_s, x_s x_0^{\Psi_s} * h_s\}.$$

5. Découplage par bouclage statique par retour d'état

Soit $X = \{x_0, x_1, \ldots, x_m\}$ (resp. $\hat{X} = \{\hat{x}_0, \hat{x}_1, \ldots, \hat{x}_{\hat{m}}\}$) un alphabet fini. Nous posons $X_0 = X \setminus \{x_0\}$ (resp. $\hat{X}_0 = \hat{X} \setminus \{\hat{x}_0\}$).

5.1. Bouclage statique par retour d'état

Lorsque (S) ne présente pas les propriétés qu'on souhaite (dans notre étude c'est le découplage de H_1 par rapport à X_1), on lui applique une transformation du type **bouclage statique par retour d'état** :

$$(*) \quad \forall x \in X, \quad a^x = \sum_{\hat{x} \in \hat{X}} a^{\hat{x}} \Phi_{\hat{x}}^x,$$

où : - Pour tout $x \in X$ et pour tout $\hat{x} \in \hat{X}$, $\Phi_{\hat{x}}^x$ est une fonction analytique de $C^\omega(Q)$.

- Pour tout $x \in X$, $\Phi_{\hat{x}_0}^x$ est constante et égale à 0, et $\Phi_{\hat{x}_0}^{x_0}$ est égale à 1.

- Pour tout $\hat{x} \in \hat{X}$, $a^{\hat{x}}$ est une application continue par morceaux sur $I\!R_+$. En particulier, $a^{\hat{x}_0}$ est une application constante et égale à 1. On note $\hat{a}$ le vecteur $\begin{pmatrix} a^{\hat{x}_0} & a^{\hat{x}_1} & \ldots & a^{\hat{x}_{\hat{m}}} \end{pmatrix}$.

En posant $\Phi = \begin{pmatrix} 1 & \alpha^{x_1} & \dots & \alpha^{x_m} \\ 0 & \beta^{x_1}_{\hat{x}_1} & \dots & \beta^{x_m}_{\hat{x}_1} \\ \vdots & \vdots & \ddots & \vdots \\ 0 & \beta^{x_1}_{\hat{x}_{\hat{m}}} & \dots & \beta^{x_m}_{\hat{x}_{\hat{m}}} \end{pmatrix} = \begin{pmatrix} 1 & \alpha \\ 0 & \beta \end{pmatrix}$, on a $a = \hat{a}\Phi$ et $\hat{A} = \Phi A$. Le système bouclé obtenu $(\hat{S})$ est encore un système de (NL) :

$$(\hat{S}) \quad \begin{cases} \dot{q}(t) = \hat{a}(t)\hat{A}(q) \\ y(t) = h(q(t)) \end{cases}$$

Définition 5.1.1. : *On appelle* Φ *la matrice de bouclage.*
Le bouclage par retour d'état sera dit **régulier** *si et seulement si la matrice de bouclage* Φ *est de rang maximal, ou de manière équivalente* β *est de rang maximal.*

Il est montré que les nombres caractéristiques ne peuvent pas décroître par bouclage statique par retour d'état (voir par exemple [4]). Ils sont **invariants dans le cas d'un bouclage statique par retour d'état régulier.**

Nous notons $\{\hat{\Psi}_s\}_{1 \le s \le p}$ les nombres caractéristiques de $(\hat{S})$. Lorsqu'ils sont tous définis, on note $\hat{\Omega}$ la matrice de bouclage de $(\hat{S})$, $\begin{pmatrix} \hat{x}_1 * \hat{\Delta} \\ \hat{x}_2 * \hat{\Delta} \\ \vdots \\ \hat{x}_{\hat{m}} * \hat{\Delta} \end{pmatrix}$, et $\hat{\Gamma}$ la matrice $\begin{pmatrix} \hat{x}_0 * \hat{\Delta} \\ \hat{\Omega} \end{pmatrix}$,

où $\hat{\Delta}$ est la matrice $\begin{pmatrix} \hat{x}_0^{\hat{\Psi}_1} * h_1 & \hat{x}_0^{\hat{\Psi}_2} * h_2 & \dots & \hat{x}_0^{\hat{\Psi}_p} * h_p \end{pmatrix}$, et pour toute lettre $\hat{x} \in \hat{X}$, $\hat{x} * \hat{\Delta}$ représente la matrice $\begin{pmatrix} \hat{x}\hat{x}_0^{\hat{\Psi}_1} * h_1 & \hat{x}\hat{x}_0^{\hat{\Psi}_2} * h_2 & \dots & \hat{x}\hat{x}_0^{\hat{\Psi}_p} * h_p \end{pmatrix}$. Avec ces notations, nous avons (voir par exemple [4]) :

Théorème 5.1. : *Si les nombres caractéristiques* $\{\Psi_s\}_{1 \le s \le p}$ *de* (S) *sont définis, et si le bouclage par retour d'état est régulier alors* $\hat{\Gamma} = \Phi\Gamma$.

5.2. Découplage par bouclage statique par retour d'état

Définition 5.2.1. : *Un système* (S) *de* (NL) *est* **découplable par bouclage statique par retour d'état** *si et seulement si il existe un bouclage régulier par retour d'état tel que le système* $(\hat{S})$ *bouclé admette l'ensemble* $\{y_s\}_{s \in \hat{\pi}}$ *de sorties découplé par rapport à l'ensemble* $\{a^{\hat{x}}\}_{\hat{x} \in \hat{X}_1}$ *d'entrées.*

Si les nombres caractéristiques $\{\Psi_s\}_{1 \le s \le p}$ de (S) sont définis, et si le bouclage par retour d'état est régulier alors nous avons :

$$\begin{pmatrix} \hat{x}_0 * \hat{\Delta} \\ \hat{\Omega} \end{pmatrix} = \begin{pmatrix} 1 & \alpha \\ 0 & \beta \end{pmatrix} \begin{pmatrix} x_0 * \Delta \\ \Omega \end{pmatrix}.$$

Par conséquent, lorsque le bouclage est régulier on a le système d'équations suivant :

$$(**) \quad \begin{cases} \hat{x}_0 * \hat{\Delta} - x_0 * \Delta = \alpha\Omega \\ \hat{\Omega} = \beta\Omega. \end{cases}$$

Théorème 5.2.1. : *Un système (S) de (NL) est* **découplable par bouclage statique par retour d'état** *si et seulement si le système d'équations $(**)$ admet une solution en α et β tel que le système bouclé $(\hat{S})$ admet l'ensemble $\{y_s\}_{s \in \hat{\pi}}$ de sorties découplé par rapport à l'ensemble $\{a^{\hat{x}}\}_{\hat{x} \in \hat{X}_1}$ d'entrées. Une telle solution existe en particulier si Ω admet un inverse à droite noté par Ω^d. Dans cette condition, nous avons :*

$$\begin{cases} \alpha = (\hat{x}_0 * \hat{\Delta} - x_0 * \Delta)\Omega^d \\ \beta = \hat{\Omega}\Omega^d \end{cases}$$

6. Bibliographies

[1] **D. Claude & P. Dufresne**, *An Application of MACSYMA to Nonlinear Systems Decoupling*, Lecture Notes in Computer Science, 144, 1982.

[2] **D. Claude**, *Découplage et Linéarisation des systèmes non linéaires par bouclages statiques*, Thèse d'état, Université de Paris Sud, Centre d'Orsay, 1986.

[3] **M. Fliess**, *Fonctionnelles causales non linéaires et indéterminées non commutative*, Bull. Soc. Math France, 109, 1981, pp. 3-40.

[4] **Hoang Ngoc Minh**, *Contribution au développement d'outils informatiques pour résoudre des problèmes d'automatique non linéaire*, Thèse, Université Lille I, 1990.

[5] **C. Moog**, *Inversion, Découplage, Poursuite de modèle des Systèmes Non Linéaires*, Thèse d'état, Université Nantes, 1987.

[6] **A. Isidori**, *Nonlinear Control Systems : an Introduction*, Springer-Verlag, 1989.

[7] **G. Viennot**, *Algèbres de Lie libres et Monoïdes libres*, Lecture Notes in Mathematics, 691, 1978.

Annexe

Exemple : Soit S le système dynamique suivant :

$$(S) \begin{cases} \dot{q}_1(t) = q_2(t) + 2q_2(t)q_3(t) + 2q_2(t)a(t) + 2q_2(t)b(t) \\ \dot{q}_2(t) = q_3(t) + a(t) + b(t) \\ \dot{q}_3(t) = q_4(t) - q_3^2(t) - q_5^2(t) + a(t) \\ \dot{q}_4(t) = q_5(t) + 2q_3(t)(q_4(t) - q_3^2(t) - q_5^2(t)) + 2q_3(t)a(t) + 2q_5(t)b(t) \\ \dot{q}_5(t) = b(t) \\ y_1(t) = h_1(q) = q_1(t) - q_2^2(t) \\ y_2(t) = h_2(q) = q_3(t). \end{cases}$$

Soit $X = \{x_0, x_1, x_2\}$ l'alphabet de codage :

_ x_0 code la partie autonome :

$$Y_{x_0} = (q_2 + 2q_2 q_3)\frac{\partial}{\partial q_1} + q_3 \frac{\partial}{\partial q_2} + (q_4 - q_3^2 - q_5^2)\frac{\partial}{\partial q_3} + (q_5 + 2q_3(q_4 - q_3^2 - q_5^2))\frac{\partial}{\partial q_4},$$

_ x_1 code l'entrée $a(t)$:

$$Y_{x_1} = 2q_2 \frac{\partial}{\partial q_1} + \frac{\partial}{\partial q_2} + \frac{\partial}{\partial q_3} + 2q_3 \frac{\partial}{\partial q_4}$$

_ x_2 code l'entrée $b(t)$:

$$Y_{x_2} = 2q_2 \frac{\partial}{\partial q_1} + \frac{\partial}{\partial q_2} + 2q_5 \frac{\partial}{\partial q_4} + \frac{\partial}{\partial q_5}.$$

On souhaite rendre ce système non interactif par bouclage par retour d'état (voir 5.). Et on souhaite aussi que le système non interactif obtenu (voir 4.4.) ait la matrice de commande étendue $\hat{\Gamma}$ sous la forme $\begin{pmatrix} 0 & 0 \\ k_1 & 0 \\ 0 & k_2 \end{pmatrix}$ (k_1 et k_2 sont des constantes arbitraires).

Pour cela, nous allons calculer les lois de bouclage Φ correspondantes (voir définition 5.1.1. et théorème 5.2.1.) à l'aide de MACSYMA.

Nous allons utilisé la fonction "**Action(P,f)**" calculant l'action (voir 1.1.) d'un polynôme **P** sur la fonction analytique **f** pour écrire la procédure "**Control_matrix()**" contruisant la matrice de commande Ω et la matrice Γ d'un système dynamique (voir définition 1.3.3.). Cette procédure utilise la liste **Psi** contenant les nombres caractéristiques $\{\Psi_s\}_{1 \le s \le p}$ d'un système dynamique (voir définition 1.3.2.), et aussi p listes **sgn** de lettres significatives des fonctions d'observation $\{h_s\}_{1 \le s \le p}$, et enfin la liste **Gamma** représentant la première ligne de Γ : $\left(z_0^{\Psi_1 + 1} * h_1 \quad z_0^{\Psi_2 + 1} * h_2 \quad \dots \quad z_0^{\Psi_p + 1} * h_p \right)$, tout cela étant fourni par la procédure "**Characteristic_numbers()**". Cette dernière procédure calcule (comme son nom l'indique) les nombres caractéristiques d'un système dynamique (voir définition 1.3.1.).

This is Macsyma 415.69 for SUN-3 Series Computers.
Copyright (c) 1982 Massachusetts Institute of Technology.
All Rights Reserved.
Enhancements (c) 1982, 1989 Symbolics, Inc. All Rights Reserved.
Type "DESCRIBE(TRADE_SECRET);" to see important legal notices.
Type "HELP();" for more information.

```
(C1) Action(P,f):=if atom(P) then ev(Y[P](f),diff)
             else if inpart(P,0)="." then Action(inpart(P,1),
                                        Action(inpart(P,allbut(1)),f))
                 else if inpart(P,0)="+" then Action(inpart(P,1),f)
                                            +Action(inpart(P,allbut(1)),f)
                     else if inpart(P,0)="*" then
                          Action(inpart(P,allbut(1)),f)*inpart(P,1)
                         else if inpart(P,0)="^" then
                             Action(inpart(P,1),
                             Action(inpart(P,1)^(inpart(P,2)-1),f))
                             else print("Error in polynomial :",P)$
```

```
(C2) Characteristic_numbers():=block([s,i,j,z,nu],
Gamma:[],
for s:1 thru p do
    (hs:h[s],sgn[s]:[],nu:-1,
    while (nu<N-2) and (sgn[s]=[]) do
         (nu:nu+1,i:0,G[s]:hs,
         while (i<m) and (sgn[s]=[]) do
              (i:i+1,z:concat('z,i),
              if Action(z,hs)#0 then (sgn[s]:[z],Psi[s]:nu)),
         hs:Action(z0,hs)),
    if sgn[s]=[] then Psi[s]:INF
    else for j:i+1 thru m do
         (z:concat('z,j),if Action(z,G[s])#0 then sgn[s]:endcons(z,sgn[s])),
    Gamma:endcons(hs,Gamma)))$
```

```
(C3) Control_matrix():=block([s,i,z],
Gamma:matrix(Gamma),
for i:1 thru m do
    (z:concat('z,i),
    Gamma:addrow(Gamma,makelist(if member(z,sgn[s]) then Action(z,G[s])
                                            else 0,s,1,p))),
Omega:submatrix(1,Gamma))$
```

(C4) p:2$

(C5) m:2$

(C6) N:5$

(C7) depends(h,[q1,q2,q3,q4,q5])$

```
(C8) Y[z0](h):=(q2+2*q2*q3)*diff(h,q1)+q3*diff(h,q2)+(q4-q3^2-q5^2)*diff(h,q3)
       +(q5+2*q3*(q4-q3^2-q5^2))*diff(h,q4)$
```

(C9) Y[z0];

$$
(D9)\ \text{LAMBDA}([H],\ \frac{dH}{dQ3}\ (-\ Q5^2\ +\ Q4\ -\ Q3^2)\ +\ \frac{dH}{dQ4}\ (2\ Q3\ (-\ Q5^2\ +\ Q4\ -\ Q3^2)\ +\ Q5)
$$

$$
+\ \frac{dH}{dQ1}\ (2\ Q2\ Q3\ +\ Q2)\ +\ \frac{dH}{dQ2}\ Q3)
$$

(C10) Y[z1](h):=2*q2*diff(h,q1)+diff(h,q2)+diff(h,q3)+2*q3*diff(h,q4)$

(C11) Y[z1];

(D11)   LAMBDA([H], 2 $\frac{dH}{dQ4}$ Q3 + 2 $\frac{dH}{dQ1}$ Q2 + $\frac{dH}{dQ3}$ + $\frac{dH}{dQ2}$)

(C12) Y[z2](h):=2*q2*diff(h,q1)+diff(h,q2)+2*q5*diff(h,q4)+diff(h,q5)$

(C13) Y[z2];

(D13)   LAMBDA([H], 2 $\frac{dH}{dQ4}$ Q5 + 2 $\frac{dH}{dQ1}$ Q2 + $\frac{dH}{dQ5}$ + $\frac{dH}{dQ2}$)

(C14) h[1]:q1-q2^2$

(C15) h[2]:q3$

(C16) Characteristic_numbers()$

(C17) sgn[1];

(D17)   [Z1, Z2]
(C18) psi[1];

(D18)   1
(C19) sgn[2];

(D19)   [Z1]
(C20) psi[2];

(D20)   0
(C21) Control_matrix();

(D21)
$$\begin{bmatrix} 1 & 1 \\ 1 & 0 \end{bmatrix}$$
(C22) Omega_chapeau:entermatrix(2,2)$

What kind of matrix?
   1. Diagonal   2. Symmetric   3. Antisymmetric   4. General
Selection: 1;
Row 1 Column 1:   k1;
Row 2 Column 2:   k2;
Matrix entered.
(C23) Alpha:-submatrix(2,3,Gamma) . Omega^^-1;

(D23)
$$\begin{bmatrix} 2 & 2 & 2 & 2 \\ Q5 - Q4 + Q3 & - Q5 + Q4 - Q3 & - Q3 \end{bmatrix}$$
(C24) Beta:Omega_chapeau . Omega^^-1;

(D24)
$$\begin{bmatrix} 0 & K1 \\ K2 & - K2 \end{bmatrix}$$
(C25) Phi:addcol(ematrix(3,1,1,1,1),addrow(Alpha,Beta));

(D25)
$$\begin{bmatrix} 1 & Q5^2 - Q4 + Q3^2 & - Q5^2 + Q4 - Q3^2 & - Q3 \\ 0 & 0 & & K1 \\ 0 & K2 & & - K2 \end{bmatrix}$$
(C26) quit();

UTILISATION DU CALCUL FORMEL
POUR L'ETUDE ET LA COMMANDE DES SYSTEMES
LINEAIRES : APPROCHE TRANSFERT.

A. BENAMARA, J. P. GUERIN

Laboratoire d'Automatique de Grenoble (U.R.A. CNRS 228)

ENSIEG -INPG- BP 46.

38402 Saint Martin d'Hères FRANCE.

Résumé : Des outils de base pour l'étude et la commande des systèmes linéaires par l'approche transfert sont développés en MACSYMA ; on obtient alors des calculs exacts des formes de Smith et de Smith-MacMillan ainsi que des factorisations premières à droite et à gauche des matrices de transfert. Ces résultats sont utilisés pour l'étude structurelle des systèmes linéaires mutivariables et la synthèse de correcteurs réalisant des objectifs de commande. Les structures de résolution des problèmes de poursuite parfaite avec stabilité et de rejet de perturbation par mesure de sorties avec stabilité sont présentées.

1. Introduction

L'introduction de l'anneau des fractions rationnelles propres et stables ($\mathcal{R}_\infty$) dans l'étude des systèmes linéaires a permis d'aborder sous un nouvel angle les problèmes de l'automatique linéaire. L'idée principale de cette approche - dite approche transfert - est la factorisation des matrices de transfert des systèmes en un quotient de deux matrices à éléments dans $\mathcal{R}_\infty$ [6][15]. Cette factorisation conduit alors à un résultat essentiel qui est la paramétrisation de tous les correcteurs qui stabilisent un procédé donné. On choisit parmi cette infinité de correcteurs ceux qui réalisent divers objectifs que l'automaticien peut se fixer. Les objectifs du type stabilisation, poursuite parfaite, rejet de perturbation ou découplage sont réalisés en utilisant uniquement des outils algébriques (forme de Smith, forme de Smith-MacMillan, factorisations premières à droite et à gauche, ...). Par ailleurs, les propriétés analytiques et topologiques de $\mathcal{R}_\infty$ (en plus de ses propriétés algébriques) permettent d'aborder les problèmes qui n'ont pas de solutions exactes ; ce sont des problèmes liés à des critères de minimisation : poursuite approchée, minimisation de sensibilité par apport aux perturbations, robustesse... .

Dans ce papier nous nous intéressons aux objectifs du premier type et que l'on peut appeler objectifs algébriques. Les outils de base (forme de Smith, forme de Smith-MacMillan et factorisations premières) ont déjà fait l'objet de développements algorithmiques et ont été programmés dans l'environnement de calcul formel MACSYMA [1][3][10]. Ces procédures de base reposent essentiellement sur un calcul de pgcd (plus grand diviseur commun) sur $\mathcal{R}_\infty$ évitant les difficultés numériques (séparation des polynômes en parties "stable" et "instable") de la division euclidienne sur $\mathcal{R}_\infty$ [1][8]. Cette approche est différente de celle qui utilise les matrices polynomiales [2][16] - dite approche polynomiale - ; en effet, en travaillant directement sur l'anneau $\mathcal{R}_\infty$, on intègre de façon naturelle les problèmes liés à la stabilité. Les correcteurs ainsi obtenus garantissent la stabilité de la boucle fermée. Nous nous proposons, ici, de montrer comment ces outils de base permettent de résoudre - de façon simple - les problèmes de poursuite parfaite avec stabilité et de rejet de perturbation par mesure de sorties avec stabilité.

Nous adopterons le plan suivant : dans une première partie nous rappelons les formes de Smith et de Smith-MacMillan ainsi que les factorisations premières à droite et gauche des matrices de transfert. Nous présentons ensuite la paramétrisation des correcteurs stabilisant un procédé donné. Nous terminons en présentant l'utilisation de cette paramétrisation pour réaliser la poursuite parfaite et le rejet de perturbation avec mesure de la sortie. Les structures des algorithmes de résolution sont présentés en montrant les outils algébriques utilisés. Quelques indications sur l'utilisation du calcul formel (MACSYMA) pour le développement de ces algorithmes sont données.

2. Préliminaires

On notera par $\mathcal{R}_p$ l'ensemble des fractions rationnelles propres et $\mathcal{R}_\infty$ l'anneau des fractions rationnelles propres et stables. $\mathcal{M}_p$ et $\mathcal{M}_\infty$ désigneront les ensembles des matrices à éléments dans $\mathcal{R}_p$ et dans $\mathcal{R}_\infty$ respectivement. On omet les dimensions par souci de simplicité des notations.

Rappelons qu'une matrice U est une *unimodulaire* de $\mathcal{M}_\infty$ si det U est une unité de $\mathcal{R}_\infty$ (i.e : U est inversible dans $\mathcal{M}_\infty$). Nous rappelons quelques résultats classiques sur les matrices de $\mathcal{M}_\infty$ et de $\mathcal{M}_p$.

Forme de Smith [9][11] : Soit $A \in \mathcal{M}_\infty$ avec rang(A) = k. Alors il existe U et V unimodulaires de $\mathcal{M}_\infty$ telles que :

$$U \cdot A \cdot V = \begin{bmatrix} a_1 & & 0 \\ & \cdot & \\ & & a_k \\ 0 & & 0 \end{bmatrix}$$

où $a_i \in \mathcal{R}_\infty$ (i = 1, ..., k) et a_i divise a_{i+1} pour i = 1, ..., k-1.

Les a_i sont uniques à des unités prés de $\mathcal{R}_\infty$ et sont appelés facteurs invariants de la matrice A.

Forme de Smith-MacMillan [14][15] : Soit $P \in \mathcal{M}_p$ avec rang(P) = k. Alors il existe U et V unimodulaires de $\mathcal{M}_\infty$ telles que :

$$U \cdot P \cdot V = \begin{bmatrix} a_1 . b_1^{-1} & & 0 \\ & a_k . b_k^{-1} & \\ 0 & & 0 \end{bmatrix}$$

Avec a_i , b_i $\in \mathcal{R}_\infty$, De plus a_i divise a_{i+1} , b_i divise b_{i-1} et les a_i et b_i sont premiers entres eux (i = 1, ..., k).

Notons que la forme de Smith-MacMillan est très utile pour l'étude de la structure du système linéaire. En effet les zéros instables des a_i sont les zéros instables du système, les zéros instables des b_i sont ses pôles instables. Les zéros à l'infini des a_i sont aussi ceux du système.

Soient A et B deux matrices de $\mathcal{M}_\infty$. A et B sont dites *premières à droite* (resp. à gauche) ssi il existe deux matrices X et Y de $\mathcal{M}_\infty$ telles que : X.A + Y.B = I (resp. A.X + B.Y = I).
Soit $P \in \mathcal{M}_p$. Une paire (N_d, D_d) $(N_d \in \mathcal{M}_\infty, D_d \in \mathcal{M}_\infty)$ est une *factorisation première à droite* (**FPD**) de P si : D_d est carrée avec det $D_d \neq 0$, $P = N_d . D_d^{-1}$ et N_d et D_d sont premières à droite. De même, une paire (N_g, D_g) $(N_g \in \mathcal{M}_\infty, D_g \in \mathcal{M}_\infty)$ est une *factorisation première à gauche* (**FPG**) de P si : D_g est carrée avec det $D_g \neq 0$, $P = D_g^{-1} . N_g$ et N_g et D_g sont premières à gauche.

Théorème 1 [15] : *Soit $P \in \mathcal{M}_p$. Alors il existe N_d, D_d, N_g, D_g, X_d, Y_d, X_g et Y_g dans $\mathcal{M}_\infty$ telles que : (N_d, D_d) et (N_g, D_g) sont des factorisations premières à droite et à gauche de P et :*

$$X_d . N_d + Y_d . D_d = I \text{ et } N_g . X_g . + D_g . Y_g = I .$$

3. Correcteurs stabilisants [6][15]

Etant donné un procédé caractérisé par sa matrice de transfert $G(s) \in \mathcal{M}_p$, on cherche l'ensemble des correcteurs C(s) qui stabilisent le procédé P.

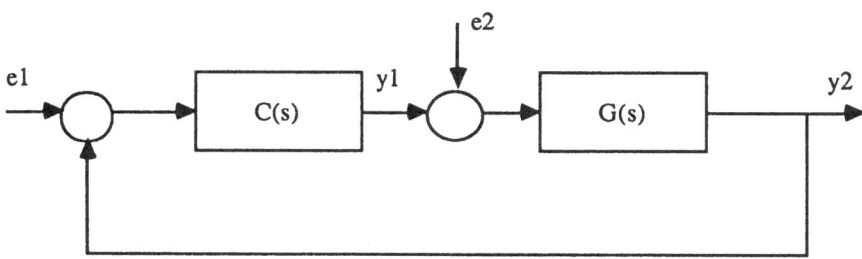

Figure 1

e_1 et e_2 sont les entrées externes ; y_1 et y_2 sont les sorties du correcteur et du procédé. Soit G(s) le transfert entre (e_1, e_2) et (y_1, y_2) :

$$\begin{bmatrix} y1 \\ y2 \end{bmatrix} = G(s) \cdot \begin{bmatrix} e1 \\ e2 \end{bmatrix}.$$

Le correcteur C est dit *stabilisant* ou *stabilise* G si le système de la figure 1 est stable, i.e : la matrice de transfert G(s) $\in \mathcal{M}_\infty$.

En utilisant les factorisations premières à droite et à gauche de G, on peut paramétriser tous les correcteurs stabilisant G.

Théorème 2 [6][15] : Soit P$\in \mathcal{M}_p$. Soient (N_d, N_d) une FPD et (N_g, N_g) une FPG de P. On choisit X_d, Y_d, X_g et Y_g dans $\mathcal{M}_\infty$ tels que :

$$X_d.N_d + Y_d.D_d = I \ \text{ et } \ N_g.X_g + D_g.Y_g = I .$$

Alors l'ensemble des correcteurs stabilisants P est donné par les C(s) qui s'écrivent :

ou

$$C(s) = (Y_d - R . N_g)^{-1} (X_d + R . D_g)$$

$$C(s) = (X_g + D_d . Q).(Y_g - N_d . Q)^{-1}$$

Avec : $R\in \mathcal{M}_\infty$, $Q\in \mathcal{M}_\infty$, $Det(Y_d - R . N_g) \neq 0$ et $Det(Y_g - N_d . Q) \neq 0$.

Ainsi nous avons la paramétrisation (par R ou Q) de tous les correcteurs qui stabilisent un procédé donné. Parmi ceux là, nous choisirons ceux qui réalisent des objectifs divers tels que la poursuite de modèle ou le rejet de perturbation.

4. Poursuite parfaite avec stabilité [15]

Soit un procédé caractérisé par son transfert P(s). Le problème de la poursuite parfaite avec stabilité consiste à trouver un compensateur C(s) stabilisant P tel que le comportement entrée-sortie du système de la figure 2 soit identique à celui de la figure 3.

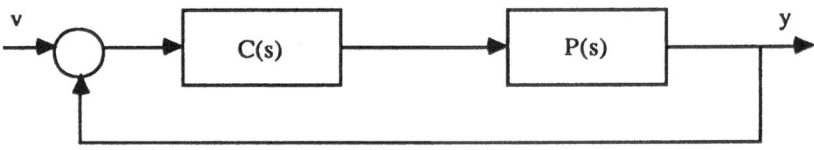

Figure 2

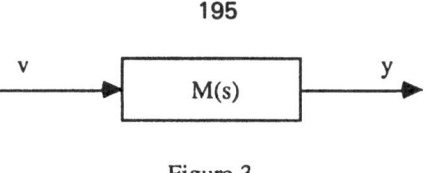

Figure 3

Ceci revient à trouver C(s) dans $\mathcal{H}_p$ qui stabilise P et tel que :

$$P(s).C(s).[\ I + P(s).C(s)\ ]^{-1} = M(s)$$

où M(s) est la matrice de transfert du modèle (M(s)∈$\mathcal{H}_\infty$).

En utilisant le théorème 2, on montre que le problème se ramène à la résolution d'une équation matricielle sur $\mathcal{H}_\infty$ du type :

$$A \ . \ X = B \ .$$

qui se résout en utilisant la forme de Smith de A dans $\mathcal{H}_\infty$ [9][15].

Nous pouvons résumer la résolution du problème de la poursuite parfaite avec stabilité en donnant la structure globale de la procédure (fig. 4) qui met en relief les deux modules de base que sont la forme de Smith et les factorisations premières. Ces deux modules ont été programmés sous MACSYMA et on voit que la résolution de la poursuite parfaite en découle de façon immédiate.

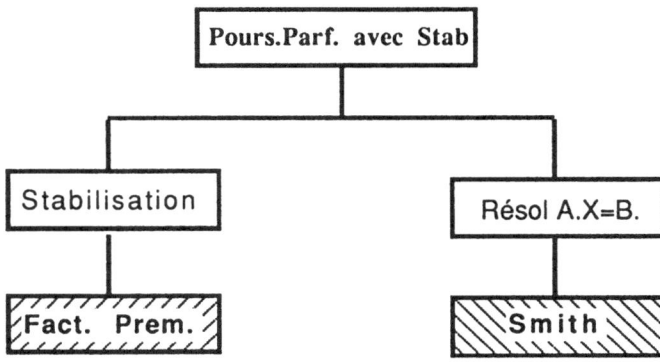

Figure 4

5. Rejet de perturbation avec mesure de sortie [4][5][6][12][13]

Soit le système représenté par le schéma suivant :

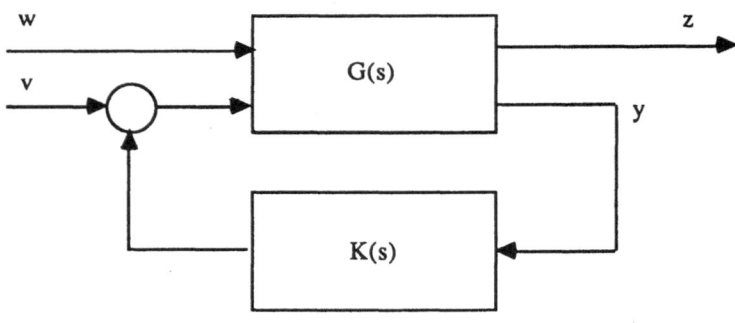

Figure 5

Où w est la perturbation non mesurée, u la commande, z la sortie à réguler, y la sortie mesurée, et v l'entrée externe.

Le procédé est représenté par sa matrice de transfert G(s) telle que :

$$\begin{bmatrix} z \\ y \end{bmatrix} = \begin{bmatrix} G_{11} & G_{12} \\ G_{21} & G_{22} \end{bmatrix} \cdot \begin{bmatrix} w \\ u \end{bmatrix}$$

On se propose de trouver une commande du type : u = K . y + v telle que :

> 1/ K stabilise le procédé G
>
> 2/ K rejette la perturbation w.

La résolution de ce problème nécessite l'introduction de la notion d'admissibilité des procédés.

Définition [4][12] : Soit (N,D) une FPD de G et (N_{22}, D_{22}) une FPD de G_{22}. On dira que G est *admissible* s'il existe u unité de $\mathcal{R}_\infty$ tel que Det D = u . Det D_{22}.

La notion d'admissibilité, dans l'approche transfert, est l'équivalent des notions de stabilisabilité et de détéctabilité dans l'approche des systèmes linéaires par l'état.

Sous l'hypothése d'admissibilité du procédé et en utilisant la paramétrisation des correcteurs stabilisant G_{22} (Théorème 2), on montre que le problème se ramène à la résolution d'une équation matricielle sur $\mathcal{M}_\infty$ du type [5]:

$$A \cdot X \cdot B = C$$

qui se résout en utilisant les formes de Smith de A et de B [13].

Une structure algorithmique globale (fig. 6) résume la résolution du rejet de perturbation avec stabilité. Les deux modules de base de la forme de Smith et des factorisations premières sont mis en évidence.

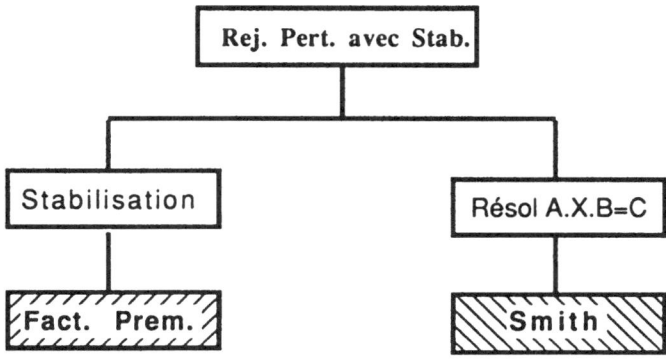

Figure 6

6. Utilisation du calcul formel

L'orientation vers le calcul formel (MACSYMA) pour le développement des différents algorithmes a été motivée par différentes raisons : obtenir des calculs exacts, traiter des problèmes dépendants de paramètres, éviter des calculs fastidieux, ...

Les différents modules de base (forme d'Hermite, forme de Smith, forme de Smith-MacMillan, factorisations premières, résolutions de A.X=B et A.X.B=C sur $\mathcal{M}_\infty$) reposent essentiellement sur le calcul de pgcd sur $\mathcal{R}_\infty$. Un algorithme original évitant les difficultés numériques de la division euclidienne sur $\mathcal{R}_\infty$ a été développé [1]. De ces modules de base découlent des fonctions directement applicables en automatique des systèmes linéaires. Les principales sont les suivantes :

- Factorisation première à droite   Entrée : $P(s) \in \mathcal{M}_p$ (transfert du système)

         Sortie : N_d, D_d, X_d, Y_d dans $\mathcal{M}_\infty$ tels que :

$$P = N_d . D_d^{-1} \quad \text{et} \quad X_d . N_d + Y_d . D_d = I$$

- Factorisation première à gauche   Entrée : $P(s) \in \mathcal{M}_p$ (transfert du système)

         Sortie : N_g, D_g, X_g et Y_g dans $\mathcal{M}_\infty$ tels que :

$$P = D_g^{-1} . N_g \quad \text{et} \quad N_g . X_g + D_g . Y_g = I$$

- Paramétrisation des correcteurs stabilisant   Entrée : $P(s) \in \mathcal{M}_p$

         Sortie : $C(s)$ stabilisant P

- Poursuite parfaite avec stabilité

Entrée : $P(s) \in \mathcal{M}_p$ et $M(s) \in \mathcal{M}_\infty$ (modèle)

Sortie : $C(s)$ réalisant la poursuite avec stabilité.

- Rejet de perturbation avec stabilité

Entrée : $P(s) \in \mathcal{M}_p$ et $P(s)$ admissible.

Sortie : $C(s)$ réalisant le rejet avec stabilité.

D'autres algorithmes feront l'objet de développements sous MACSYMA et seront intégrés à ce logiciel : factorisations antipremières, découplage... . Par ailleurs nous ne disposons pas encore de module pour tester l'appartenance des fractions rationnelles à $\mathcal{R}_\infty$, mais ce problème a déjà fait l'objet de travaux de recherche [7]. Ce module permettra de vérifier la validité de certains théorèmes : admissibilité des procédés, existence des solutions des équations matricielles... .

7. Conclusion

Un calcul de pgcd sur l'anneau des fractions rationnelles propres et stables évitant les difficultés numériques de la division sur cet anneau a été programmé sur MACSYMA. Ce module de base a permis de développer, sous MACSYMA, les outils algébriques nécessaires à l'approche transfert : formes de Smith et de Smith-MacMillan, factorisations premières à gauche et à droite. Dans ce papier, nous avons mis en relief le rôle fondamental de ces outils pour l'étude structurelle des systèmes linéaires mutivariables, pour la stabilisation et la réalisation d'objectifs de commande (poursuite parfaite, rejet de perturbation).

Le calcul formel est particulièrement adapté pour ce genre de calcul et nous mettons ainsi à la disposition de l'automaticien des outils qui permettent la synthèse - de manière exacte - de régulateurs pour la commande des systèmes linéaires.

Références

[1] **Benamara A., Guerin J.P.**, "Quelques outils pour l'étude des systèmes linéaires. Utilisation du calcul formel", Note interne L.A.G. No 90/89.

[2] **Chenin P.**, "AUTORED : un logiciel pour l'automatique linéaire", Outils et modèles mathématiques pour l'automatique, l'analyse des systèmes et le traitement du signal, ed. C.N.R.S., Paris, 1983.

[3] **Davenport J., Siret Y., Tournier E.**, "Calcul formel : systèmes et algorithmes de manipulations algébriques", Masson, Paris, 1987.

[4] **Desoer C.A., Gündes A.N.**, "Algebraic theory of linear time-invariant feedback systems with two-input two-output plant and compensator.", Int. J. Control, Vol. 47, No. 1, 1988.

[5] **Ferreira P.M.G.**, "DDP and model following problem revisited", Systems & Control Letters 12, 1989.

[6] **Francis B.A.**, "A course in control theory", Lecture notes in control and information science, No. 88, Springer, Berlin - New-York, 1987.

[7] **Gleyse B.**, "Calcul formel et nombre de racines d'un polynôme dans le disque unité : application en automatique et biochimie", Th. 3e. cycle, Math. Appl. Grenoble, 1986.

[8] **Hung N.T., Anderson B.D.O.**, "Triangularization technique for the design of multivariable control systems", IEEE Trans. Automat. Contr., Vol. AC-24, No. 3, June 1978.

[9] **MacDuffee C.C.**, "The theory of matrices", Chelsea, New-York, 1956.

[10] **MACSYMA** reference manual, Symbolics inc., 1988.

[11] **Marcus M.**, "Introduction to modern algebra", Marcel Decker inc., New-York, 1978.

[12] **Nett C.N.**, "Algebraic aspects of linear control system stability", IEEE trans. automat. contr., Vol. AC-31, No. 10, October 1986.

[13] **Özgüler A.B., Eldem V.**, "Disturbance decoupling problems via dynamic output feedback", IEEE Trans. Automat. Contr., Vol. AC-30, No. 8, August 1985.

[14] **Vardulakis A.I.G., Karcanias N.**, "Structure, Smith-MacMillan form and coprime MFDs of a rational matrix inside a region $\mathcal{P} = \Omega \cup \{\infty\}$ ", Int. J. Control, Vol. 38, No.5, 1983.

[15] **Vidyasagar M.**, "Control system synthesis : a factorization approach", Cambridge, M.I.T. Press, 1985.

[16] **Wolovich W. A.**, "Linear multivariable systems", Springer-Verlag, New-York, 1974.

FAST ALGORITHMS FOR STRUCTURED MATRICES AND INTERPOLATION PROBLEMS

Georg Heinig

Universität Leipzig, Sektion Mathematik, D-O-7010 Leipzig

1. INTRODUCTION

Many problems in systems and control theory are actually interpolation or approximation problems for rational functions or other special function classes. For example, partial realization for linear time-invariant systems is equivalent to a special case of Padé approximation. Furthermore, many considerations in H_∞ control and the theory of passive systems and scattering theory lead to the Nevanlinna-Pick or related problems, which are interpolation problems with metric constraints. For application the rational solutions are of special interest. In view of the importance of rational interpolation and approximation for application there exists a vast literature in this field. We quote only, as examples, [3,5] and especially the recent monograph [4] by Ball/Gohberg/Rodman, where a fairly complete theory is developed and many applications to systems theory are considered. However, in [4] there is no treatment of the algorithmical side of the problems. It would be a task for the future to write a systematic account of algorithms for rational interpolation and approximation covering the many different results contained in the literature.

In the present contribution we want to point out only one but we think important view-point in connection with the theory and algorithms of interpolation which consists in the following. Interpolation and approximation problems are often related with linear systems of equations in such a way that the solution of the system provides a solution of the original problem. The coefficient matrix of such a system is usually not sparse but possesses a certain structure such as the structure of a Vandermonde, Cauchy, Hankel, Toeplitz, Loewner, Pick matrix or generalization of these types.

The complexity of the classical solution algorithms for systems of n linear equations equals $O(n^3)$. For many systems with a structured coefficient matrix there exist solution algorithms with $O(n^2)$ or even lower complexity. Algorithms with complexity less than $O(n^3)$ are said to be fast. Fast algorithms exist for arbitrary systems (see [1]). An essential gain, however, is obtained only for structured or sparse systems. Classical fast algorithms for structured systems are the Levinson algorithm, originally developed for special Toeplitz systems, and the Schur algorithm, which is in its original form an algorithm in complex function theory. A survey of the state of the art of the theory of structured matrices in the mid-eighties is the monograph [13]. The Schur algorithm has the advantage that it can be completely parallized to an $O(n)$ complexity algorithm at an $O(n)$ processors computer (see [7]). Algorithms with the complexity $O(n \log^2 n)$ for the solutions of Toeplitz systems were presented by Brent/Gustavson/Yun, de Hoog, Ammar and others (see [11]). Such algorithms are called assymptotically fast or superfast. So far these algorithms are only partly transferred to other classes of matrices.

A natural idea is to utilize the theory and algorithms for structured matrices in order to solve interpolation problems. It turned out that there is actually a two-sided interplay between interpolation and structured matrices. Many structured systems can be interpreted as interpolation problems and this interpretation leads to a better understanding of the matrix structure and to new algorithms.

In the present contribution we want to give a flavour of this interplay. Furthermore, we present a general approach of the construction of fast algorithms for solving different classes structure systems including parallel and superfast algorithms.

Especially we pay attention to the following problem. Many algorithms presented in the literature work only under certain conditions such as definiteness or strong regularity. Therefore there is the task to construct algorithms without additional assumptions. This is just what we do.

2. A COLLECTION OF STRUCTURED MATRICES IN INTERPOLATION

Speaking about interpolation we have in mind the following type of problems:

Given a class $\mathcal{F}$ of complex functions, points c_i of the extended complex plane and numbers $y_{ik} \in \mathbb{C}$; find $f \in \mathcal{F}$ satisfying $\frac{1}{k!} f^{(k)}(c_i) = y_{ik}$ ($k=0,\ldots,\nu_i-1, i=1,\ldots,n$).

The case when all ν_i equal one is called the non-confluent case.

We give some examples how interpolation problems lead to structured systems. First of all let us note that the corresponding matrix essentially depends on the parametrization of $\mathcal{F}$, if $\mathcal{F}$ is a linear space it depends on the basis in $\mathcal{F}$. The parametritzation does not only influence the obtained matrix but, what is still more important in practice, the stability of the corresponding numerical procedures.

The simplest case of rational interpolation is the non-confluent polynomial interpolation with respect to the standard basis $\{\lambda^k\}$ ($k=0,\ldots,n-1$). This problem leads to a system $Ax=y$ the coefficient matrix of which is the Vandermonde matrix $A=[c_i^{j-1}]_1^n$. In the comfluent case one has a generalized Vandermonde matrix with the entries $\binom{j}{k} c_i^{j-k}$ ($k=0,\ldots,\nu_j-1$).

Vandermonde matrices are, in general, ill-conditioned which causes numerical instabilities. The situation is, as a rule, better if one takes orthogonal polynomials as the basis of $\mathcal{F}$. Especially Chebyshev polynomials turned out to be very convenient. This leads to a certain class of Vandermonde-like matrices that can be treated similarly as the classical Vandermonde matrices (see [10]).

Polynomial interpolation is a special case of rational interpolation with prescribed poles. The simplest non-polynomial problem of this kind is that for $\mathcal{F} = \text{span} \{\frac{1}{\lambda-d_1}, \ldots, \frac{1}{\lambda-d_n}\}$. This leads to systems with a so-called Cauchy coefficient matrix $[\frac{1}{c_i-d_j}]_1^n$.

If the number of interpolation conditions is larger than the dimension of $\mathcal{F}$ then the interpolation must be replaced by an approximation problem, for example by the problem to find $f \in \mathcal{F}$ such that $\sum |f(c_i)-y_i)|^2$ becomes minimal. This or more general least-square approximation problems lead to systems of equations the coefficient matrix of which is the Gram matrix of the columns of the corresponding interpolation matrix. For $\mathcal{F}$ being a polynomial space and the standard basis this matrix equals $[s_{i+j}]$, where $s_i = \sum_{k=1}^{n} c_k^i$, i.e. it is a Hankel matrix. If one chooses Chebyshev polynomials as a basis one will obtain a coefficient matrix which is the sum of a Toeplitz and a Hankel matrix, i.e. it is of the form $[s_{i+j}+t_{i-j}]$. If $\mathcal{F}$ is the span of functions $(\lambda-d_j)^{-1}$, then the coefficient matrix will be a symmetric Loewner matrix.

Now we discuss the problem of rational interpolation without prescribed poles. Here $\mathcal{F}$ is the class of all rational functions with a fixed degree of the numerator and denominator. $\mathcal{F}$ is no linear space, hence the problem is nonlinear. However the solution set can be imbedded in a slightly larger solution set of a close linear problem, which is obtained after the multiplication by the denominator.

In the case $n=1$, $c_i=0$ or $c_i=\infty$ the problem is the classical Padé approximation problem. It is well-known that the coefficient vectors of the denomiantors of the Padé approximations are solutions of homogeneous systems with Hankel coefficient matrices of the form $[y_{i+j}]$. In the non-confluent case the linearized rational interpolation problem is related to matrices of the form $[\frac{y_i-y_j}{c_i-c_j}]$, where i and j run over disjoint index sets. Matrices of this kind are referred to as (nonsymmetric) Loewner matrices., They are closely related to the Pick matrices occurring in the Nevanlinna-Pick problem.

From the kernel of the Loewner matrices corresponding to given interpolation data one gets all denominator and numerator polynomials of the interpolation solutions. More general interpolation conditions lead to matrices that we shall call matrices of Hankel-Loewner type.

3. INVERSION FORMULAS

The first step of our approach are formulas for the inverse matrix involving $O(n)$ parameters. As we shall see in the case of Cauchy-Vandermonde type matrices from these formulas algorithms can immediately deduced.

Cauchy-Vandermonde matrices

It is a consequence of Lagrange's interpolation formula that the inverse of a Vandermonde matrix V corresponding to the data c_i is given by

$$V^{-1} = HV^T D,$$ (3.1)

where $D = \text{diag}(1/h'(c_i))$, $h(\lambda) = \prod_{i=1}^{n} (\lambda - c_i)$ and H is a triangular Hankel matrix generated by $h(\lambda)$. The parameters occurring in the formula can easily be computed, if FFT is utilized even with $O(n \log^2 n)$ complexity (see [1]). For our approach it is important to note that the coefficient vector of $h(\lambda)$ is a solution of the homogeneous Vandermonde system obtained by adding another row to V.

From Hermite's interpolation formula one can deduce a similar formula for the inverse of a generalized Vandermonde matrix, involving the transposed matrix.

For Cauchy matrices C one can evaluate a formula

$$C^{-1} = D_1 C^T D_2,$$ (3.2)

where D_1 and D_2 are diagonal matrices involving values of the polynomials $\prod(\lambda - c_i)$ and $\prod(\lambda - d_i)$ and their derivatives. It was recently observed by the author that a formula of the form (3.2) also holds for general Cauchy-Vandermonde matrices. Here D_1 and D_2 are block diagonal matrices the blocks of which are triangular Hankel. This result will be published elsewhere. The parameters can be computed directly from the data by fast and superfast algorithms or immediately from the solution of a certain homogeneous Cauchy-Vandermonde system.

The multiplication of a vector by a transposed Vandermonde or Cauchy matrix can be carried out with $O(n \log^2 n)$ complexity if FFT is utilized (see [1] for Vandermonde and [6] for Cauchy). In that way one can construct fast and superfast algorithms for the solution of Cauchy-Vandermonde systems and, therefore, for rational interpolation with prescribed poles.

Matrices of Hankel-Loewner type

For matrices of Hankel-Loewner type the inversion problem is some more difficult. We consider here only the cases of a pure Hankel and Loewner matrix. Let $H = [s_{i+j}]_0^{n-1}$ be regular. We introduce the $(n-1) \times (n+1)$ matrix $H' = [s_{i+j}]_0^{n-2} {}_0^n$. Clearly ker H' is two-dimensional. Any basis $\{u_1, u_2\}$ of ker H' will be called fundamental system, since the inverse matrix can be represented with the help of such a system via a Gohberg-Semencul type formula (see [10]):

$$H^{-1} = c(B(u_1)T(u_2) - B(u_2)T(u_1)) , \qquad (3.3)$$

where

$$B(u_i) = \begin{bmatrix} u_{i1} & \cdots & u_{in} \\ \cdot & & \cdot \\ \cdot & & \cdot \\ \cdot & \cdot & \\ u_{in} & & 0 \end{bmatrix} \qquad T(u_i) = \begin{bmatrix} u_{i0} & \cdots & u_{in-1} \\ & & \cdot & \vdots \\ & & \cdot & \vdots \\ 0 & & u_{i0} \end{bmatrix}$$

$(i = 1, 2)$ and c is a constant. In that way the solution of $Hx = b$ is reduced to some convolutions and can be carried out with $O(n \log n)$ flops if FFT is applied. Let us note that recently Ammar/Gader [2] observed that other formulas involving circulant matrices are still more effective for computation. A general account of the different inversion formulas and generalizations is contained in [9]. Let us still note a fact that is known to specialists but usually not explicitly stated in the literature. Formula (3.3) can be transformed into a form that is similar to the formulas for Cauchy-Vandermonde matrices:

$$H^{-1} = DH_1 D. \tag{3.4}$$

Here D is a triangular Hankel matrix and H_1 is another Hankel matrix. Different to Cauchy-Vandermonde matrices, the parameters occurring in this formula cannot be computed directly from the data but only, as described above, via the solution of a homogeneous Hankel system. Let us note that the rational functions corresponding to the matrices H and H_1 have the same denominator and the numerators are connected via a polynomial congruence modulo the denomiantor. We think that it is worth to test and estimate also formulas of this type.

For Loewner matrices the situation is completely analogous. Let the Loewner matrix $L = [1_{ij}]$, $i = n+1,\ldots,2n$; $j = 1,\ldots,n$, $1_{ij} = (y_i - y_j)/(c_i - c_j)$ be regular. L' is by definition the $(n-1) \times (n+1)$ matrix $[1_{ij}]$, $i=n+2,\ldots,n$; $j=1,\ldots,n$. We also introduce $\hat{L}' = (L^T)'$. We have again dim ker L' = dim ker $\hat{L}'$ = 2. Any basis of ker L' will be called right and any basis of ker$\hat{L}'$ left fundamental system. Now L^{-1} possesses entries of the form $(x_i \hat{v}_j - v_i \hat{x}_j)/(c_{i+n} - c_j)$, where the $x_i, v_i, \hat{x}_i, \hat{v}_i$ can simply obtained from the fundamental system. It follows from the arguments in [6] that the multiplication of matrices L^{-1} by a vectors requires only $O(n \log^2 n)$ flops. That means Lx = b can be solved with this amount, provided that a fundamental system is given. Analogously to (3.3) there is a representation $L^{-1} = D_1 L_1 D_2$, where D_1 and D_2 are diagonal and L_1 is Loewner.

Let us finally consider Toeplitz-plus-Hankel matrices. Suppose that $A = [s_{i+j} + t_{i-j}]$ is nxn and regular. Define $A' = [s_{i+j} + t_{i-j}]_1^{n-2} {}_{-1}^n$. Then ker A' has dimension 4. Any basis of ker A' will be called right fundamental system. Replacing A by its transposed we get the definition of a left fundamental system. In [14] it is shown that A^{-1} can be represented with the help of triangular Toeplitz matrices involving the vectors of the fundamental system. Applying FFT one will obtain an O(n log n) complexity algorithm for solving a system Ax = b for a Toeplitz-plus-Hankel matrix A, provided that a fundamental system is given.

4. KERNEL STRUCTURE

We saw in the previous section that the inverse of a matrix with a structure of any type under consideration is determined by a fundamental system, which is the basis of the kernel of a certain related structured matrix. In order to solve a structured system of one of the forms considered above it remains to construct algorithms for the fundamental systems. This can be done recursively. Since we want to avoid any restriction we have to define the concept of a fundamental system also for nonregular matrices. This can be done using some kernel structure properties of these matrices.

In the sequel we restrict ourselves to Hankel, Loewner and Toeplitz-plus-Hankel matrices. Let us point out that all consider- ations can also be carried out for Cauchy-Vandermonde matrices, but in view of the more explicit inversion formulas this consideration are not so essential than for the other matrix classes.

(a) <u>Hankel matrices.</u> A given sequence $s=(s_0, \ldots, s_{N-1})$ will be associated with the family of Hankel matrices $H_k = [s_{i+j}]_0^{N-k} \, _0^{k-1}$, $k=1, \ldots, N$. We study the subspaces ker H_k in "polynomial language", i.e. we look for polynomials $x(\lambda) = x_0 + \ldots + x_{k-1}\lambda^{k-1}$ for which $x = (x_i)$ belongs to ker H_k. It turnes out that there exist two polynomials $u_1(\lambda), u_2(\lambda)$ such that any $x(\lambda)$ can be represented as linear combination of polynomials $\lambda^k u_1(\lambda)$ and $\lambda^k u_2(\lambda)$ ($k = 0, \ldots, \nu_i-1$, $i = 1, 2$). The pair $(u_1(\lambda), u_2(\lambda))$ will called fundamental system.

(b) <u>Loewner matrices.</u> Two sequences $y=(y_1, \ldots, y_N)$ and $c=(c_1, \ldots, c_N)$ will be associated with the Loewner matrices $L_k = [l_{ij}]_{k+1}^N \, _1^k$, $l_{ij} = (y_i - y_j)/(c_i - c_j)$ ($k=1, \ldots, N$). Analogously to the Hankel case the kernels of the L_k are generated by two vectors x_1 and x_2. It is convenient to utilize the connection between the x_1 and x_2 and the denominator polynomials $u_1(\lambda)$ and $u_2(\lambda)$ of the solution of the corresponding rational interpolation problem for, viz.

$$u_i(\lambda) = h(\lambda) \sum_{j=1}^{k_i} x_{ij}/(\lambda - c_j), \quad h(\lambda) = \prod_{j=1}^{n} (\lambda - c_j), \quad x_i = (x_{ij})_1^{k_i}.$$

(i=1,2). The pair $(u_1(\lambda), u_2(\lambda))$ will also be called fundamental system. We shall also work with the corresponding to $u_1(\lambda)$ and $u_2(\lambda)$ numerator polynomials denoted by $p_1(\lambda)$ and $p_2(\lambda)$.

(c) Toeplitz-plus-Hankel matrices. For two given sequences s and t of length N we define $A_k = H_k + T_k$, where H_k is the (N+1-k)×k Hankel matrix corresponding to s and T_k is the (N+1-k)×k Toeplitz matrix related to t. We consider the two families $\{A_{2j}\}$ and $\{A_{2j+1}\}$ seperately. For each of the two families there exist 4 polynomials $u_i(\lambda)$ such that each $x(\lambda)$ with $x \in \ker A_k$ is a linear combination of polynomials $(\lambda^2+1)^j u_i(\lambda)$. Clearly the collection of the $u_i(\lambda)$ will be called fundamental system.

5. TYPE I ALGORITHMS

The next step is to study how fundamental systems change after the extension of the parameter set. The results will lead to $O(N^2)$ algorithms for computing a fundamental system, which are, in a sense, generalizations of the Levinson algorithm.

Let us start with Hankel matrices. If $\{u_1^n, u_2^n\}$ is a fundamental system of $s(n)=(s_0, \ldots, s_{n-1})$ then a fundamental system of $s(n+1)$ can be obtained via a formula

$$[u_1^{n+1} \quad u_2^{n+1}] = [u_1^n \quad u_2^n]\, \Phi_n(\lambda) \qquad (4.1)$$

where $\Phi_n(\lambda)$ is a unimodular 2×2 degree one polynomial matrix, which can be obtained via 2 inner product calculations (for details see [10]).

(b) A formula of the form (1) also holds for the fundamental system of Loewner matrices after extensions $c \to (c, c_{n+1})$, $y \to (y, y_{n+1})$ (only $\Phi_n(\lambda)$ is not unimodular). In terms of the kernel vectors x_1, x_2 in the formula (1) the polynomial product has to be replaced by the Hadamard product of the corresponding vectors.

(c) For Toeplitz-plus-Hankel matrices the situation is some more involved, but the principle is the same: The fundamental system after extension is obtained via multiplication by a 4×4 polynomial matrix $\Phi_n(\lambda)$. More over $\Phi_n(\lambda)$ is obtained via inner product calculations.

6. TYPE II ALGORITHMS

The algorithms in Section 4 are not very suited for parallel computations since inner products are involved. The idea for another class of algorithms is to study the recursions for the so called "residuals". The corresponding formulas are very similar to the classical Schur algorithm, which was originally developed to answer the question whether a given function maps the unit disk into itself.

(a) First we consider Hankel matrices. Suppose that s with length N is given and $\{u_1^n, u_2^n\}$ is a fundamental system of $s(n)$. Let $\bar{H}_k$ denote the matrix with k columns corresponding to the $N+k-1$ dimensional vector $(0, \ldots, 0, s)$. The vectors $\bar{H}_{k_1} u_1^n = r_1^n$, $\bar{H}_{k_2} u_2^n = r_2^n$ (for suitable k_1, k_2) are called residuals of u_1^n and u_2^n. The polynomials cooresponding to the residuals fulfil a recursion

$$[r_1^{n+1} \quad r_2^{n+1}] = [r_1^n \quad r_2^n] \Psi_n(\lambda),$$

where $\Psi_n(\lambda)$ involves only components of the residuals, and Ψ_n is closely related to Φ_n in (4.1). The recursion of the residual can be done parallelly. This leads to algorithms for computing fundamental systems which require $O(N)$ steps on a N processor computer (for details see [10])

(b) For Loewner matrices the residuals are defined by

$$r_{1i}^n = p_1^n(c_i) - w_i u_1^n(c_i), \quad r_{2i}^n = p_2^n(c_i) - w_i u_2^n(c_i).$$

The recursion is given by

$$[r_1^{n+1} \quad r_2^{n+1}] = [r_1^n \quad r_2^n] \circ \phi_n,$$

where "$\circ$" denotes the Hadamard product and ϕ_n is obtained from $\Phi_n(\lambda)$ in (4.1) replacing λ by the vector$(c_i)$and 1 by the vector $(1, \ldots, 1)$. The inner products involved in $\Phi_n(\lambda)$ are in fact components of the residuals. For this reason the fundamental system can be computed parallelly and fast.

(c) In order to deal with Toeplitz-plus-Hankel matrices one has to apply a simple idea: Consider separately the residuals of the Hankel and Toeplitz parts. The formulas however are more complicated in this case.

7. SUPERFAST ALGORITHMS

The type II algorithms have another advantage comparing with the type I algorithms. They can be organized.in such a way that divide and conquer strategies can be applied. The reason for the possibility of a speed up consists in the fact that the algorthms involve only polynomial multiplications and Hadamard products. The latter ones require only $O(n)$ flops and polyomial multiplication can be carried out with $O(n \log n)$ complexity if FFT is employed. This leads to $O(N \log^2 N)$ algorithms for the computation of fundamental systems. Together with the inversion formulas this will provide superfast algorithms for the solution structured systems with a Hankel, a Loewner or a Toeplitz-plus-Hankel coefficient matrix.

8. FINAL REMARKS

The algorithms described above for Hankel matrices have been implemented and tested by P.Jankowski. He did this also for the analogous algorithms for Toeplitz matrices. It turned out that our approach has an advantage also in the classical case of a strongly regular matrix, since it permits a scaling during the computation, which improves the stability. If a certain section of the matrix is almost singular (this can be checked during the computation), then we propose to consider the section like a singular matrix (practically this means that a small number is replaced by zero). In that way we obtained even quite reasonable numerical results for the Hilbert matrix $\left[\frac{1}{i+j+\alpha} \right]_1^n$.

For general Toeplitz-plus-Hankel matrices we implemented only the type I algorithm. Various algorithms working ·under additional assumptions are constructed and discussed in our paper [12].

Let us finally note that the approach described above works also for other classes of structured matrices. This concerns first of all the generalizations to the block case (for Hankel matrices see[8]), but also various kinds of Toeplitz-like or Vandermonde-like matrices, mosaics of structured matricesand others.

REFERENCES

[1] A. V. Aho, J. E. Hopcroft, J. D. Ullman, The design and analysis of computer algorithms. Addison-Wesley 1976.

[2] G. Ammar, P. Gader, New decompositions of the inverse of a Toeplitz matrix. In: Progress in Systems and Control Theory, vol. 5, Birkhäuser Boston 1990, 421-428.

[3] A. C. Antoulas, B. D. O. Anderson, On the problem of atable rational interpolation. Linear Algebra Appl. 122-24 (1989), 301-329.

[4] J. Ball, I. Gohberg, L. Rodman, Rational interpolation of matrix-valued functions. Birkhäuser Boston 1990.

[5] M. van Barel, A. Bultheel, A new formal approach to the rational interpolation problem. Report TW 134 (1990), Katholieke Universiteit Leuven.

[6] A. Gerasoulis, A fast algorithm for the multiplication of generalized Hilbert matrices with vectors. Math. Comp. 50 (1987), 179-188.

[7] I. Gohberg, T. Kailath, I. Koltracht, P. Lancaster, Linear complexity parallel algorithms for linear systems of equations with recursive structure. Linear Algebra Appl. 88/89 (1987), 271-315.

[8] G. Heinig. Formulas and algorithms for block Hankel matrix inversion and partial realization. In: Progress in Systems and Control Theory, vol. 5, Birkhäuser Boston 1990, 79-90.

[9] G. Heinig, Matrix representations of Bezoutians. Linear Algebra Appl. (to appear).

[10] G. Heinig, W. Hoppe, K. Rost, Structuredmatrices in interpolation and approximation problems. Wissensch. Zeitschr. d. TU Karl-Marx-Stadt 31, 2 (1989), 196-202.

[11] G. Heinig, P. Jankowski, Parallel and superfast algorithms for Hankel systems of equations. Numer. Math. 58 (1990), 109-127.

[12] G. Heinig, P. Jankowski, K. Rost, Fast algorithms for Toeplitz-plus-Hankel matrices. Numer. Math. 52 (1988), 665-682.

[13] G. Heinig, K. Rost. Algebraic methods for Toeplitz-like matrices and operators. Birkhäuser Boston 1984.

[14] G. Heinig, K. Rost, Matrix representations of Toeplitz-plus-Hankel matrix inverses. Linear Algebra Appl. 113 (1989), 65-78.

POLYNOMIAL DYNAMICAL SYSTEMS OVER FINITE FIELDS

Michel Le Borgne, Albert Benveniste, Paul Le Guernic
IRISA, Campus de Beaulieu, 35042 RENNES CEDEX (FRANCE)

Abstract

This paper is a short introduction to the theory of Polynomial Dynamical Systems over Galois fields. These dynamical systems have been found to be a usefull tool in the study of synchronisation relations specified by a declarative style real time programming langage, the SIGNAL language.

1 Introduction and motivations

Discrete Event Dynamical Systems (DEDS) have been introduced as a theoretical framework for the study of flexible manufacturing and related systems by Wonham and Ramadge [1] [2], and have been widely studied since their introduction. Roughly speaking, DEDS are finite state transition systems which are observed and can be controlled by the language generated by the labels that are attached to each transition, regardless of the precise meaning of these labels. On the other hand, *Hybrid Dynamical Systems (HDS)* theory has been introduced in [3] and [4] to handle synchronization, logic, and their interconnections to numerics in dynamical systems. HDS theory covers areas such as real-time process control, real-time signal processing systems, and more generally, C^3-systems.

In a previous paper [7] we have established a connection between these two theories and the study of non linear dynamical systems over finite fields, and we have shown how polynomial ideal theory techniques can be used to solve various questions that are fundamental in DEDS supervision in the sense of Ramadge and Wonham, and in the compilation of the SIGNAL language.

In this paper we show how simple algorithmic techniques using classical tools in computational algebra, can solve various control problems. In the area of real-time programming it is essential to prove that programs meet some vital specifications. With this objective in mind we also introduce several tools which we think will be useful for proofs on SIGNAL programs.

2 Algebraic tools

2.1 Algebraic sets and polynomial ideals.

The theory of polynomial dynamical systems over finite fields uses classical tools in elementary algebraic geometry. Since the fields considered are Galois fields, ideals and algebraic sets have special properties. Every algebraic variety is of dimension zero and every subset of an affine space $\mathcal{F}_p^n$ is an algebraic set. In order to present some more special properties of finite fields, let us introduce some notations.

Notations: rings of polynomials, ideals. The notation X will denote a collection of formal variables $X(i)$, and similarly for Y, etc... The notation $Q(X, Y)$ will denote a *collection* of polynomials in the variables $X(i), Y(j)$. The ring of the polynomials in these variables with coefficients in $\mathcal{F}_p$ will be denoted by $\mathcal{F}_p[X, Y]$, this notation extends to any set $Z, T, \ldots$ of formal variables. Let us denote by $< Q(X, Y) >$ the ideal spaned by the mentionned collection of polynomials, and by $\mathcal{V}(Q(X, Y))$ the algebraic set associated with the set of equations $Q(x, y) = 0$. We have:

$$\mathcal{V}(Q(X, Y)) = \mathcal{V}(< Q(X, Y) >) \tag{1}$$

On the other hand, given a subset S of $\mathcal{F}_p^{dim(x)} \times \mathcal{F}_p^{dim(v)}$, the set of polynomials R in $\mathcal{F}_p[X, Y]$ such that $R(s) = 0$ for all $s \in S$, is an ideal of $\mathcal{F}_p[X, Y]$ denoted $\mathcal{I}(S)$.

Thanks to properties of Galois fields, the equivalent of the Nullstellensatz is very simple and can be stated as follows:

$$\mathcal{I}(\mathcal{V}(\underline{a})) = < \underline{a}, X^p - X, Y^p - Y > \tag{2}$$

where $X^p - X$ (resp. $Y^p - Y$) stands for the set of polynomials $X^p(i) - X(i)$ (resp. $Y^p(i) - Y(i)$). This result means that we have only to add to every ideal the set of everywhere null polynomials to obtain the equivalent of the radical ideal of an ideal. So, from now on, $< Q(X, Y) >$ will denote $< Q(X, Y), X^p - X, Y^p - Y >$ and *all the ideals we shall consider in the sequel will be completed with the associated polynomials $X^p - X$, etc ....*

On the other hand, for all set S we have, as usual:

$$\mathcal{V}(\mathcal{I}(S)) = S$$

A nice consequence of the former results is that any ideal associated with algebraic sets may be considered as an ideal of the quotient ring

$$\mathcal{F}_p[X, Y] \; / < X^p - X, Y^p - Y >$$

Since this ring is a finite dimensional vector space over $\mathcal{F}_p$ with dimension $p^{(dim(x)+dim(v))}$, *strictly increasing chains of ideals are finite with length bounded by $p^{(dim(x)+dim(v))}$.* This fact also indicates that algorithms from linear algebra may be used and this gives some indications about the complexity of the algorithms introduced below.

2.2 Morphisms.

Consider multivariables Z and T. A morphism $\theta : \mathcal{F}_p^{dimZ} \to \mathcal{F}_p^{dimT}$ is a mapping such that each component is a polynomial function. In fact, due to the finiteness of the involved spaces, every mapping $\mathcal{F}_p^{dimZ} \to \mathcal{F}_p^{dimT}$ is a morphism. This can be easily proved using the Lagrange interpolation polynomials.

Given a morphism θ as above with components $(\theta_1(Z), \ldots, \theta_{dimT}(Z))$, we define, as usual, the comorphism: $\theta^* : \mathcal{F}_p[T] \rightarrow \mathcal{F}_p[Z]$ by:

$$p \in \mathcal{F}_p[T] : \quad \theta^*(p(T)) = \theta^*(p(T_1, \ldots, T_{dimT})) \triangleq p(\theta_1(Z), \ldots, \theta_{dimT}(Z)) \qquad (3)$$

The proofs of the following relations are straighforward:

$$\mathcal{I}(\theta(\mathcal{V}(\underline{a}))) = \theta^{*-1}(\underline{a})$$

$$\mathcal{V}(\theta^*(\underline{b})) = \theta^{-1}(\mathcal{V}(\underline{b})) \qquad (4)$$

for all completed ideals $\underline{a}$ in $\mathcal{F}_p[Z]$ and all completed ideals $\underline{b}$ in $\mathcal{F}_p[T]$.

2.3 Gröbner bases

One of our favorite algorithmic tools in polynomial dynamical systems are Gröbner bases. Gröbner bases are now widely known in the algebraic community so we will only make a few remarks about their use in our particular context.

One of the important issue, when using Gröbner bases, is the complexity issue. We don't know yet the complexity of the Buchberger algorithm for finite fields. As we deal exclusively with ideals containing the polynomials $X^p - X$, the degrees of polynomials appearing in the computation of Gröbner bases is alway bounded by p. Moreover, the different computations done in Buchberger algorithm, when projected on the vector space $\mathcal{F}_p[X,Y] / < X^p - X, Y^p - Y >$, look like Gaussian elimination. Hence our conjecture is that the complexity of the Buchberger algorithm is simply exponential in the number of variables in our case. Notice also that in the case $p = 2$, deciding whether a set of equation has solutions or not is exactly, after coding, the problem of the satisfiability of a Boolean formula. It is well known to be the prototype of NP-complete problems.

3 Polynomial dynamical systems

3.1 Definitions

A polynomial dynamical system over a field $\mathcal{F}_p$ may be specified via a set of polynomial equations:

$$\begin{cases} p_1(x_{n+1}, x_n, y_n) &= 0 \\ \quad \cdots \\ p_k(x_{n+1}, x_n, y_n) &= 0 \end{cases}$$

where x_{n+1}, x_n, y are collections of variables and $dim(x_{n+1}) = dim(x_n)$. More briefly, such a system will be denoted as:

$$P(x_{n+1}, x_n, y_n) = 0$$

The variables x represent the state of the system and the variables y are called *events*. Such a dynamical system is generally implicit and no distinction is made between inputs and outputs. As for algebraic sets, the representation of a dynamical system as a set of equations is not unique and any set of generators of the ideal associated with the equations is another valid representation. A case of particular interest is when it is possible to find a set of generators of the form:

$$\begin{cases} x_{n+1} & = & P(x_n, y_n) \quad \cdot \\ Q(x_n, y_n) & = & 0 \end{cases} \tag{5}$$

Such a dynamical system is said to be *state explicit*. In the form (5), the first equation is termed the *state transition equation* and the second one the *constraints equations*. In this paper we will only consider *state explicit dynamical systems* but the following definitions and concepts can be extended to more general ones.

The ideal $\underline{c}$ generated by the constraints equations is the ideal of constraints and the corresponding algebraic set is the set of constraints. An element (x_n, y_n) in this set is termed *admissible*. An event y is admissible in the state x if (x, y) is admissible. A trajectory $(x_n, y_n)_{n \geq 0}$ initialized in (x_0, y_0) is a sequence of admissible pairs (x_n, y_n), such that $x_{n+1} = P(x_n, y_n)$.

An exemple of polynomial dynamical system over the field $\mathcal{F}_3$ of the integers modulo 3 is:

$$\begin{aligned} \mathcal{L}_{dyncl} \quad = \quad < & (-x_1^2 + x_1)y + x_1^2 - x_1 \,, \\ & (x_2^2 - x_2)y - x_2^2 + x_2, -y^2 + 1 \,, \\ & (x_1 - 1)x_2^2 + (-x_1^2 + 1)x_2 + x_1^2 - x_1 \,, \\ & x_1^3 - x_1, x_2^3 - x_2 > \end{aligned} \tag{6}$$

$$\begin{aligned} p_{dyncl}^1 & = & x_1 x_2 y + x_2 y - y - x_2^2 - x_1^2 x_2 - 1 \\ p_{dyncl}^2 & = & -x_1 x_2 y + x_2 y + x_1 y + y + x_2^2 - 1 \end{aligned}$$

This dynamical system may be pictured as a finite automaton as in the figure (1). Conversely, any finite state automaton may be described as a polynomial dynamical sys-

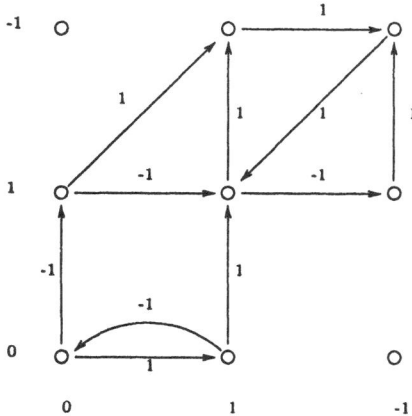

Figure 1: The dynamical system (6)

tem over a finite field.

Since we are only interested in admissible pairs (x, y), the state transition equations of a dynamical system are not unique. Any polynomial belonging to the ideal of constraints may be added to the right side of any state transition equation, giving the same constrainted dynamical system. As a consequence, the set of dynamical systems with $\underline{c}$ as ideal of constraints, is isomorphic to $(\mathcal{F}_p[X, Y]/\underline{c})^{dim(X)}$.

3.2 Basic Properties

3.2.1 Liveness

Among the most basic properties, the liveness one is of great interest. This kind of property is also encountered in Petri nets theory and in the theory of discrete events systems. Given a state explicit polynomial dynamical system (5), a state x is termed *dead* if there is no admissible event y in the state x (it is termed *alive* otherwise). A trajectory with a dead state is a finite trajectory.

A system is termed *alive* if for each state x and each y such that $(x, y) \in \mathcal{V}(Q)$, $P(x, y)$ is a live state.

In other words, a state is termed alive if it belongs to $proj_X(\mathcal{V}(Q))$. A system is alive iff $P(\mathcal{V}(Q)) \subseteq proj_X(\mathcal{V}(Q))$. Translating into corresponding ideals properties with the help of relations (4), we get:

Proposition 1 *A dynamical system is alive if an only if*

$$P^*(<Q>) \subseteq \ <Q> \cap \mathcal{F}_p[X]$$

where P^* is the comorphism associated with P.

Hence, checking the liveness property needs 1/ the computation of $<Q> \cap \mathcal{F}_p[X]$ which can be done by Gröbner bases method, 2/ the computation of its image by P^* which is done by simple substitution, and then 3/ checking the inclusion of two ideals again via Gröbner bases methods.

3.2.2 Invariance

A second important basic property is the invariance property. This kind of property is know under different names in the litterature, depending whether dynamical systems, Petri nets, automata and even Markov chains are under consideration.

Definition 1 *A subset E of states of a dynamical system is invariant if for every x in E and every event y admissible in the state x, the state $x' = P(x, y)$ is in E.*

Theorem 1 *A subset E is invariant for the dynamical system (5) if and only if:*

$$P^*(I(E)) \subseteq \underline{c} + \mathcal{I}(E)\mathcal{F}_p[X, Y]$$

where $\underline{c}$ is the ideal of constraints of the dynamical system and $\mathcal{I}(E)\mathcal{F}_p[X, Y]$ is the ideal in $\mathcal{F}_p[X, Y]$ generated by $\mathcal{I}(E)$.

Proof The definition is clearly equivalent to:

$$P(\mathcal{V}(\underline{c}) \cap (E \times \mathcal{F}_p^{dim(Y)})) \subseteq E$$

Ideal, morphisms and comorphisms properties allow us to derive the following equivalences:

$$
\begin{aligned}
& P(\mathcal{V}(\underline{c}) \cap (E \times \mathcal{F}_p^{dim(Y)})) \subseteq E \\
\Leftrightarrow\ & \mathcal{I}(E) \subseteq \mathcal{I}(P(\mathcal{V}(\underline{c}) \cap (E \times \mathcal{F}_p^{dim(Y)}))) \\
\Leftrightarrow\ & \mathcal{I}(E) \subseteq \mathcal{I}(P(\underline{c} + \mathcal{I}(E)\mathcal{F}_p[X,Y])) \\
\Leftrightarrow\ & \mathcal{I}(E) \subseteq P^{*-1}(\underline{c} + \mathcal{I}(E)\mathcal{F}_p[X,Y]) \\
\Leftrightarrow\ & P^*(\mathcal{I}(E)) \subseteq \underline{c} + \mathcal{I}(E)\mathcal{F}_p[X,Y]
\end{aligned}
$$

□

The greatest invariant subset The union of two invariant subsets of states is again an invariant subset. Hence, given a subset of states E, there exists a largest subset of E which is invariant for a given dynamical system. It is of great interest to compute this set. Let us consider first the case of a dynamical system *without* constraints. Such a system is given by a set of state transition equations alone:

$$x_{n+1} = P(x_n, y_n)$$

We need the:

Definition 2 *If $\underline{a}$ is an ideal of $\mathcal{F}_p[X,Y]$, $Coef_X(\underline{a})$ is the ideal in $\mathcal{F}_p[X]$, generated by the coefficients of the polynomials in $\underline{a}$ when they are written as polynomials with variables Y and coefficients in $\mathcal{F}_p[X]$.*

It is easy to check that this ideal is generated by $\{Coef_X(\alpha_1), \ldots, Coef_X(\alpha_k)\}$ for all set of generators $\{\alpha_1(X,Y), \ldots, \alpha_k(X,Y)\}$ of $\underline{a}$. With this definition we get the straightforward proposition:

Proposition 2 *Given an ideal $\underline{a}$ in $\mathcal{F}_p[X]$ and an ideal $\underline{b}$ in $\mathcal{F}_p[X,Y]$, the following assertions are equivalent:*

i) $\underline{b} \subseteq \underline{a}\mathcal{F}_p[X,Y]$

ii) $Coef_X(\underline{b}) \subseteq \underline{a}$

With this property, it is then possible to get an algorithm to compute the largest invariant subset included in a given set E of states. Consider the following sequence of ideals:

$$
\begin{aligned}
\underline{a}_0 &= \mathcal{I}(E) \\
\underline{a}_1 &= \underline{a}_0 + Coef_X(P^*(\underline{a}_0)) \\
&\cdots \\
\underline{a}_k &= \underline{a}_{k-1} + Coef_X(P^*(\underline{a}_{k-1}))
\end{aligned}
$$

This is an increasing , thus stationary chain of ideals. Let $\underline{a}(E, P)$ the largest ideal of this chain.

Theorem 2 *With the previous notations, if $\underline{a}(E, P)$ is not the entire ring then $\mathcal{V}(\underline{a}(E, P))$ is the largest invariant subset of E. If $\underline{a}(E, P)$ is the entire ring $\mathcal{F}_p[X]$ then E contains no invariant subset.*

The proof is straightforward.

The absence of invariant subset is also an interesting property. It means that every trajectory entering E *may* eventually leave E. The computation of the largest invariant subset of a set E for a dynamical system with constraints can be done by replacing it by another one, which is free from constraints, but possesses the same behaviour with respect to invariance properties. If the ideal of constraints of the dynamical system $(P, \underline{c})$ is generated by:

$$q_1(X, Y), \ldots, q_k(X, Y)$$

the constraint-free dynamical system given by:

$$p_L^i(x, y) = (\prod_{j=1}^{k}(1 - q_j^2(x, y)))p^i(x, y) + (1 - \prod_{j=1}^{k}(1 - q_j^2(x, y)))x_i$$

is such that $P_L(x, y) = P(x, y)$ if (x, y) is admissible and $P_L(x, y) = x$ on the contrary. A set E is then invariant for $(P, \underline{c})$ if and only if it is invariant for the dynamical system P_L.

Example The set $E = \{ (0, 0), (0, 1), , (1, 0), (1, 1) \}$ doesn't possess any invariant subset for the constraint-free dynamical system derived from (6) by removing the constraints. The previous algorithm gives the following ideals:

$$
\begin{aligned}
\underline{a}_0 &= \; < -x_1^2 + x_1, \; -x_2^2 + x_2 > \\
\underline{a}_1 &= \; < \underline{a}_0, \; \mathrm{Coef}_X(P^*(\underline{a}_0)) > \; = \; < -x_1^2 + x_1, \; x_2 > \\
\underline{a}_2 &= \; < \underline{a}_1, \; \mathrm{Coef}_X(P^*(\underline{a}_1)) > \; = \; < 1 >
\end{aligned}
$$

E-stabilisation The invariance property described above is a property of a set with respect to a dynamical system. The property we are interested in now is a property of a dynamical system with respect to a set E of states.

Definition 3 *A dynamical system $(P, \underline{c})$ is E-stabilisable if for all state x in E there exists an event y which is admissible in the state x and such that $P(x, y)$ is still an element of E.*

Theorem 3 *A dynamical system $(P, \underline{c})$ is E-stabilisable if and only if:*

$$(\underline{c} \; + \; P^*(\mathcal{I}(E))) \cap \mathcal{F}_p[X] \subseteq \mathcal{I}(E)$$

Proof The definition states that for each x in E the fiber of the constraints set over x contains a y such that $P(x, y) \in E$. But for all ideal $\underline{a}$, $\mathcal{V}(P^*(\underline{a})) = \{(x, y) \; / \; P(x, y) \in \mathcal{V}(\underline{a})\}$. The E-stabilisability is then equivalent to:

$$\mathcal{F}_x(V(P^*(\mathcal{I}(E)))) \; \cap \; \mathcal{F}_x(\mathcal{V}(\underline{c})) \neq \emptyset$$

for all $x \in E$ and with the obvious notation for fibers over x. This is then equivalent to:

$$E \subseteq Pr_x(V(P^*(\mathcal{I}(E))) \cap \mathcal{V}(\underline{c}))$$

which, when translated into the corresponding ideal properties, gives the theorem. □

It is also clear from the proof that $\underline{c} + P^*(\mathcal{I}(E))$ defines a *relational* feedback law E-stabilising the system, or if preferred, the set E is invariant for the dynamical system $(P, \underline{c} + P^*(\mathcal{I}(E)))$.

Example Consider the dynamical system of example (6). This system is certainly E-stabilisable for $E = \{ (0,0), (1,0) \}$. The ideal associated with this set is $\{ x_2, x_1(x_1-1) \}$. The Gröbner basis of $\underline{c} + P^*_{dyncl}(\mathcal{I}(E))$ with respect to the lexicographic order $x_1 \prec x_2 \prec y$ is:

$$\{ y - x_1 - 1, \ x_2, \ x_1(x_1 - 1) \}$$

showing that for each state in E there exists only one admissible event $y = x_1 + 1$ such that $P(x_1, x_2, y) \in E$. In general there are several admissible events with that property and we get a *relational* feedback instead of a functional one.

As for the largest invariant subset of a given set E, it is possible to compute the largest subset F of a set E for which the dynamical system $(P, \underline{c})$ is F-stabilisable. The algorithm is as follows:

$$
\begin{aligned}
\underline{a}_0 &= \mathcal{I}(E) \\
\underline{a}_1 &= \underline{a}_0 + (\underline{c} + P^*(\underline{a}_0)) \cap \mathcal{F}_p[X] \\
&\cdots \\
\underline{a}_{k+1} &= \underline{a}_k + (\underline{c} + P^*(\underline{a}_k)) \cap \mathcal{F}_p[X]
\end{aligned}
$$

and the proof of correctness is similar.

It is also of interest to know that a set of states E doesn't contain any subset for which the dynamical system is stabilisable. This means that all trajectory entering E *must* exit E after a finite delay.

Example A visual inspection of figure (1) reveals that the largest stabilisable subset of $E = \{(0,0), (1,0), (0,1), (1,1)\}$ is the subset $F = \{(0,0), (1,0)\}$. The above algorithm gives:

$$
\begin{aligned}
\underline{a}_0 &= \ < x2^2 - x2, \ x1^2 - x1 \ > \\
\underline{a}_1 &= \ < -x1^2 + x1, \ x1x2, \ -x2^2 + x2 \ > \\
\underline{a}_2 &= \ < x2, \ -x1^2 + x1 \ > \\
\underline{a}_3 &= \ \underline{a}_2
\end{aligned}
$$

and it is easy to check that $\mathcal{I}(\underline{a}_2) = F$.

3.3   Some control problems

Among the applications of polynomial dynamical systems belong Discrete Events Dynamical Systems (DEDS) control problems. A lift is a simple example of a plant that can be modeled as a DEDS. The control of such systems is achieved by inhibating some events. For example it is desirable that the doors of a lift are kept closed when the lift is moving.

This is obtained by inhibating the events corresponding to the opening of the doors when the system is in a state corresponding to a move. The translation into the framework of dynamical systems over finite fields results in equations involving states and events.

In a quality sofware production approach, after designing a controller for a DEDS it is desirable to check if the controlled system meets the specifications. Does the lift always move with the doors closed? Again the specifications tranlate into constraints equations that must be satisfied for all possible trajectories. These specifications may involve dynamics. The question is then: does the system behave like a given dynamical system taken as a model? If not, how is it possible to influence its behaviour to meet the specifications. The static case corresponds to problems of stabilisation and the dynamic case belong to the well known family of model following control problems.

3.3.1 Control problems static in nature

Given a dynamical system:

$$\begin{cases} x_{n+1} &= P(x_n, y_n) \\ Q(x_n, y_n) &= 0 \end{cases} \tag{7}$$

and a control objective expressed as a set of equations $Q_c(x, y)$, the control objective is fullfiled if for all trajectory $(x_n, y_n)_{n \geq 0}$, of the controlled system, we have $Q_c(x_n, y_n) = 0$. (The initial conditions must also meet the specifications). These trajectories are simply trajectories of the dynamical system obtained by adding the control constraints $Q_c(x, y) = 0$ to the constraints of the original system. In general, the dynamical system obtained in this way is not alive. Some transitions have to be inhibited to make the new system alive and this is where control takes place.

Given a dynamical system $(P, \underline{c})$, an algorithm can be derive from the condition for liveness proved in (1). Compute the following sequence of ideals:

$$\begin{aligned} c_0 &= \underline{c} \\ c_1 &= < c_0, P^*(c_0) \cap \mathcal{F}_p[X, X_m]) > \\ &\quad \cdots \\ c_i &= < c_{i-1}, P^*(c_{i-1} \cap \mathcal{F}_p[X, X_m]) > \end{aligned} \tag{8}$$

This chain must be constant after some $k \geq 0$. It is then straightforward to check that $P^*(c_k) \subseteq c_k \cap \mathcal{F}_p[X, X_m]$. If the largest ideal of the chain is the whole ring, *it is impossible* for the dynamical system to meet the specifications while having infinite trajectories. In the other case the largest ideal c_k gives the constraints we have to add in order to force it to behave as wished.

Example Suppose that for some reason we want to force the dynamical system (6) to behave in such a way that $x_1 x_2 y = 0$ along every trajectory. The previous algorithm, applied to (6) with this new constraint, gives:

$$< -x_2, \ x_1^2 - x_1, \ y - x_1 - 1 \ >$$

as maximal ideal. So to say this is the largest sub-automaton of (6) having a behaviour compatible with the added constraints.

Static objectives on states with initial conditions. In some cases we want only to have a dynamical system meeting some specifications for a given set of initial conditions. Moreover the specifications are expressible as equations involving only states and no events. The initial conditions are also defined by a set of polynomial equations $Q_0(x) = 0$. The control objective may be stated as follows: is it possible to find constraints involving states and events such that for the dynamical system subject to these constraints, all the trajectories starting in $\mathcal{V}(< Q_0(x) >)$ meet the specification $Q_c(x) = 0$.

The union of all trajectories originating from a point x_0, (the orbit of x_0) is clearly an invariant subset of the state space of the dynamical system. Hence, checking that the specification $Q_c(x) = 0$ is fulfilled is equivalent to check that the orbit of x_0 (in fact checking this for x_0 itself is enough) is contained in the largest invariant subset contained in $\mathcal{V}(Q_c(x))$. For all orbits with given initial conditions $Q_0(x_0) = 0$ it is equivalent to check that $\mathcal{V}(Q_0(x))$ is included is the largest invariant subset of $\mathcal{V}(Q_c(x))$.

For the control issue, it is enough to notice that a state trajectory is a stabilisable subset of states. Hence any trajectory starting inside $\mathcal{V}(Q_0(x))$ and included in $\mathcal{V}(Q_c(x))$ is contained in the largest subset of $\mathcal{V}(Q_c(x))$ for which the dynamical system is stabilisable. Thus it is possible to meet the specification if and only if $\mathcal{V}(Q_0(x))$ is contained in this set. If $\underline{c}_s$ is the ideal associated with this stabilisable subset, from the remark following theorem (3), adding the constraint $P^*(\underline{c}_s)$ to the dynamical system forces it to behave as desired.

3.3.2 Control problems dynamic in nature

The classical problem of exact model following control is: given the dynamical system:

$$\begin{aligned} x_{t+1} &= P(x_t, y_{1,t}, y_{2,t}) \\ 0 &= Q(x_t, y_{1,t}, y_{2,t}) \end{aligned} \tag{9}$$

and a second one considered as the reference model,

$$\begin{aligned} x_{m,t+1} &= P_m(x_{m,t}, y_{1,t}) \\ 0 &= Q_m(x_{m,t}, y_{1,t}) \end{aligned} \tag{10}$$

is it possible to find a relation $K(x_m, x, y_1, y_2) = 0$ such that all admissible sequences $(y_{1,t})$ of the system obtained by combining (9) and this relation, are admissible sequences of the reference model (10). The liveness concept gives a tool to solve the exact model following control problem. Consider the dynamical system obtained by combining (9) and (10). The exact following model control problem is solved if it is possible to add constraints on x, y_1, y_2 such that the resulting dynamical system is alive. The algorithm described in (8) gives the solution to this problem.

4   Conclusion

This paper gives a flavor of the theory of Dynamical Systems over finite fields. Although we have use classical concepts from dynamical systems theory and control theory, the results obtained are in a form which is not quite familiar in the control community. The best illustration is probably the notion of a *relational feedback*. The situation is similar in the study of observability and external representations. Some dynamical systems do not possess a transfer function, but rather a transfer *relation*. The nature of a realisation theory in dynamical systems over finite fields is not yet clear. There is no satisfactory notion of minimality. Moreover, since our dynamical systems come from the coding of automata, it is essential to compare dynamical systems with different spaces of events. Realisation theory may then be inspired from equivalences between transition systems [8].

References

[1] P.J. Ramadge, W.M. Wonham. *Supervisory Control of a class of Discret Events Processes* , SIAM J. Control and Opt., 25, 1, 1987, 206-230.

[2] P.J. Ramadge, W.M. Wonham. *Controllable Sublanguage of a Given Language*, SIAM J. Control and Opt., 25, 3, 1987, 637-659.

[3] A. Benveniste, P. Le Guernic. *Hybrid Dynamical Systems theory and nonlinear dynamical systems over finite fields,* Proc. of CDC 88, 209-213, Austin 7-9 Dec 1989.

[4] A. Benveniste, B. Le Goff, P. Le Guernic. *Hybrid Dynamical Systems theory and the language SIGNAL.* INRIA research report No 838, 1988.

[5] P.J.G. Ramadge, *Some tractable supervisory control problems for discrete event systems modelled by Büchi automata,* IEEE-AC-34 No 1, 10-20, 1989.

[6] B. Buchberger. *Gröbner Bases: An Algorithmic Method in Polynomial Ideal Theory* N.K. Bose (ed.), Multidimensional Systems Theory, 184-232, D. Reidel Publishing Company.

[7] M. Le Borgne, A. Benveniste, P. Le Guernic. *Polynomial ideal theoretic methods in Diecrete Events and Hybrid Dynamical Systems* Proceeding of the 28th Conference on Decision and Control, IEEE Control Systems Society, Volume 3 of 3, 1989, 2695-2700.

[8] A. Arnold. *Systèmes de transitions finis et sémantique des processus communicants* T.S.I vol 9, No 3, 1990, 193-216.

Computing with P.B.W.
in enveloping algebras

Gérard DUCHAMP – Daniel KROB

Université de Rouen, Université Paris 6 and CNRS [1]

0 Introduction

Let M be a connected finite dimensional **R**-analytic manifold, let $(Y_i)_{i=0,n}$ be a family of analytic vector fields over M, let $(u_i)_{i=1,n}$ be a family of input functions in $\mathcal{C}^0_{pw}(\mathbf{R}, \mathbf{R})$ and let h be an analytic output function from M into **R**. Then we can consider the analytic system (Σ) :

$$(\Sigma) \qquad \left\{ \begin{array}{c} \dot{q}(t) = Y_0(q(t)) \; + \; \sum_{i=1}^{n} Y_i(q(t)) \, u_i(t) \\ y(t) = h(q(t)) \end{array} \right.$$

where q belongs to M, the initial state $q(0)$ being given. Fliess showed (see [14]) that non-commutative formal power series over the alphabet $A = \{a_i, \; i \in [0, n]\}$ can be used to describe the input/output behaviour of (Σ). More precisely, if $\mathcal{Y}(w)$ denotes for every $w \in A^*$ the differential operator defined by the following relations :

$$\mathcal{Y}(w) = \left\{ \begin{array}{cc} Id & \text{if } w = 1 \\ Y_i \circ \mathcal{Y}(v) & \text{if } w = a_i v \end{array} \right.$$

and if $\mathcal{E}_a(w) = \int_0^t \delta_a(w)$ denotes the iterated integral of order w associated with the intput functions (see [14] or [15]), the output of (Σ) is given by Fliess' main formula (cf [14]) :

$$(\mathcal{FL}) \qquad y(t) = \sum_{w \in A^*} \mathcal{E}_a(w) \, \mathcal{Y}(w) \circ h|_{q(0)} \; = \; \mathcal{H} \circ h|_{q(0)}$$

where $\mathcal{H}$ is the differential operator (called also transport operator) :

$$(\mathcal{TO}) \qquad \mathcal{H} = \sum_{w \in A^*} \mathcal{E}_a(w) \, \mathcal{Y}(w) = (\mathcal{E}_a \otimes \mathcal{Y})(\sum_{w \in A^*} w \otimes w)$$

Thus, Fliess' formula can be expressed using the above "double" formal power series. But an algebraic result of Mélançon and Reutenauer (see [17]) shows that we have the

---

[1] Mailing adress : Laboratoire d'Informatique de Rouen (LITP) – Université de Rouen – Faculté des Sciences – 76134 Mont Saint-Aignan Cedex – FRANCE

following relation in $\mathbf{Q} \ll A \gg \otimes \mathbf{Q} < A >$ if we equip $\mathbf{Q} \ll A \gg$ with the Cauchy product and $\mathbf{Q} < A >$ with the shuffle product :

$$(\mathcal{SC}) \qquad \sum_{w \in A^*} w \otimes w = \prod_{l \in L_y} exp\left(P_l \otimes [l]\right)$$

where $([l])_{l \in L_y}$ denotes the Lyndon basis of the free Lie algebra $L(A)$ and where $(P_l)_{l \in L_y}$ is a certain effectively computable family of non-commutative polynomials in $\mathbf{Q} < A >$. Since $\mathcal{E}_a$ is a morphism for the shuffle product and since $\mathcal{Y}$ is a morphism for the Cauchy product, $(\mathcal{SC})$ used in connection with $(\mathcal{TO})$ provides a new way for computing the output of (Σ) which allows to construct structural nilpotent [2] approximations of the output of (Σ) (see [15]).

However all the above work was made in free algebras and the real structure of the Lie algebra $\mathfrak{g}$ generated by the vector fields $(Y_i)_{i=0,n}$ was never used. In this paper, we will give some indications on how to obtain some algebraic results that would permit to work in a non-free framework. In a first part, we will prove the general version of relation $(\mathcal{SC})$ for a Lie K-algebra $\mathfrak{g}$ that is a free K-module. More precisely, we will show that if $g = (g_i)_{i \in I}$ is a K-basis of $\mathfrak{g}$ and if $((g^\nu)^*)_{\nu \in \mathbf{N}^{(I)}}$ is the dual family of the P.B.W. [3] basis of $\mathcal{U}(\mathfrak{g})$ associated with g, we have the relation :

$$\sum_{\nu \in \mathbf{N}^{(I)}} (g^\nu)^* \otimes g^\nu = \prod_{i \in I} exp\left(g_i^* \otimes g_i\right)$$

which involves a new multiplication on $\mathcal{U}(\mathfrak{g})^* \otimes \mathcal{U}(\mathfrak{g})$ (see th. 1.6 below).

In a second part, we will only consider the free partially commutative Lie algebra which is the Lie algebra $L(A, \theta)$ defined by the presentation :

$$L(A, \theta) = < A ; [a, b] = 0 \text{ for } (a, b) \in \theta >_{\text{Lie}}$$

Then, we will give the partially commutative generalizations of some basic tools that are needed for computing nilpotent structural approximations of the output of an analytic system (see [15]). In fact, we will show here how to obtain a partially commutative Lazard's elimination process that permit to compute bases of the free module $L(A, \theta)$, to get an effective partially commutative version of relation $(\mathcal{SC})$ and to obtain P.B.W. decompositions in $K < A, \theta >$.

1 Some consequences of P.B.W.

1.1 General conventions

First, we will refer to [2] for all the properties of tensor products that will be used in the sequel. We will also refer to [5] and [6] for all general properties and definitions concerning Lie algebras and enveloping algebras. In the same way, we refer to [2] and [6] for the properties of coalgebras and bialgebras that we will use.

Let us introduce now some general conventions that will be used throughout this paper. First, let us recall that a *multidegree* ν on a set I is just an element of $\mathbf{N}^{(I)}$.

---

[2] It means that the computations are made in a free nilpotent Lie algebra.

[3] In all this paper, P.B.W. will mean Poincaré-Birkhoff-Witt.

The *support* of ν is then the finite subset of I, denoted supp(ν), which is defined by : supp(ν) = $\{\, i \in I,\, \nu(i) \neq 0 \,\}$ and the length $|\nu|$ of ν is the integer : $|\nu| = \sum \nu(i)$. In the same way, we will also denote by $\nu!$ the integer $\prod \nu(i)!$.

For every totally ordered finite set $I = \{\, i_1 < \ldots < i_n \,\}$, for every monoid M and for every family $g = (g_i)_{i \in I}$ of M^I, we will denote :

$$\prod_{i \in I} g_i = g_{i_1} g_{i_2} \cdots g_{i_n}$$

Moreover, we will also use the same denotation whenever the set supp(g) = $\{\, i \in I, g \neq 1 \,\}$ is finite, even if I is infinite. We will also denote for every multidegree $\nu \in \mathbb{N}^{(I)}$:

$$g^{\nu} = g_{i_1}^{\nu(i_1)} g_{i_2}^{\nu(i_2)} \cdots g_{i_n}^{\nu(i_n)} \qquad (*)$$

Let K be a ring, let H be a free K-module and let $(b_i)_{i \in I}$ be a K-basis of H (cf [2]). Then every element $m \in H$ can be written uniquely as follows :

$$m = \sum_{i \in I} b_i^{\bullet}(m)\, b_i \qquad \text{where } b_i^{\bullet}(m) \in K$$

Hence, these relations define linear forms $b_i^{\bullet} \in H^* = \mathrm{End}_K(H, K)$ which will be called the *coordinate linear forms*. The family $(b_i^{\bullet})_{i \in I}$ will also be called the *dual family* of the family $(b_i)_{i \in I}$. It is easy to see that the dual family is uniquely defined by the relations : $b_i^{\bullet}(b_j) = \delta_{i,j}$ for every $i, j \in I$, where $\delta_{i,j}$ denotes the usual Kronecker symbol.

Finally, we will denote by $X \subseteq_f Y$ whenever X is a finite subset of the set Y.

1.2 Coordinates in an enveloping algebra

Let $\mathfrak{g}$ be a Lie algebra over a ring K. It is well-known (cf [6]) that the enveloping algebra $\mathcal{U} = \mathcal{U}(\mathfrak{g})$ of $\mathfrak{g}$ can be equipped with a unique structure of bialgebra where the comultiplication c from $\mathcal{U} \otimes \mathcal{U}$ into $\mathcal{U}$ satisfies to :

$$\forall\, g \in \mathfrak{g}, \quad c(g) = g \otimes 1 + 1 \otimes g$$

Moreover this comultiplication is coassociative and cocommutative. Let us first recall that coassociativity means that, whenever $c(w) = \sum u_i \otimes v_i$ for $w \in \mathcal{U}$, then the two "second fold expansions" : $\sum c(u_i) \otimes v_i$ and $\sum u_i \otimes c(v_i)$ are equal in $\mathcal{U} \otimes \mathcal{U} \otimes \mathcal{U}$. In the same way, cocommutativity means that, for every w in $\mathcal{U}$, $c(w) = \sum u_i \otimes v_i$ is invariant by the twist symmetry T of $\mathcal{U} \otimes \mathcal{U}$ defined by $T(x \otimes y) = y \otimes x$.

By duality, it is now possible to equip $\mathcal{U}^* = \mathrm{End}_K(\mathcal{U}, K)$ with a structure of K-algebra. The corresponding law, denoted $*$, is defined for every $f, g \in \mathcal{U}^*$ as follows :

$$f * g = {}^t c\,(f \otimes g) \iff \forall\, w \in \mathcal{U}, \; (f * g)(w) = (f \otimes g)(c(w))$$

The above law $*$ can be precised. Indeed, for every $f, g \in \mathcal{U}^*$, $f * g$ is the linear form of $\mathcal{U}^*$ which is defined for every $w \in \mathcal{U}$ as follows :

$$f * g\,(w) = \sum f(u_i)\, g(v_i) \quad \text{if we have : } c(w) = \sum u_i \otimes v_i$$

From the coassociativity and the cocommutativity of c, we can easily deduce that the law $*$ is associative and commutative. When $\mathfrak{g}$ is the free partially commutative Lie algebra (cf section 2.3 below), the product $*$ is called the *shuffle product* (cf [16] for the classical free case and [18] for the general partially commutative case).

Example : When $\mathfrak{g}$ is the free Lie K-algebra $L(A)$, we can identify $\mathcal{U}(L(A))$ with $K\!<\!A\!>$ and $\mathcal{U}^*$ with $K\!<\!<\!A\!>\!>$, the duality being defined by :

$$\forall\, S \in K\!<\!<\!A\!>\!>,\ \forall\, P \in K\!<\!A\!>,\ <S,P> = \sum_{w\in A^*} (S|w)\,(P|w)$$

(see [6] and [1]). It can be shown that the shuffle product (denoted here ⊔⊔) is then given on A^* (and hence everywhere by bilinear and continuous extensions) by the laws :

$$\forall\, u,v,w \in A^*,\ \forall\, x,y \in A, \quad \left\{ \begin{array}{c} w \sqcup\!\sqcup 1 = 1 \sqcup\!\sqcup w = w \\ ux \sqcup\!\sqcup vy = (ux \sqcup\!\sqcup v)y + (u \sqcup\!\sqcup vy)x \end{array} \right.$$

The law $*$ appears as a way to construct new linear forms on $\mathcal{U}$. We will now study its relations with P.B.W. decompositions. Therefore, let K be a ring, let $\mathfrak{g}$ be a Lie K-algebra which is a free K-module and let $g = (g_i)_{i\in I}$ be a K-basis of $\mathfrak{g}$ indexed by a totally ordered set I. The P.B.W. theorem ensures that the family $(g^\nu)_{\nu\in\mathbb{N}^{(I)}}$ is a basis of the enveloping algebra $\mathcal{U} = \mathcal{U}(\mathfrak{g})$ of $\mathfrak{g}$. Then, according to our general conventions, let us denote by $((g^\nu)^*)_{\nu\in\mathbb{N}^{(I)}}$ its dual family. The following result precises now this last family :

THEOREM 1.1 : (Mélançon-Reutenauer; [17]) Under the previous hypotheses, let us denote $h = (g_i^*)_{i\in I}$.[4] Then, we have for every multidegree $\nu \in \mathbb{N}^{(I)}$:

$$\nu!\,(g^\nu)^* = h^\nu$$

where the product in the above second member is taken in the sense of $*$.

Proof : First, let us establish a binomial formula for the comultiplication :

LEMMA 1.2 : Under the previous denotations, we have :

$$c(g^\nu) = \sum_{\alpha+\beta=\nu} \binom{\nu}{\alpha}\, g^\alpha \otimes g^\beta \qquad \text{where} \qquad \binom{\nu}{\alpha} = \frac{\alpha!}{\nu!\,(\nu-\alpha)!} \in \mathbb{N}$$

Proof of the lemma : First, the above property is obvious for $\nu = 0$. Otherwise, let $i_1 < i_2 < \ldots < i_n$ be the support of a non-zero multidegree ν. Then, we can write :

$$c(g^\nu) = c(g_{i_1}^{\nu(i_1)})\, c(g_{i_2}^{\nu(i_2)}) \ \ldots \ c(g_{i_n}^{\nu(i_n)}) \qquad (*)$$

$$\forall\, i \in I,\ c(g_i^{\nu(i)}) = (g_i \otimes 1 + 1 \otimes g_i)^{\nu(i)} = \sum_{k+l=\nu(i)} \binom{\nu(i)}{k}\, g_i^k \otimes g_i^l \qquad (**)$$

The lemma is now an immediate consequence of relations $(*)$ and $(**)$. ∎

Let us now come to the proof of our result that we will establish by induction on $|\nu|$. First, observe that 1^*, which is the coordinate form on $g^0 = 1_\mathcal{U}$, is also the unity for $*$ as it is easily checked. Hence, we immediately obtain the formula : $(g^0)^* = 0!\, h^0$ which proves our theorem for $|\nu| = 0$. Otherwise, let us suppose our result proved for every multidegree of length $< N$ where N is ≥ 1. Then, let us consider a multidegree ν such that $|\nu| = N$ and let us define $m = \min(\mathrm{supp}(\nu))$. Let now (m) be the multidegree of support $\{m\}$ whose unique non-zero entry of order m is 1. Then, according to the previous lemma, we can write for every multidegree α :

---

[4] For every $i \in I$, g_i^* just denotes the element of the dual family of $(g^\nu)_{\nu\in\mathbb{N}^{(I)}}$ corresponding to the element g_i of the K-basis g of $\mathfrak{g}$.

$$h^\nu(g^\alpha) = (g_m^* * h^{\nu-(m)})(g^\alpha) = (g_m^* \otimes h^{\nu-(m)})(c(g^\alpha))$$

$$= (g_m^* \otimes h^{\nu-(m)})(\sum_{\gamma+\delta=\alpha} \binom{\alpha}{\gamma} g^\gamma \otimes g^\delta) = \sum_{\gamma+\delta=\alpha} \binom{\alpha}{\gamma} g_m^*(g^\gamma) \otimes h^{\nu-(m)}(g^\delta)$$

Two cases are now to be considered :

- if $\alpha(m) = 0$, then $\gamma(m) = 0$ for every γ such that $\gamma + \delta = \alpha$. It follows that $g_m^*(g^\gamma) = 0$ for all such γ. Hence, we immediately obtain that : $h^\nu(g^\alpha) = 0$ here.

- if $\alpha(m) \neq 0$, we can use the induction hypothesis to conclude that we have : $h^{\nu-(m)}(g^\delta) = 0$ when $\delta \neq \nu-(m)$ and $h^{\nu-(m)}(g^\delta) = (\nu-(m))!$ when $\delta = \nu-(m)$. It follows now easily from these results and from the above formula that we have : $h^\nu(g^\alpha) = 0$ when $\alpha \neq \nu$ and $h^\nu(g^\nu) = \binom{\nu}{(m)} g_m^*(g_m) (\nu-(m))! = \nu!$.

Hence, we can summarize the above study by the following property :

$$\forall \alpha \in \mathbf{N}^{(I)}, \ h^\nu(g^\alpha) = \nu! \, \delta_{\alpha,\nu} \iff h^\nu = \nu! \, (g^\nu)^*$$

Thus, this ends our induction and proves our theorem. ∎

Note : (Mélançon-Reutenauer; [17]) Let us consider a totally ordered alphabet A and let $g = ([l])_{l \in L_y}$ be the associated Lyndon basis of the free Lie $\mathbf{Q}$-algebra $\mathfrak{g} = L_{\mathbf{Q}}(A)$ over A, totally ordered by the "opposite lexicographic order" (see [16], [17] or [20]). Then, using the one-to-one correspondance given by the fact that the Lyndon words form a complete factorization of A^* (see [16] or [17]), we can index the family $(g^\nu)_{\nu \in \mathbf{N}^{(L_y)}}$ by the words of A^*, denoting it $([w])_{w \in A^*}$. More precisely, we will have for every $w \in A^*$:

$$[w] = [l_1]^{i_1} \dots [l_n]^{i_n}$$

where $w = l_1^{i_1} \dots l_n^{i_n}$ is the Lyndon decomposition of the word w. Hence, the dual family $((g^\nu)^*)_{\nu \in \mathbf{N}^{(L_y)}}$ of the P.B.W. basis of $\mathbf{Q}\langle A\rangle$ associated with the Lyndon basis of $L_{\mathbf{Q}}(A)$ can now be denoted $(P_w)_{w \in A^*}$. Then, according to th. 1.1, this family satisfies to :

$$P_1 = 1 \quad \text{and} \quad P_w = \frac{1}{i_1! \dots i_n!} (P_{l_1})^{i_1} \sqcup \dots \sqcup (P_{l_n})^{i_n}$$

where $l_1^{i_1} \dots l_n^{i_n}$ is again the Lyndon decomposition of w. Moreover, if $l = am$ is a Lyndon word with $a \in A$, we also have : $P_l = aP_m$ (see [17]). It should be noted that all the above properties permit to recursively compute every element of the dual basis of g.

Example : Let $A = \{a, b\}$ with $a < b$. Then $aabab$ is a Lyndon word and we have :

$$P_{aabab} = aP_{(ab)^2} = a\frac{1}{2!}(P_{ab} \sqcup P_{ab}) = a\frac{1}{2!}(aP_b \sqcup aP_b)$$

$$= a\frac{1}{2!}(abP_1 \sqcup abP_1) = a\frac{1}{2!}(ab \sqcup ab) = aabab + 2aaabb$$

1.3 Expressions of the identity of $\mathcal{U}$

Let $\mathfrak{g}$ be a Lie algebra which is a *free* K-module. Then, according to the P.B.W. theorem, $\mathcal{U} = \mathcal{U}(\mathfrak{g})$ is also a free K-module. Therefore, the natural mapping $\theta_\mathcal{U}$ from $\mathcal{U}^* \otimes \mathcal{U}$ into $\text{End}_K(\mathcal{U})$, defined by the following relations :

$$\forall\, f \in \mathcal{U}^*,\ \forall\, u \in \mathcal{U},\ \theta_\mathcal{U}(f \otimes u) : v \in \mathcal{U} \longrightarrow f(v)\, u \in \mathcal{U}$$

is injective (cf [2]) and we can identify $\mathcal{U}^* \otimes \mathcal{U}$ with a dense subset [5] of $\mathrm{End}_K(\mathcal{U})$, equipped with the topology of pointwise convergence [6] for which it is Haussdorf and complete. The following result, which follows easily from the definition of pointwise convergence, characterizes some summable families of $\mathcal{U}^* \otimes \mathcal{U}$:

PROPOSITION 1.3 : A family $(f_i \otimes a_i)_{i \in I}$ of $\mathcal{U}^* \otimes \mathcal{U}$ is summable iff it is locally finite, i.e. iff for every $w \in \mathcal{U}$, the family $(f_i \otimes a_i(w))_{i \in I} = (f_i(w)\, a_i)_{i \in I}$ has finite support. In this case, we have :

$$\forall\, w \in \mathcal{U},\ \big(\sum_{i \in I} f_i \otimes a_i \big)(w) \ =\ \sum_{i \in I} f_i(w)\, a_i$$

Remark : For every K-basis $(b_n)_{n \in N}$ of $\mathcal{U}$ and for every endomorphism ϕ of $\mathrm{End}_K(\mathcal{U})$, the family $((b_n^* \circ \phi) \otimes b_n)_{n \in N}$ is summable and we have :

$$\phi = \sum_{n \in N} (b_n^* \circ \phi) \otimes b_n$$

Observe that this relation allows to prove easily that $\mathcal{U}^* \otimes \mathcal{U}$ is dense in $\mathrm{End}(\mathcal{U})$. It follows also immediately from this last formula that we have for every K-basis $(b_n)_{n \in N}$ of $\mathcal{U}$:

$$\sum_{n \in N} b_n^* \otimes b_n \ =\ Id_\mathcal{U}$$

$\mathrm{End}_K(\mathcal{U})$ can be now equipped with a new product, inherited from the structure of $\mathcal{U}$ and of its dual. To this purpose, let us consider first the K-algebra $\mathcal{U}^* \otimes \mathcal{U}$ which is the tensor product of the K-algebras $\mathcal{U}^*$ (with the product $*$) and $\mathcal{U}$ (with the usual product $\times$). Then we have the following result :

LEMMA 1.4 : The law $* \otimes \times$ on $\mathcal{U}^* \otimes \mathcal{U}$ is bilinear and continuous.

Proof : The bilinearity of $* \otimes \times$ is straightforward. Let us now show that this product is continuous. Therefore, let $(b_n)_{n \in N}$ be a K-basis of $\mathcal{U}$. Then, every element R of $\mathcal{U}^* \otimes \mathcal{U}$ can be uniquely decomposed as follows : $R = \sum R(n) \otimes b_n$. Let now P and Q be two elements of $\mathcal{U}^* \otimes \mathcal{U}$ and let $\mathcal{V}$ be a neighbourhood of $P * \otimes \times Q$ for the pointwise convergence topology. By definition of this topology, there exists a finite set $F \subset \mathcal{U}$ such that :

$$W = \{\, v \in \mathcal{U}^* \otimes \mathcal{U},\ \forall\, f \in F,\ v(f) = (P * \otimes \times Q)(f)\,\} \ \subset\ \mathcal{V}$$

For every $f \in F$, we can write : $c(f) = \sum f_n \otimes b_n$ where the support of this sum is a finite set $J_f \subset N$. Hence, we can define the two following finite sets :

$$G = \bigcup_{f \in F} \{\, f_n, n \in J_f \,\} \ \subset \mathcal{U} \qquad \text{and} \qquad H = \bigcup_{f \in F} \{\, b_n, n \in J_f \,\} \ \subset \mathcal{U}$$

It follows immediately that the two following sets :

$$\mathcal{P} = \{\, p \in \mathcal{U}^* \otimes \mathcal{U}, \; \forall \, g \in G, \; p(g) = P(g) \,\}$$

$$\mathcal{Q} = \{\, q \in \mathcal{U}^* \otimes \mathcal{U}, \; \forall \, h \in H, \; q(h) = Q(h) \,\}$$

are neighbourhoods of P and Q for the pointwise convergence topology on $\mathcal{U}^* \otimes \mathcal{U}$. Let us now show that we have the following inclusion :

$$\mathcal{P} * \otimes \times \mathcal{Q} \subset \mathcal{W} \qquad (1)$$

which will clearly prove the continuity of the product $* \otimes \times$. Thus, let f be in F and let p, q be respectively in $\mathcal{P}$ and $\mathcal{Q}$. Then, we have :

$$
\begin{aligned}
(p * \otimes \times q)(f) &= \Big(\sum_{n,m} p(n) * q(m) \otimes b_n \, b_m \Big)(f) = \sum_{n,m} (p(n) * q(m))(f) \, b_n \, b_m \\
&= \sum_{n,m} (p(n) \otimes q(m))(c(f)) \, b_n \, b_m = \sum_{n,m} (p(n) \otimes q(m)) \Big(\sum_k f_k \otimes g_k \Big) b_n \, b_m \\
&= \sum_{n,m,k} p(n)(f_k) \, q(m)(g_k) \, b_n \, b_m = \sum_k \Big(\sum_n p(n)(f_k) \, b_n \Big) \Big(\sum_m q(m)(g_k) \, b_m \Big) \\
&= \sum_k p(f_k) \, q(g_k) = \sum_k P(f_k) \, Q(g_k) = (P * \otimes \times Q)(f)
\end{aligned}
$$

the last equality being clearly a consequence of the previous computations. Hence, we proved that inclusion (1) holds. Therefore this ends our proof. ∎

It is now possible to extend the law $* \otimes \times$ to $\operatorname{End}_K(\mathcal{U})$. The following result makes more precise the definition of this new product :

PROPOSITION 1.5 : There is a unique extension of the product $* \otimes \times$ of $\mathcal{U}^* \otimes \mathcal{U}$ to the K-algebra $\operatorname{End}_K(\mathcal{U})$ in such a way that $(\operatorname{End}_K(\mathcal{U}), +, * \otimes \times)$, equipped with the pointwise convergence topology, becomes a topological K-algebra which is the completion of $(\mathcal{U}^* \otimes \mathcal{U}, +, * \otimes \times)$. Moreover this new product can be computed for every f, g in $\operatorname{End}_K(\mathcal{U})$ when they are given by the following sums of summable families :

$$f = \sum_{i \in I} \alpha_i \otimes a_i \qquad \text{and} \qquad g = \sum_{j \in J} \beta_j \otimes b_j$$

where $\alpha_i, \beta_i \in \mathcal{U}^*$ and where every a_i and every b_i of $\mathcal{U}$ has no torsion. [7] Indeed the family $((\alpha_i \otimes a_i) * \otimes \times (\beta_j \otimes b_j))_{i,j}$ is then locally finite and we have :

$$f * \otimes \times g = \sum_{i,j} (\alpha_i \otimes a_i) * \otimes \times (\beta_j \otimes b_j)$$

Proof : The first assertion of our proposition follows from [3] III. 6.5 th. 1 since $\operatorname{End}_K(\mathcal{U})$ is complete for the pointwise convergence topology, since $\mathcal{U}^* \otimes \mathcal{U}$ is dense in $\operatorname{End}_K(\mathcal{U})$ and since $* \otimes \times$ is bilinear and continuous in $\mathcal{U}^* \otimes \mathcal{U}$ according to lemma 1.4. To prove the second assertion, it suffices clearly to only show the local finiteness of the family $((\alpha_i \otimes a_i) * \otimes \times (\beta_j \otimes b_j))_{i,j}$ under the above hypotheses. To this purpose, let w be in $\mathcal{U}$. Then there exists a finite set F such that we have :

$$c(w) = \sum_{j \in F} u_j \otimes v_j$$

---

[7] This is always possible to realize since we can take the a_i's and b_j's in a K-basis of $\mathcal{U}$.

It follows then from an easy computation that we have for every $i \in I$ and $j \in J$:

$$((\alpha_i \otimes a_i) * \otimes \times (\beta_j \otimes b_j))(w) = (\sum_{f \in F} \alpha_i(u_f)\beta_j(v_f)) \, a_i b_j \qquad (1)$$

But, according to prop. 1.3, the families $(\alpha_i \otimes a_i)_i$ and $(\beta_j \otimes b_j)_j$ are locally finite. Therefore, since every a_i and b_j has no torsion, the families $(\alpha_i)_i$ and $(\beta_j)_j$ are also locally finite. Hence, since F is finite, the family $(\sum \alpha_i(u_f)\beta_j(v_f))_{i,j}$ has clearly a finite support. Our proposition follows now immediately from relation (1). ■

In a topological K-algebra $\mathcal{A}$, a totally ordered family $(t_i)_{i \in I}$ of $\mathcal{A}$ is said to be *multipliable* with product $P \in \mathcal{A}$ iff for every neighborhood $\mathcal{V}$ of P, we have :

$$\exists \, J \subseteq_f I, \; \forall \, K, \; J \subseteq K \subseteq_f I \implies \prod_{i \in K} t_i \in \mathcal{V}$$

Note also that we will assume from now to the end of this section that K is a **Q**-algebra. In this context, we can deduce (see also [17]) from th. 1.1 that we have :

THEOREM 1.6 : Under the previous hypotheses and the denotations of th. 1.1, we have the following relations :

$$Id_{\mathcal{U}} = \sum_{\nu \in \mathbf{N}^{(I)}} (g^\nu)^* \otimes (g^\nu) = \prod_{i \in I} exp(g_i^* \otimes g_i)$$

where sums and products are computed in $(\text{End}(\mathcal{U}), +, * \otimes \times)$.

Proof : Let us first give the following general lemma :

LEMMA 1.7 : Let $\mathcal{A}$ be a topological Hausdorff complete K-algebra over a ring K of characteristic 0, let I be a totally ordered set and let $t = (t_i)_{i \in I}$ be a family of $\mathcal{A}$ such that $(\frac{1}{\nu!} t^\nu)_{\nu \in \mathbf{N}^{(I)}}$ is summable. Then the family $(exp(t_i))_{i \in I}$ exists, is multipliable and the following relation holds :

$$\sum_{\nu \in \mathbf{N}^{(I)}} \frac{1}{\nu!} t^\nu = \prod_{i \in I} exp(t_i)$$

Proof of the lemma : It follows from th. 2 of [3] III. 5.3 that all the subfamilies of $(\frac{1}{\nu!} t^\nu)_{\nu \in \mathbf{N}^{(I)}}$ are summable. In particular, the subfamily indexed by $P(i) = \{ n.(i) \}_{n \in \mathbf{N}}$ (where (i) denotes the multidegree of support i whose only non-zero entry (of order i) is 1) is summable for every $i \in I$. Hence, this proves clearly the existence of $exp(t_i)$. Let us set now $S = \sum \frac{1}{\nu!} t^\nu$ and let $\mathcal{V}$ be a closed neighbourhood of S. Then, there is a finite subset K of $\mathbf{N}^{(I)}$ such that :

$$\forall \, K', \; K \subseteq K' \subseteq_f \mathbf{N}^{(I)} \implies \sum_{\nu \in K'} \frac{1}{\nu!} t^\nu \in \mathcal{V}$$

But, since $\mathcal{V}$ is closed, the above property holds even if K' is not finite. Hence, if J denotes the finite set which is the union of the supports of all ν in K, we will have for every finite subset I' of I containing J :

$$\prod_{i \in I'} (\sum_{n \in \mathbf{N}} \frac{1}{n!} t_i^n) = \sum_{\text{supp}(\nu) \subseteq I'} \frac{1}{\nu!} t^\nu \in \mathcal{V}$$

The lemma follows now immediately since $\mathcal{A}$ is Hausdorff. ■

Let us now come back to our theorem. First, according to th. 1.1, $(\frac{1}{\nu!}(h^\nu))_{\nu \in \mathbb{N}^{(I)}}$ is the dual basis of $(g^\nu)_{\nu \in \mathbb{N}^{(I)}}$. Hence it follows clearly that we have :

$$Id_{\mathcal{U}} = \sum_{\nu \in \mathbb{N}^{(I)}} (g^\nu)^* \otimes (g^\nu) = \sum_{\nu \in \mathbb{N}^{(I)}} \frac{1}{\nu!} (h^\nu) \otimes (g^\nu)$$

But, from the previous lemma, we also have :

$$\sum_{\nu \in \mathbb{N}^{(I)}} \frac{1}{\nu!} (h^\nu) \otimes (g^\nu) = \prod_{i \in I} exp(g_i^* \otimes g_i)$$

Thus, our theorem follows immediately from these last two relations. ∎

Note : Mélançon-Reutenauer's formula $(\mathcal{SC})$ given in the introduction of this paper appears now clearly as a particular case of th. 1.6. Indeed, taking again the denotations of the note ending section 1.2, th. 1.6 gives us the following relation :

$$Id_{K\langle A \rangle} = \sum_{w \in A^*} P_w \otimes [w] = \sum_{l \in L_y} exp(P_l \otimes [l])$$

Hence, relation $(\mathcal{SC})$ follows immediately since we can also obviously write :

$$Id_{K\langle A \rangle} = \sum_{w \in A^*} w \otimes w$$

2 The partially commutative case

The free partially commutative Lie algebra is the natural algebraic object to consider when some of the vector fields of an analytic system commute. We will give here the partially commutative versions of some basic methods that are used for computing structural nilpotent approximations of the output of an analytic system (cf [15]).

2.1 The free partially commutative monoid

Let us first define the objects with commutations freely generated in the categories of monoids and algebras. Hence let A be an alphabet and let θ be a symmetric and irreflexive subset of $A \times A$. [8] We can represent θ, that will be called a *partial commutation relation*, by its *commutation graph* which is the non-directed graph with vertices in A and where an edge is drawn between two letters $a, b \in A$ iff $(a, b) \in \theta$ (see [7], [8]).

With the pair (A, θ), called an *alphabet with commutations*, we can associate several objects freely generated by it. First, the *free partially commutative monoid* is the monoid $M(A, \theta)$ defined by the following monoid presentation :

$$M(A, \theta) = \langle A; (ab = ba)_{(a,b) \in \theta} \rangle_{\text{Mon}}$$

Explicitely, $M(A, \theta)$ is the quotient monoid $M(A, \theta) = A^* / \equiv_\theta$ where $\equiv_\theta$ denotes the finest congruence on A^* such that : $ab \equiv_\theta ba$ for every $(a, b) \in \theta$. This monoid solves by definition an obvious universal problem (see [7]).

Let now K be a ring. Then, we can consider the associative K-algebra $K\langle A, \theta \rangle$ which is defined by the following algebra presentation :

---

[8] Symmetric means that we have : $(a, b) \in \theta \implies (b, a) \in \theta$ for every $a, b \in A$ and irreflexive means that we have : $(a, a) \notin \theta$ for every $a \in A$.

$$K<A,\theta> = \; < A; \; (ab = ba)_{(a,b)\in\theta} >_{\text{K-Alg}}$$

This means that $K<A,\theta>$ is the quotient of the free associative K-algebra $K<A>$ by its ideal I_θ generated by the differences $ab - ba$ for $(a,b) \in \theta$. Since the relators $ab = ba$ are monoidal, it turns that $K<A,\theta> \simeq K[M(A,\theta)]$ (cf [9]). Hence $K<A,\theta>$ can be called the algebra of *partially commutative polynomials*.

2.2   Lazard's factorization of M(A,θ)

We will now give a process that will permit us to factorize $M(A,\theta)$ into free monoids. For this purpose, let us first give the following definition :

DEFINITION 2.1 :   Let (A,θ) be an alphabet with commutations, let Z be a non-commutative subalphabet [9] of A and let $B = A - Z$. Then we will call Z-*code* the subset of $M(A,\theta)$, denoted $C_Z(B)$, and defined by :

$$C_Z(B) = \{\, w.z, \; w \in \,<B>, \; z \in Z, \; TA(wz) = \{z\}\,\}$$

where $<B> \simeq M(B,\theta)$ denotes the submonoid of $M(A,\theta)$ generated by B and where $TA(wz)$ is the set of the terminal letters of wz (i.e. $a \in TA(wz)$ iff $wz \in M(A,\theta)\,a$).

Note :   There are effective methods for computing Z-codes. We refer the reader to [11] where two such methods are presented.

We can now give the following result (see [10] or [12] for more details) :

THEOREM 2.1 :   Let (A,θ) be an alphabet with commutations, let Z be a non-commutative subalphabet of A and let $B = A - Z$. Then the submonoid of $M(A,\theta)$ generated by $C_Z(B)$ is free of basis $C_Z(B)$ and $M(A,\theta)$ can be factorized as follows :

$$M(A,\theta) = C_Z(B)^* \, M(B,\theta) \qquad (\mathcal{LZ})$$

Note :   The previous theorem means exactly that every word w of $M(A,\theta)$ can be written in a unique way as follows :

$$w = c_1 \ldots c_n \, \beta$$

where every c_i belongs to $C_Z(B)$ and where β belongs to $M(B,\theta)$.

Examples :   1) In the free case, every subset Z of A is non-commutative and the associated Z-code is clearly equal to $C_Z = (A-Z)^* \, Z$. Hence th. 2.1 gives us just the classical Lazard's bisection of A^* (cf [20]) :

$$A^* = ((A-Z)^* \, Z)^* \, (A-Z)^*$$

2) Let $A = \{a,b,c\}$ and let θ be defined by the following graph :

$$a \;\text{------}\; b \qquad\qquad c$$

Then $Z = \{c\}$ is non-commutative and the associated Z-code is : $C_Z = a^*c \cup b^*c$. Hence it follows from th. 2.1 that $M(A,\theta)$ can be factorized here in free monoids as follows :

---

[9]   This means exactly that we have $(a,b) \notin \theta$ for every $a,b \in Z$.

$$M(A, \theta) = ((a^* + b^*)c)^* \, b^* \, c^*$$

From now to the end of this section, we will suppose that A is a *finite* alphabet. [10] Then the factorization $(\mathcal{LZ})$ given by th. 2.1 can clearly be iterated on $M(B, \theta)$ in order to obtain a finite factorization of $M(A, \theta)$ into free monoids. But let us first give the following definition that will permit us to precise this fact :

DEFINITION 2.2 : Let (A, θ) be an alphabet with commutations. Then we will call *chromatic partition* of A any partition $(Z_i)_{i=1,n}$ of A such that Z_i is a non-commutative subalphabet for every $i \in [1, n]$.

Note : A chromatic partition is clearly equivalent to a colouring of the commutation graph of (A, θ) in such a way that every pair of letters that are related by an edge are coloured in two distinct colours.

Let now $Z = (Z_i)_{i=1,n}$ be a chromatic partition of A. Then it follows clearly from th. 2.1 that $M(A, \theta)$ can be factorized into free monoids as follows :

$$M(A, \theta) = C_{Z_1}(B_1)^* \, C_{Z_2}(B_2)^* \, \ldots \, C_{Z_n}(B_n)^* \qquad (\mathcal{FM})$$

where $B_i = Z_{i+1} \cup \ldots \cup Z_n$ for every $i \geq 1$.

Let us introduce also a lexicographic ordering $<_i$ of the free monoid $C_{Z_i}(B_i)^*$ for every $i \in [1, n]$. Then there is a unique associated family $Ly(Z_i)$ of Lyndon words on the free monoid $C_{Z_i}(B_i)^*$ and it is possible to give the definition :

DEFINITION 2.3 : We will call *family of coloured Lyndon words* of $M(A, \theta)$ and denote by $Ly(Z, \theta)$ the union of the families $Ly(Z_i)$ for every $i \in [1, n]$.

Notes : 1) If $u \in C_{Z_i}(B_i)^*$ and $v \in C_{Z_j}(B_j)^*$ are two coloured Lyndon words, we will say that $u < v$ either if $i < j$ or if $v <_i u$ when $i = j$. Then the coloured Lyndon words, ordered by this total order, form clearly a complete factorization of $M(A, \theta)$ (cf [20]). This means that every word w of $M(A, \theta)$ can be written in a unique way as follows :

$$w = l_1^{i_1} \ldots l_m^{i_m}$$

where $(l_k)_{k \in [1,m]}$ are coloured Lyndon words such that $l_1 < \ldots < l_m$.

2) In the free case, the coloured Lyndon words can be a different family of words than the usual Lyndon ones. However, it suffices to only consider minimal chromatic partitions in order to have a good partially commutative generalization of the usual Lyndon words.

2.3 The free partially commutative Lie algebra

Let K be a ring and let (A, θ) be an alphabet with commutations. Then the *free partially commutative Lie K-algebra* is the quotient Lie K-algebra $L(A, \theta)$ defined by :

$$L(A, \theta) = L(A)/I_\theta$$

---

[10] In fact, all the following results can be stated for an infinite alphabet A with some minor adaptations (see [10] for more details).

where I_θ denotes the Lie ideal of $L(A)$ generated by $\{[a, b],\ (a, b) \in \theta\}$. It is easy to see that $L(A, \theta)$ satisfies to the following universal problem : for every mapping f from A into a Lie K-algebra $\mathfrak{g}$ such that :

$$\forall\ (a, b) \in \theta,\ [f(a), f(b)] = 0$$

there is a unique Lie K-algebra morphism $\overline{f}$ from $L(A, \theta)$ into $\mathfrak{g}$ which extends f. It can be shown that $L(A, \theta)$ is isomorphic to the Lie K-subalgebra of $K\!<\!A, \theta\!>$ generated by the letters of A (see [9]). We have also the following result concerning its enveloping algebra (see [19] or [9]) :

PROPOSITION 2.2 : The enveloping algebra of $L(A, \theta)$ is $K\!<\!A, \theta\!>$.

Observe now that we can define an intrinsic notion of adjoint in $L(A, \theta)$ by virtue of the following property (see [9] for more details) :

$$u \equiv_\theta v \quad \Longrightarrow \quad ad(u) = ad(v) \ \text{ in } L(A, \theta)$$

Let us now consider a non-commutative subalphabet Z of A and let us denote $B = A - Z$. Then we can define the following family of $L(A, \theta)$:

$$T_Z(B) = \{\, ad(w).z,\ wz \in C_Z(B)\,\}$$

since w and z are unique in $wz \in C_Z(B)$. This family will permit us to give the following Lie algebra version of th. 2.1 that generalizes the classical Lazard's elimination process to the partially commutative case (see also [12]) :

THEOREM 2.3 : (Duchamp-Krob; [9] or [10]) Let (A, θ) be an alphabet with commutations, let Z be a non-commutative subset of A and let $B = A - Z$. Then the Lie subalgebra of $L(A, \theta)$ generated by $T_Z(B)$ is a free Lie algebra of basic family $T_Z(B)$ such that $L(A, \theta)$ can be decomposed as follows :

$$L(A, \theta) = L(T_Z(B)) \oplus L(B, \theta)$$

Note : $L(T_Z(B))$ is a Lie ideal of $L(A, \theta)$. It is in fact the graded Lie ideal of $L(A, \theta)$ generated by the homogeneous Lie polynomials whose partial degree in Z is not 0.

Example : Let us consider again the second example of section 2.2. With the same choices, $L(A, \theta)$ can be decomposed into the following direct sum of free Lie algebras :

$$L(A, \theta) = L(\{ad\, a^n.c \cup ad\, b^n.c,\ n \in \mathbb{N}\}) \oplus K.a \oplus K.b$$

We will suppose from now to the end of the next section that A is a *finite* alphabet. Let then $Z = (Z_i)_{i=1,n}$ be a chromatic partition of A. Then we can iterate the decomposition given by th. 2.3 as in the monoid case in order to obtain the following direct sum decomposition of $L(A, \theta)$ into free Lie algebras :

$$L(A, \theta) = \bigoplus_{i=1}^{n} L(T_{Z_i}(B_i)) \qquad (\mathcal{FLA})$$

where B_i has the same meaning as in relation $(\mathcal{FM})$ of section 2.2. It follows now clearly from this decomposition and from the fact that every free Lie algebra is a free K-module that $L(A, \theta)$ is also a free K-module. Moreover, since there are several classical methods for constructing bases of a free Lie algebra (see [20]), the above decomposition gives us in fact an effective way for obtaining bases of $L(A, \theta)$.

2.4   The P.B.W. Lyndon basis of K(A,θ)

Let (A, θ) be an alphabet with commutations, let K be a ring which is also a **Q**-algebra, let $Z = (Z_i)_{i=1,n}$ be a chromatic partition of A, let $<_i$ be a lexicographic ordering of the free monoid $C_{Z_i}(B_i)^*$ and let $Ly(Z_i)$ be the corresponding family of Lyndon words on $C_{Z_i}(B_i)^*$ for every $i \in [1, n]$. Then let us recall that the Lyndon basis $[Ly](Z_i) = ([l])_{l \in Ly(Z_i)}$ of the free Lie algebra $L(T_{Z_i}(B_i))$ is defined for every $i \in [1, n]$ as follows : for every coloured Lyndon word $l \in Ly(Z_i)$, we define $[l]$ by the following rules :

- if $l = w.z \in C_{Z_i}(B_i)$, we have : $[l] = ad(w).z$

- if $l \notin C_{Z_i}(B_i)$, l can be decomposed in a unique way as : $l = m.n$ where $m <_i n$ are two Lyndon words of $Ly(Z_i)$ and we have :

$$[l] = [m][n] - [n][m]$$

Decomposition ($\mathcal{FLA}$) gives then sense to the following definition :

DEFINITION 2.4 :   We will call Lyndon basis of $L(A, \theta)$ the union of all Lyndon bases $[Ly(Z_i)]$ for every $i \in [1, n]$.

Let us consider a fixed Lyndon basis $[Ly] = [Ly](Z_1) \cup \ldots \cup [Ly](Z_n)$ of $L(A, \theta)$. Then since the corresponding coloured Lyndon words form a complete factorization of $M(A, \theta)$, we can index the associated P.B.W. basis of $K<A, \theta>$ by the words of $M(A, \theta)$. Indeed, if we define for every word $w \in M(A, \theta)$:

$$[w] = [l_1]^{i_1} \ldots [l_m]^{i_m}$$

where $w = l_1^{i_1} \ldots l_m^{i_m}$ is the factorization of w in coloured Lyndon words given by the first note ending section 2.2, then the family $PBW([Ly]) = ([w])_{w \in M(A, \theta)}$ is the P.B.W. basis of $K<A, \theta>$ associated with $[Ly]$. Let us now introduce the dual family of $PBW([Ly])$ that will be denoted $(P_w)_{w \in M(A, \theta)}$ [11] In order to study how to compute this dual family, let us first give the following immediate consequence of th. 1.1 :

PROPOSITION 2.4 :   For every $w \in M(A, \theta)$, we have :

$$P_w = P_{w_1} \;\sqcup\!\sqcup\; \ldots \;\sqcup\!\sqcup\; P_{w_n}$$

where $w = w_1 \ldots w_n$ denotes the unique decomposition of w corresponding to the factorization ($\mathcal{FM}$) (i.e. such that $w_i \in C_{Z_i}(B_i)^*$ for every $i \in [1, n]$).

According to this last result, it suffices to know how to compute for every $i \in [1, n]$ the elements $(P_w)_{w \in C_{Z_i}(B_i)^*}$ [12] in order to be able to compute all the dual family of $PBW([Ly])$. But it follows also from th. 1.1 that we have :

PROPOSITION 2.5 :   For every $w \in C_{Z_i}(B_i)$, we have :

---

[11]   Note that as in the free case, we can identify in a natural way the dual of $K<A, \theta>$ with the K-algebra $K<<A, \theta>>$ of the partially commutative series.

[12]   Unfortunately, this last problem can not be solved with Mélançon-Reutenauer's method. This difficulty comes from the fact that the natural scalar product in the free K-algebra $K<T_{Z_i}(B_i)>$ obtained by considering the words over the alphabet $T_{Z_i}(B_i)$ as an orthonomal basis is in general different from the trace on $K<T_{Z_i}(B_i)>$ of the natural scalar product on $K<A, \theta>$.

$$P_w = \frac{1}{i_1! \ldots i_n!} \; P_{l_1}^{i_1} \; \sqcup\!\sqcup \; \ldots \; \sqcup\!\sqcup \; P_{l_n}^{i_n}$$

where $w = l_1^{i_1} \ldots l_n^{i_n}$ denotes the Lyndon factorization of w in the free monoid $C_{Z_i}(B_i)^*$.

Hence our problem can be reduced to the computation of P_l for every coloured Lyndon word $l \in C_{Z_i}(B_i)^*$ when $i \in [1, n]$. Let us now consider a fixed integer $i \in [1, n]$. Then we have the following result which precises the value of the linear form P_l :

PROPOSITION 2.6 : Let cu be a coloured Lyndon word of $C_{Z_i}(B_i)^*$ which has $c \in C_{Z_i}(B_i)$ as first letter. Then P_{cu} is equal to the unique linear form of the free K-algebra $K{<}T_{Z_i}(B_i){>}$ [13] which maps 1 onto 0 and satisfies to the relation :

$$\forall \, d \in C_{Z_i}(B_i), \; \forall \, v \in T_{Z_i}(B_i)^*, \; P_{cu}([d]\,v) = \delta_{c,d}\, P_u(v)$$

Proof : Since $T_{Z_i}(B_i)^*$ form a K-basis of $K < T_{Z_i}(B_i) >$, the above relations defines clearly a unique linear form f of $K{<}T_{Z_i}(B_i){>}$. Moreover it follows easily from the proof of th. 2 (ii) of [17] that f satisfies to the following relation :

$$\forall \, \nu \in \mathbb{N}^{(Ly)}, \; f([Ly](Z_i)^\nu) = \delta_{\nu,(l)}$$

where (l) denotes the multidegree of support l whose unique l-entry is 1. In the same way, an easy argument of multihomogeneous degree permits to see that we have :

$$(\, f \,|\, [Ly](Z_1)^{\nu_1} \ldots [Ly](Z_n)^{\nu_n}\,) \; = \; 0 \qquad \text{if there is } \nu_j \neq 0 \text{ with } j \neq i$$

It follows now immediately from the above relations that $f = P_{cu}$. Therefore this ends the proof of our proposition. ∎

Using the two last propositions, it is now easy to compute by induction P_l for every Lyndon word $l = cu \in C_{Z_i}(B_i)^*$. Indeed, let us suppose that P_u is a polynomial which is already known. Let us then introduce the set H of the words over the alphabet $T_{Z_i}(B_i)$ which have a multihomogeneous degree of the form : $\mathrm{md}(c) + \mathrm{md}(s)$ where $\mathrm{md}(c)$ (resp. $\mathrm{md}(s)$) denotes the multihomogeneous degree of c (resp. of some element s in the support of P_u) when all these elements are considered as polynomials of $K{<}A, \theta{>}$. Then, since the family $T_{Z_i}(B_i)$ is multihomogeneous, it follows clearly from prop. 2.6 that P_l is also a polynomial of $K{<}T_{Z_i}(B_i){>}$ whose support is contained in H. Hence we can write :

$$P_l = \sum_{t \in H} k_t\, t$$

Since the family $T_{Z_i}(B_i)$ is multihomogeneous, an easy dimension argument permits now to show that prop. 2.6 is in fact equivalent to a square linear system where every equation has the following form : [14]

$$\sum_{t \in H} k_t\, (t \,|\, [d]\,v) \; = \; \delta_{c,d}\, (P_u|v)$$

and which involves every element $[d]\,v$ that has the same multihomogeneous degree as the words of H. Note first that the matrix of this system has obviously all its entries in $\mathbb{Q}$. Hence, since K is a $\mathbb{Q}$-algebra and since P_l is uniquely defined by these relations, the

---

[13] We identify here a linear form of $K{<}T_{Z_i}(B_i){>}$ to a formal power series of $K{<<}T_{Z_i}(B_i){>>}$.

[14] Note that the scalar product which is involved in these equations is the trace on $K{<}T_{Z_i}(B_i){>}$ of the natural scalar product in $K{<}A, \theta{>}$.

above system must be a Cramer system. It suffices now to solve it in order to obtain the value of the partially commutative polynomial P_l.

Notes : 1) It follows clearly from the above results that the dual basis $(P_w)_{w \in M(A,\theta)}$ of the P.B.W. Lyndon basis of $K<A,\theta>$ consists in fact of polynomials.

2) The shuffle product that is involved in prop. 2.4 and 2.5 is the shuffle product in $K \ll A, \theta \gg$ (in $K<A,\theta>$ in fact according to the previous note) which was defined in section 1.2 (see also [18]). To compute this product, it suffices just to know how to compute the shuffle product of two words of $M(A,\theta)$ and this can be made by recursive formulas that generalize the classical free ones (see [13] for more details).

Example : Let us take again the second example of section 2.2 and let us consider the chromatic decomposition of A defined by : $Z_1 = \{c\}$, $Z_2 = \{b\}$, $Z_3 = \{a\}$. Let us then order the alphabet $C_{Z_1}(B_1)$ as follows :

$$c < bc < ac < \ldots < b^n c < a^n c < \ldots$$

This defines a unique corresponding family of coloured Lyndon words. Then using the above results, it is easy to see that we have :

$$P_c = c, \ P_{bc} = \tfrac{1}{2}(bc - cb), \ P_{ac} = \tfrac{1}{2}(ac - ca), \ P_{b^2 c} = \tfrac{1}{6}(b^2 c - 2bcb + cb^2), \ \ldots$$

Let us now take as an example the coloured Lyndon word $bcac$. Then it follows from prop. 2.6 and from a previous computation that we have :

$$P_{bcb^2 c}([d_1] \ldots [d_n]) = \delta_{bc,d_1} \tfrac{1}{6} \left(b^2 c - 2bcb + cb^2 \,|\, [d_2] \ldots [d_n]\right)$$

for every letters $(d_i)_{i=1,n}$ in $a^* c \cup b^* c$. Hence it follows easily from the general analysis that we must here have :

$$
\begin{aligned}
P_{bcb^2 c} \ = \ & k_1 \left(bc - cb\right)(b^2 c - 2bcb + cb^2) + k_2 \left(b^2 c - 2bcb + cb^2\right)(bc - cb) \\
& + k_3 \, c \left(b^3 c - 3b^2 cb + 3bcb^2 - cb^3\right) + k_4 \left(b^3 c - 3b^2 cb + 3bcb^2 - cb^3\right) c
\end{aligned}
$$

where $(k_i)_{i=1,4}$ are unknown constants in K. Then an easy computation based on the previous general discussion shows that $(k_1, \ldots, k_4)$ satisfies to the linear system :

$$
\begin{cases}
12k_1 - 8k_2 - 10k_3 + 4k_4 & = 1 \\
-8k_1 + 12k_2 + 3k_3 - 9k_4 & = 0 \\
-10k_1 + 3k_2 + 20k_3 - k_4 & = 0 \\
4k_1 - 9k_2 - k_3 + 20k_4 & = 0
\end{cases}
\implies (k_1, k_2, k_3, k_4) = \tfrac{1}{1610}(507, 329, 207, 57)
$$

whose solution reported in the above relation gives the exact value of $P_{bcb^2 c}$.

Remark : According to th. 1.6 and to the definition of the Lyndon basis of $L(A,\theta)$, we clearly have the following effective formula :

$$Id_{K<A,\theta>} \ = \ \sum_{w \in M(A,\theta)} w \otimes w \ = \ \prod_{i=1}^{n} \prod_{l \in \mathbf{Ly}(Z_i)} exp\left(P_l \otimes [l]\right)$$

that involves only computations in free K-algebras.

2.5 Tensor decomposition of $K{<}A,\theta{>}$

Let (A,θ) be an alphabet with commutations, let Z be a non-commutative subalphabet of A and let $B = A - Z$. Then there is a tensor decomposition of the enveloping algebra $K{<}A,\theta{>}$ that corresponds to the decomposition $L(A,\theta) = L(T_Z(B)) \oplus L(B,\theta)$ given by th. 2.3 (see [5] or [10]). More precisely, we have : $K{<}A,\theta{>} \simeq K{<}T_Z{>} \otimes K{<}B,\theta{>}$. We will now see how to obtain effectively this decomposition.

Hence let $P = P_1 P_2 \ldots P_n$ be a word of $(T_Z(B) \cup B)^*$. We will then say that an integer $i \in [1, n-1]$ is an *inversion* of P iff $P_i \in B$ and $P_{i+1} \in T_Z(B)$. We can now introduce the K-module endomorphism ρ of the free associative algebra $K{<}T_Z(B) \cup B{>}$ which is defined on every word $P = P_1 P_2 \ldots P_n$ of $(T_Z(B) \cup B)^*$ by the rules :

$(R1)$ If P has no inversion, $\rho(P) = P$

$(R2)$ If P has an inversion, let i be the index of the leftmost inversion of P. Then we can write : $P_i = b$ and $P_{i+1} = ad(w).z$ with $b \in B$ and $w.z \in C_Z(B)$ and $\rho(P)$ is defined as follows :

$(R2.1)$ If $b.wz = wz.b$ in $M(A,\theta)$, then : $\rho(P) = P_1 \ldots P_{i-1} P_{i+1} P_i P_{i+2} \ldots P_n$

$(R2.2)$ If $b.wz \neq wz.b$ in $M(A,\theta)$, then $bw.z \in C_Z(B)$ and we can define :

$$\rho(P) = P_1 \ldots P_{i-1} (ad(bw).z) P_{i+2} \ldots P_n + P_1 \ldots P_{i-1} P_{i+1} P_i P_{i+2} \ldots P_n$$

Let now α be the canonical morphism from $K{<}T_Z(B) \cup B{>}$ into $K{<}A,\theta{>}$ defined by $\alpha(P) = P_1 \ldots P_n$ for every word $P = P_1 \ldots P_n$ of $(T_Z(B) \cup B)^*$ where the product in the second member is taken in the sense of $K{<}A,\theta{>}$. Then, it is easy to check that the following diagram is commutative :

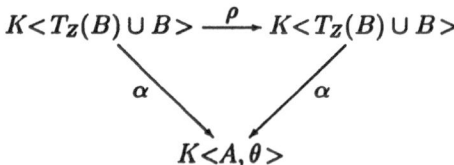

Hence this shows that every formal transformation that can be made with ρ on an element of $K{<}T_Z(B) \cup B{>}$ will not change the meaning of this element interpreted in $K{<}A,\theta{>}$. We can now give the following main property of ρ :

PROPOSITION 2.7 : Let n be an integer ≥ 1 and let Q be an homogeneous polynomial of degree n of $K{<}T_Z(B) \cup B{>}$. Then there exists an integer $N \leq n^2/4$ such that $\rho^N(Q)$ has no inversion.

Proof : Note first that the above result can be clearly reduced to prove the same result for words of length n in $(T_Z(B) \cup B)^*$. Let us now associate the following integer $N(P)$ with every word $P = P_1 \ldots P_n$ of $(T_Z(B) \cup B)^*$:

$$N(P) = \mathrm{Card}\,\{\,(i,j) \in [1,n] \times [1,n],\ i < j,\ P_i \in B,\ P_j \in T_Z(B)\,\}$$

An easy computation left to the reader permits to show that : $N(P) \leq n^2/4$ for every word $P \in (T_Z(B) \cup B)^*$ of length n. Hence our proposition will be proved if we can show that the following relation holds :

$$\forall\, P \in (T_Z(B) \cup B)^*,\ \rho^{N(P)}(P) = \rho^{N(P)+1}(P) \qquad (1)$$

since this property is obviously equivalent to the fact that $\rho^{N(P)}$ has no inversion for every word $P \in (T_Z(B) \cup B)^*$. But we get immediately from the defining rules of ρ that :

$$\rho(P) = u + v \quad \text{where } N(u) = N(P)-1 \text{ and } (v = 0 \text{ or } N(v) \leq N(P)-1)$$

Using an induction on $N(P)$, it is now easy to conclude that relation (1) holds. Hence this ends our proof. ∎

Since every letter of A belongs to $T_Z(B)$ or to B, we can consider in a natural way every word of $M(A, \theta)$ as a word of $(T_Z(B) \cup B)^*$. Hence, we can identify every polynomial $Q \in K{<}A, \theta{>}$ with an element of $K{<}T_Z(B) \cup B{>}$. The previous theorem gives us then an algorithm for computing the decomposition of an element of $K{<}A, \theta{>}$ according to the tensor product decomposition $K{<}T_Z(B){>} \otimes K{<}B, \theta{>} \simeq K{<}A, \theta{>}$. Indeed, let Q be a polynomial of $K{<}A, \theta{>}$. Then, according to prop. 2.7 and to the fact that $\alpha = \alpha \circ \rho$, it suffices clearly to compute successively $\rho(Q)$, $\rho^2(Q)$, ... (using the previous identification) until an element without inversion occurs, in order to obtain the desired decomposition. Observe finally that this result can be used for computing P.B.W. decompositions in $K{<}A, \theta{>}$ when A is finite since it reduces this problem to the free case which is classical (see [17] for instance).

Examples : 1) Let us consider the free monoid over the alphabet $A = \{a, b\}$ and let us take $Z = \{a\}$. Then the corresponding tensor decomposition of $baba$ is given by :

$$\rho^3(baba) = \rho^2([b,a]ba + abba) = \rho([b,a][b,a] + [b,a]ab + ab[b,a] + abab)$$
$$= [b,a][b,a] + [b,a]ab + a[b,[b,a]] + 2a[b,a]b + aabb$$

2) Let $A = \{a, b, c\}$, let θ be defined by the commutation graph given in the second example of section 2.2 and let us take here $Z = \{b, c\}$. Then an easy computation shows that the associated tensor decomposition of $acabc$ is given by :

$$\rho^5(acabc) = [a,c]b[a,c] + [a,c]bca + cb[a,c]a + cbcaa + cb[a,[a,c]] + cb[a,c]a$$

Remark : The above rewriting process holds for every Lie algebra $\mathfrak{g}$ that can be decomposed as the direct sum of a Lie ideal $\mathfrak{h}$ generated by a family T and of a Lie subalgebra $\mathfrak{k}$ generated by a family B. But, it is much more efficient for $L(A, \theta)$ since B acts here by derivations within $T_Z(B) \cup \{0\}$ which is not the general case.

References

[1] BERSTEL J., REUTENAUER C., *Les séries rationnelles et leurs langages*, Masson, 1984

[2] BOURBAKI N., *Algèbre générale*, Chap. 1 à 3, CCLS, 1970

[3] BOURBAKI N., *Topologie générale*, Chap. 1 à 4, CCLS, 1971

[4] BOURBAKI N., *Topologie générale*, Chap. 5 à 10, CCLS, 1974

[5] BOURBAKI N., *Algèbres et Groupes de Lie*, Chap. 1, CCLS, 1971

[6] BOURBAKI N., *Algèbres et Groupes de Lie*, Chap. 2 à 3, CCLS, 1972

[7] CHOFFRUT C., *Free partially commutative monoids*, LITP Technical Report 86.20, Paris, 1986

[8] DUBOC C., *Commutations dans les monoïdes libres : un cadre théorique pour l'étude du parallélisme*, Thèse d'Université, Université de Rouen, LITP Technical Report 86.25, Paris, 1986

[9] DUCHAMP G., KROB D., *The partially commutative free Lie algebra : bases and ranks*, LITP Technical Report 89.93, Paris, 1989 (To appear in "Adv. in Maths.")

[10] DUCHAMP G., KROB D., *Free partially commutative structures*, LITP Technical Report 90.65, Paris, 1990

[11] DUCHAMP G., KROB D., *Lazard's factorizations of free partially commutative monoids*, LITP Technical Report 90.90, Paris, 1990

[12] DUCHAMP G., KROB D., *Factorisations dans le monoïde partiellement commutatif libre*, C. R. Acad. Sci. Paris, Série I, **312** (1), pp. 189-192, 1991

[13] DUCHAMP G., KROB D., *On the partially commutative shuffle product*, LITP Technical Report, Paris, 1991

[14] FLIESS M., *Fonctionnelles causales non linéaires et indéterminées non commutatives*, Bull. Soc. Math. France, **109**, pp. 3-40, 1981

[15] HOUANG N.M., JACOB G., OUSSOUS N.E., *Comportement Entrée/Sortie des systèmes analytiques non linéaires : approximations rationnelles, approximations structurelles nilpotentes*, Preprint LIFL, Lille, 1990

[16] LOTHAIRE M., *Combinatorics on words*, Addison-Wesley, 1983

[17] MELANCON G., REUTENAUER C., *Lyndon words, free algebras and shuffles*, Can. Journ. of Math., **XLI**, 4, pp. 577-591, 1989

[18] SCHMIDT W., *Hopf algebras and identities in partially commutative monoids*, Theor. Comput. Sc., **73**, pp. 335-340, 1990

[19] THIBON J.Y., *Intégrité des algèbres de séries formelles sur un alphabet partiellement commutatif*, Theor. Comput. Sc., **41**, 1985

[20] VIENNOT G., *Algèbres de Lie libres et monoïdes libres*, Lecture Notes in Maths., **691**, Springer Verlag, 1978

Rational computation in dioid algebra and its application to performance evaluation of Discrete Event Systems

Stéphane Gaubert
INRIA*

Carlos Klimann
INRIA†

Abstract

Modeling and analysis of a specific class of Discrete Event Systems lead to introduce an exotic algebra of formal series (*dioid* algebra). In particular, the behavior of these systems is characterized by computing *transfer matrices*. In this paper, we study the algebraic problems which arise when considering rational computations in this particular dioid. The main theorem states that rational elements are periodic, in the sense they represent the eventual periodic behavior of Timed Event Graphs. Then the algebra of periodicities is investigated. Some formulae and algorithms are presented. In particular, we show how the computation of the periodic behavior is related to the Frobenius problem for linear diophantine equations. These algorithms have been implemented in MAPLE. An application to a simple flowshop is presented.

Introduction

Our approach follows the Linear System Theory for Discrete Event Systems developed by Cohen, Dubois, Moller, Quadrat, Viot (see [2, 3]). This theory extends to a restricted class of Discrete Event Systems the main concepts of Control Theory such as transfer function and state-space representation. The first part of the paper reviews the algebraic tools needed to deal with Timed Event Graphs, summarizing the presentation given in [3]. We first explain how Timed Event Graphs can be modeled using dater functions, shift operators in dating and shift operators in counting. Then we focus on the algebraic properties of shifts. This leads to the introduction of a specific algebra

of formal series called MinMax$\langle\langle \gamma, \delta \rangle\rangle$. Solving the state equation for Timed Event Graphs results in making rational computations in this algebra, that is computing the transfer matrix of the system, $H = CA^*B$ (A^* plays a role analogous to $(sI - A)^{-1}$ in classical theory). We study the properties of rationals. The important theorem 4.11 characterizes rational series as periodic series, i.e. series corresponding to a periodic behavior of the system. Then we investigate the algebra of periodicities, i.e. the way periodicities are transformed when systems are put in parallel (proposition 5.3), in cascade (proposition 5.5) and in feedback. The results can be summarized by saying that the slowest periodicity is absorbing, which is very natural when dealing with Timed Event Graphs. This is the simple case. When the systems which are put in cascade have the same periodic slope (the same periodic throughput), some more complex "arithmetical" features appear. This corresponds to the case when the *critical subgraph* of the Timed Event Graphs is not reduced to a single circuit. In this case, the computation of the transient behavior is closely related with the "Frobenius problem" for Linear Diophantine equations. We conclude by shortly presenting an application to a simple manufacturing system (flowshop), obtained with our current implementation in MAPLE.

1 Modeling Timed Event Graphs

1.1 Dater equations

We refer the reader to [3] where it is explained in detail how Timed Event Graphs can be modeled from the dioid point of view. We just recall here the few basic facts needed for our purpose. Whith each transition i of an event graph, we associate a *dater function*, which is a map $\mathbf{Z} \to \mathbf{Z} \cup \{\pm\infty\}$, $n \mapsto x_i(n)$, defined by:

---

*Domaine de Voluceau, 78153 Le Chesnay Cedex France, e-mail gaubert@seti.inria.fr

†Domaine de Voluceau, 78153 Le Chesnay Cedex France, e-mail klimann@seti.inria.fr

$x_i(n) = t \quad \Leftrightarrow \quad$ "the firing numbered n of the transition i occurs at date t"

Dater functions are increasing. The restriction $n \in \mathbf{Z}$ is related to the discrete nature of the events, and we assume that the time only takes integer or infinite values $(x_i(n) \in \mathbf{Z} \cup \{\pm\infty\})$. This means there is a clock giving absolute times, and all the firings occur only at these times. Allowing $x_i(n) = -\infty$ or $x_i(n) = +\infty$ allows modeling situations when all the firings up to n have already occurred before we consider the system $(x_i(n) = -\infty)$, or the firing numbered n never occurs $(x_i(n) = +\infty)$. For

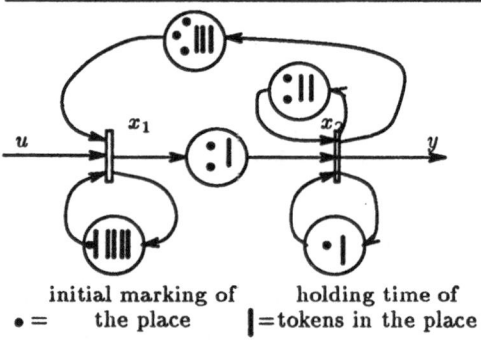

initial marking of $\quad$ holding time of
$\bullet =$ the place $\quad \mathbf{|} =$ tokens in the place

Figure 1: A simple Timed Event Graph

instance, for the event graph of Figure 1, it is immediate to obtain the following inequalities:

$x_1(n) \geq \max[5+x_1(n-1), 3+x_2(n-3), u(n)]$
$x_2(n) \geq \max[1+x_1(n-2), 2+x_2(n-2), 1+x_2(n-1)]$
$y(n) \geq x_2(n)$

$$(1)$$

Of course, the solution of (1) is not unique. We have to make the additional assumption that the transitions are fired as soon as possible (*earliest behavior*), which is equivalent to selecting the least solution of (1).

1.2 Operator representation

Let us denote by $\mathcal{T}$ the set of dater functions ("signals"). The elementary shift in dating $\delta : \mathcal{T} \to \mathcal{T}$ and the elementary shift in counting $\gamma : \mathcal{T} \to \mathcal{T}$ are defined by $\gamma x(n) = x(n-1)$ and $\delta x(n) = x(n)+1$. The set of operators $\mathcal{T}^{\mathcal{T}}$ is naturally endowed with two laws: -the addition (denoted by $\oplus$) corresponding to the max of signals, -the composition product, denoted as

usual by . or concatenation. With these notations, we write for instance $x_2(n-2) = \gamma^2 x_2(n)$ and $\max[2+x_2(n-2), 1+x_2(n-1)] = (\gamma^2\delta^2 \oplus \gamma\delta)x_2(n)$. Thus, (1) rewrites as follows:

$$\begin{cases} x_1 \geq \gamma\delta^5 x_1 \oplus \gamma^3\delta^3 x_2 \oplus u \\ x_2 \geq \gamma^2\delta x_1 \oplus (\gamma^2\delta^2 \oplus \gamma\delta)x_2 \quad (2) \\ y \geq x_2 \end{cases}$$

This system can be written in a standard state-space form:

$$x \geq Ax \oplus Bu, \quad y \geq Cx \qquad (3)$$

where A, B, C are matrices the entries of which are sum of shift operators. The main concern of this paper is to find the minimal solution of (3) in the general case by means of effective algorithms. The solution (x, y) (i.e. all the dater functions associated with transitions and outputs) will provide a complete knowledge of the earliest behavior of the Timed Event Graph.

Two examples of this approach, applied to this simple Timed Event Graph and to a more complex one corresponding to a flexible workshop can be found in Section 7.

2 Algebraic model

2.1 Absorption properties of shift operators

Since γ and δ commute, it should be clear that the only operators we need to model Timed Event Graphs can be written as finite sums of operators $\gamma^n\delta^t$ (with $n, t \in \mathbf{N}$). Conversely, given an operator $L = \bigoplus_{i=1}^{p} \gamma^{n_i}\delta^{t_i}$, we may ask whether such a decomposition with respect to γ and δ is unique. This is *not* the case because we have important absorption properties. First, if $t \geq t'$, then $(\delta^t \oplus \delta^{t'})x(k) = \max(t + x(k), t' + x(k)) = \max(t, t') + x(k) = t + x(k) = \delta^t x(k)$. Similarly, let $n \geq n'$. We have: $(\gamma^n \oplus \gamma^{n'})x(k) = \max(x(k-n), x(k-n')) = x(k-n') = \gamma^{n'}x(k)$, for the dater function $k \mapsto x(k)$ is increasing and $k - n \leq k - n'$. This implies the following fundamental rules:

$$\gamma^n \oplus \gamma^{n'} = \gamma^{\min(n,n')} \qquad (4)$$
$$\delta^t \oplus \delta^{t'} = \delta^{\max(t,t')} \qquad (5)$$

2.2 The MinMax$\langle\!\langle\gamma,\delta\rangle\!\rangle$ algebra

We now describe the algebraic structure which reflects the properties of shift operators. Let us denote by $\mathbf{B} = \{\varepsilon, e\}$ the set of Booleans, with addition $\oplus$ and product $\otimes$, ε as zero and e as the unit. $\mathbf{B}\langle\!\langle\gamma,\delta\rangle\!\rangle$ denotes the set of formal series with indeterminates γ, δ, boolean coefficients, and exponents in $\mathbf{Z}$, endowed with the usual addition $\oplus$ and product $\otimes$. Indeed, a formal series is a mapping $\mathbf{Z}^2 \to \mathbf{B}$, $(n,t) \mapsto s(n,t)$. s admits an unique expression

$$s = \bigoplus_{n,t \in \mathbf{Z}} s(n,t) \otimes \gamma^n \otimes \delta^t. \qquad (6)$$

Clearly, $\oplus$ is *idempotent* $((\forall s)\ s \oplus s = s)$. $\oplus$ and $\otimes$ both are commutative monoid laws over $\mathbf{B}\langle\!\langle\gamma,\delta\rangle\!\rangle$, $\otimes$ is distributive with respect to $\oplus$ (i.e. $(s \oplus s') \otimes r = s \otimes r \oplus s' \otimes r$) and ε is absorbing. Such a structure is known as a commutative-idempotent-semiring with absorbing zero, or simply as a commutative *dioid* (see [3]). As usual, we will omit the $\otimes$ sign and simply write $s(n,t)\gamma^n\delta^t$ instead of $s(n,t) \otimes \gamma^n \otimes \delta^t$. It is important to notice that there is a natural order associated with $\oplus$ and which is compatible with the product, namely

$$s \le r \Longleftrightarrow s \oplus r = r.$$

The ordered set $(\mathbf{B}\langle\!\langle\gamma,\delta\rangle\!\rangle, \le)$ is *complete* (i.e. every subset X admits a least upper bound, naturally denoted by $\bigoplus_{x \in X} x$). Moreover, the infinite distributivity law holds:

$$\left(\bigoplus_{x \in X} x\right) y = \left(\bigoplus_{x \in X} xy\right). \qquad (7)$$

Thus, $((\mathbf{B}\langle\!\langle\gamma,\delta\rangle\!\rangle, \oplus, \otimes)$ is a *complete* dioid [3].

The *support* of a series s is defined by $\mathrm{Supp}(s) = \{(n,t) \in \mathbf{Z}^2;\ s(n,t) \ne \varepsilon\}$. Since s can be written as $s = \bigoplus_{(n,t) \in \mathrm{Supp}(s)} \gamma^n \delta^t$, it is clear Boolean formal series are characterized by their support. When $\mathrm{Supp}(s)$ is finite, s is defined to be a *polynomial*. The subdioid of polynomials will be denoted by $\mathbf{B}\langle\gamma,\delta\rangle$.

We define the (total) *degree* of a series s as the upper bound of $\mathrm{Supp}(s)$: $\deg(s) = (\deg_\gamma(s), \deg_\delta(s)) = \sup\ \mathrm{Supp}(s)$. The *valuation* is defined dually.

Examples 2.1 $\varepsilon \oplus \gamma^n\delta^t = \gamma^n\delta^t$, $(e \oplus \gamma^3\delta^2)(\delta^{-4} \oplus \gamma^5\delta^{-4}) = \delta^{-4} \oplus \gamma^5\delta^{-4} \oplus \gamma^3\delta^{-2} \oplus \gamma^8\delta^{-2}$. Because negative exponents are allowed, γ^{-1} is a polynomial. $\deg(\gamma \oplus \delta^2) = \sup[(1,0),(0,2)] = (1,2)$.

Proposition 2.2 *The only invertible elements in $\mathbf{B}\langle\!\langle\gamma,\delta\rangle\!\rangle$ are monomials.*

Proof $\gamma^{-n}\delta^{-t}$ is the inverse of $\gamma^n\delta^t$. Conversely, noticing that $\deg(ss') = \deg(s) + \deg(s')$, $\mathrm{val}\ (ss') = \mathrm{val}\ (s) + \mathrm{val}\ (s')$ and $\deg(s) \ge \mathrm{val}\ (s)$, $ss' = e$ implies $0 = \deg(e) - \mathrm{val}\ (e) = (\deg(s) - \mathrm{val}\ (s)) + (\deg(s') - \mathrm{val}\ (s'))$, which is possible only if $\deg(s) - \mathrm{val}\ (s) = 0$ and $\deg(s') - \mathrm{val}\ (s') = 0$. This implies s is a monomial. $\blacksquare$

Before carrying on the study of formal series, we have to introduce a new operation: the star. Roughly speaking, the star plays a role analogous to the inverse in classical algebra. This is why the star is called a rational operation.

Definition 2.3 (Star operation) $a^* = e \oplus a \oplus a^2 \oplus a^3 \oplus \ldots$

Since $\mathbf{B}\langle\!\langle\gamma,\delta\rangle\!\rangle$ is a complete dioid, this infinite sum is well defined. The following properties are straightforward:

$$
\begin{array}{lll}
(i) & (a^*)^* = a^* & \\
(ii) & (a \oplus b)^* = a^*b^* & (8) \\
(iii) & a^* = a^*a^* &
\end{array}
$$

In the $\mathbf{B}\langle\!\langle\gamma,\delta\rangle\!\rangle$ algebra, we have not yet translated the absorption rules (4),(5). This can be done by taking the quotient of $\mathbf{B}\langle\!\langle\gamma,\delta\rangle\!\rangle$ by an appropriate congruence. Indeed, the first absorption rule implies $e \oplus \gamma = e$ (this is the case $n' = 0$ and $n = 1$). Writing $e \oplus \gamma = e \oplus e\gamma = e \oplus (e \oplus \gamma)\gamma$, we obtain after an immediate induction that:

$$e = e \oplus \gamma \oplus \gamma^2 \oplus \gamma^3 \oplus \ldots = \gamma^* \quad \text{(as operators)} \tag{9}$$

This suggests that e and γ^* should be identified. For the same reason, the second absorption rule requires e be identified with $(\delta^{-1})^*$. This identification is done by introducing the following quotient:

Definition 2.4 (MinMax$\langle\langle\gamma,\delta\rangle\rangle$ algebra)
The map φ : $\mathbf{B}\langle\langle\gamma,\delta\rangle\rangle$ $\rightarrow$ $\mathbf{B}\langle\langle\gamma,\delta\rangle\rangle$, $s \mapsto$ $s\gamma^*(\delta^{-1})^*$ *is a congruence. The quotient dioid* $\mathbf{B}\langle\langle\gamma,\delta\rangle\rangle/\varphi$ *is called the* MinMax$\langle\langle\gamma,\delta\rangle\rangle$ *dioid.*

Since $\gamma^*\gamma^* = \gamma^*$, we have $\varphi(e) = \gamma^*\delta^* = \gamma^*(\gamma^*\delta^*) = \varphi(\gamma^*)$. This shows that e and γ^* belong to the same equivalence class. Of course the same property holds for $(\delta^{-1})^*$, which shows that φ realizes the desired identification: rules (4),(5) are valid in MinMax$\langle\langle\gamma,\delta\rangle\rangle$. In the following, we will deal with objects in MinMax$\langle\langle\gamma,\delta\rangle\rangle$ without precising they are equivalence classes modulo φ. The equivalence classes of polynomials will be again called *polynomials*, and the subdioid of polynomials will be denoted by MinMax$\langle\gamma,\delta\rangle$. MinMax$\langle\langle\gamma,\delta\rangle\rangle$ is complete, and the lower bound will be denoted by $\wedge$.

We now present a very useful graphic representation of elements of MinMax$\langle\langle\gamma,\delta\rangle\rangle$, which makes it obvious how the simplification rules work.

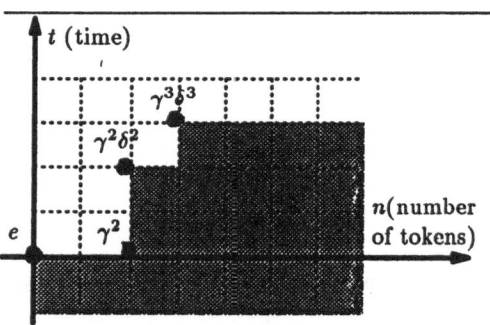

Figure 2: Graphic representation of $p = e \oplus \gamma^2\delta^2 \oplus \gamma^2 \oplus \gamma^3\delta^3$

Graphic representation It is immediate to check that given $s \in \mathbf{B}\langle\langle\gamma,\delta\rangle\rangle$, there is a *maximal representative* of the equivalence class of s modulo φ, namely $\varphi(s) = s\gamma^*(\delta^{-1})^*$. Then, a monomial $\gamma^\nu\delta^\tau \in \mathbf{B}\langle\langle\gamma,\delta\rangle\rangle$ can be represented by a point of the $\mathbf{Z}^2$-plane with coordinates (ν,τ), and the class of s by the collection of points corresponding to $\varphi(s)$, i.e. Supp $\varphi(s)$. For instance, to the equivalence class of $\gamma^2\delta^2$ corresponds the "south-east" cone of the $\mathbf{Z}^2$-plane

$\{(2 + n', 2 - t'); (n',t') \in \mathbf{N}^2\}$. More generally, an arbitrary element of MinMax$\langle\langle\gamma,\delta\rangle\rangle$ is simply represented by an union of south-east elementary cones of $\mathbf{Z}^2$. Addition corresponds to union and product to the usual addition of subsets. For instance, Figure 2 represents the polynomial $p = e \oplus \gamma^2 \oplus \gamma^2\delta^2 \oplus \gamma^3\delta^3$. The monomial γ^2 can be dropped since it lies in the "shadow" of $\gamma^2\delta^2$.

2.3 Canonical form of polynomials

In the case of polynomials, there exists a canonical form which is simpler than the maximal representative already mentioned.

Proposition 2.5 *Let* $S \in$ MinMax$\langle\gamma,\delta\rangle$. *There exists a unique minimal* $s \in \mathbf{B}\langle\gamma,\delta\rangle$ *among the representatives of* S.

Proof (i) existence: let $s \in \mathbf{B}\langle\gamma,\delta\rangle$ be a representative of S. We write s as a sum of n monomials: $s = \bigoplus_{i=1}^n a_i$. We say the monomial a_i is redundant if $\varphi(\bigoplus_{j\neq i} a_j) = \varphi(s)$. If s is not minimal, we can eliminate a redundant monomial, and so on. Hence we obviously get a minimal representative.

(ii) uniqueness: let us now assume that $p = \bigoplus_{i\in I} a_i$ and $q = \bigoplus_{j\in J} b_j$ both are minimal representatives (a_i, b_j being monomials). Let $c = \gamma^*(\delta^{-1})^*$. $\varphi(p) = \varphi(q)$ is equivalent to $(\bigoplus_{i\in I} a_i)c = (\bigoplus_{j\in J} b_j)c$, i.e. $\bigoplus_{i\in I} a_i c = \bigoplus_{j\in J} b_j c$ Since $e \leq c$, we have $a_i \leq a_i c \leq \bigoplus_{j\in J} b_j c$, and there exists $j \in J$ such that $a_i \leq b_j c$. Thus $a_i c \leq b_j c^2 = b_j c$. For the same reason, $b_j c \leq a_k c$ for some k. Indeed, we must have $k = i$ (otherwise, a_i should be redundant), hence $\varphi(a_i) = a_i c = b_j c = \varphi(b_j)$. This implies $a_i = b_j$ since the restriction of φ to the set of monomials is clearly injective (cf. the graphic interpretation of φ). This shows that $p = q$ and ends the proof. ∎

The geometric meaning of the minimal representative is obvious. In Figure 2, the minimal representative of p corresponds to the set of "north-west" corners of its maximal representative, namely $p' = e \oplus \gamma^2\delta^2 \oplus \gamma^3\delta^3$ (corresponding to the black points). The minimal representative has been obtained by eliminating the redundant monomial γ^2 (black square).

Proposition 2.5 allows extending the definition of support, degree and valuation to MinMax$\langle \gamma, \delta \rangle$. These concepts are now defined in terms of canonical representatives. The number of monomials of the canonical representative of p is called the *complexity* of p (denoted by compl(p)).

Remark 2.6 In MinMax$\langle\!\langle \gamma, \delta \rangle\!\rangle$, there is in general no canonical representative. Consider $\delta^* = e \oplus \delta \oplus \delta^2 \oplus \dots$. We also have $\delta^* = \delta^k \oplus \delta^{k+1} \oplus \dots$ for all k. This suggests that the minimal representative of δ^* should be something like $\delta^{+\infty}$ which does not exist in MinMax$\langle\!\langle \gamma, \delta \rangle\!\rangle$.

3   System Theory

We have now provided enough material to solve at least "formally" our main problem (3). The star notation which has been defined for scalars of $\mathbf{B}\langle\!\langle \gamma, \delta \rangle\!\rangle$ (and hence of MinMax$\langle\!\langle \gamma, \delta \rangle\!\rangle$) obviously extends to $\mathcal{M}_{n,n}(\text{MinMax}\langle\!\langle \gamma, \delta \rangle\!\rangle)$, the algebra of $n \times n$ square matrices with entries in MinMax$\langle\!\langle \gamma, \delta \rangle\!\rangle$. The star is related to solving equations by the following well known result [3]:

Proposition 3.1
Let $A \in \mathcal{M}_{n,n}(\text{MinMax}\langle\!\langle \gamma, \delta \rangle\!\rangle)$ and $b \in (\text{MinMax}\langle\!\langle \gamma, \delta \rangle\!\rangle)^n$. The least solution of

$$x \geq Ax \oplus b \qquad (10)$$

is given by $A^* b$.

Thus, the least solution of (3) is given by $x = A^* Bu$, $y = C A^* Bu$. $H = C A^* B$ is called the *transfer matrix* of the system.

Usually, the transfer function is characterized as the "impulse response" of the system. This still holds in the MinMax$\langle\!\langle \gamma, \delta \rangle\!\rangle$ algebra, and has important consequencies. First, we associate a formal series with a dater function $n \mapsto x(n)$ in a natural way:

$$S_x = \bigoplus_{n \in \mathbf{Z}} \gamma^n \delta^{x(n)} \qquad (11)$$

with the convention $\delta^{-\infty} = \varepsilon$ and $\delta^{+\infty} = \delta^*$. Conversely, given a series s, the only dater function $\mathcal{D}_s$ such that $s = S_x$ is given by:

$$\mathcal{D}_s = \sup\{t, \gamma^n \delta^t \leq s_x\} \qquad (12)$$

This correspondence allows identifying signals and formal series. All the notions already defined for series (support, valuation,...) are obviously extended to signals.

Let us now consider a system $u \mapsto y$ with transfer series: $H = \bigoplus_{p \in \mathbf{Z}} \gamma^p \delta^{\mathcal{D}_H(p)}$. Applying the definition of shift operators, we have $H u(n) = \max_{p \in \mathbf{Z}}[\mathcal{D}_H(p) + u(n - p)]$. Then, the input output-relation for dater function is given by the following *max-convolution* of daters:

$$y(n) \stackrel{\text{def}}{=} (\mathcal{D}_H \star u)(n) = \max_{p \in \mathbf{Z}}[\mathcal{D}_H(p) + u(n - p)] \qquad (13)$$

The translation of (13) using formal series is simply a product:

$$S_y = \bigoplus_{n \in \mathbf{Z}} \gamma^n \delta^{y(n)} =$$
$$= (\bigoplus_{n \in \mathbf{Z}} \gamma^n \delta^{\mathcal{D}_H(n)})(\bigoplus_{n \in \mathbf{Z}} \gamma^n \delta^{u(n)}) = H S_u$$

Because $e = \bigoplus_{n \geq 0} \gamma^n \delta^0$, the dater function associated with e is given by $\mathcal{D}_e(n) = -\infty$ if $n < 0$ and $\mathcal{D}_e(n) = 0$ if $n \geq 0$. This means that the events numbered $0, 1, 2, \dots$ occur at date 0. For this reason, e is called the *impulse*, it models the case when an infinite "quantity" of inputs is available at date 0. Since $S_y = H e = H$, the dater function associated with H determines the output y corresponding to an impulse (*impulse response*).

4   Rationality and periodicity

The *rational closure* of a subset $\mathcal{E}$ of a dioid is by definition the smallest subdioid $\mathcal{F}$ such that $\mathcal{E} \subset \mathcal{F}$ and $\mathcal{F}$ is rationally stable (i.e. stable for the operations $\oplus, \otimes$, and $*$). The notation $\mathcal{E}^*$ for $\mathcal{F}$ is standard. When modeling systems, only *causal polynomials* occur, that is polynomials whith nonnegative exponents in γ in δ. Let us denote by MinMax$^+\langle \gamma, \delta \rangle$ the subdioid of causal polynomials. Then, transfer matrices are objects of the form $C A^* B$ with $A^* \in [\mathcal{M}_{n,n}(\text{MinMax}^+\langle \gamma, \delta \rangle)]^*$, $C \in \mathcal{M}_{p,n}(\text{MinMax}^+\langle \gamma, \delta \rangle)$ and $B \in \mathcal{M}_{n,q}(\text{MinMax}^+\langle \gamma, \delta \rangle)$. Since

$$[\mathcal{M}_{n,n}(\text{MinMax}^+\langle \gamma, \delta \rangle)]^* = \mathcal{M}_{n,n}[(\text{MinMax}^+\langle \gamma, \delta \rangle)^*]$$

(this is simply a part of the Kleene-Schützenberger Theorem, see [3] for a proof specific to this context), we are reduced to computing the stars of scalars. Indeed, the computation of A^* provided the stars of scalars are known is no more than the Gauss elimination algorithm applied to the matrix equation $X = AX \oplus \text{Id}$. The simplest algorithm is perhaps the Jacobi-like variant developed in [6, 5]. There is no difficulty here and we shall only consider the scalar case.

Definition 4.1 (Rational series)
We call rational series, *or simply* rationals *the series belonging to the rational closure $\mathcal{R}$ of the subdioid of causal polynomials.*

Definition 4.2 (Periodicity) *Let $(\nu, \tau) \in (N \setminus \{0\} \times N \setminus \{0\}) \cup \{(1,0),(0,+\infty)\}$. A dater function $n \mapsto d(n)$ is (ν, τ)-periodic if there exists $N \in N$ such that:*

$$(\forall n \geq N) \quad d(n + \nu) = d(n) + \tau \quad (14)$$

Periodicity has a straightforward interpretation for Timed Event Graphs. (14) means that after a transient behavior of length N, *every τ units of time, ν firings occur.* The *periodic throughput* or the *periodic slope* of d is the ratio $\lambda(d) \stackrel{def}{=} \frac{\tau}{\nu}$. The minimal value of N is called the *length of the transient*, and the minimal value of (ν, τ) is called *the* periodicity of d. The *degenerate* cases $(\nu, \tau) = (1,0)$ and $(0,+\infty)$ represent respectively the situation when an infinite number of event occur in a finite time and when no events occur after the $N - 1$ th.

Remark 4.3 Indeed, definition 4.2 is not the usual one for periodic functions. Since dater functions are increasing, they cannot be periodic in the classical sense and this is not confusing. An equivalent formulation is to introduce $\tilde{d}(n) = d(n + 1) - d(n)$ and to observe that (14) implies $(\forall n \geq N) \quad \tilde{d}(n + \nu) = \tilde{d}(n)$ which means that the *increases* of d are eventually periodic in the usual sense.

Definition 4.4 (Periodic series) *A series $s \in \text{MinMax}\langle\langle \gamma, \delta \rangle\rangle$ is (ν, τ)-periodic if the dater function $\mathcal{D}_s$ associated with s is (ν, τ)-periodic.*

The following provides an algebraic characterization of periodic series:

Proposition 4.5 *s is (ν, τ)-periodic iff there exists two polynomials p, q such that*

$$s = p \oplus q(\gamma^\nu \delta^\tau)^* \quad (15)$$

This is called a periodic representation *of s, with periodicity $\gamma^\nu \delta^\tau$.*

Proof The degenerate cases being straightforward, we assume $\tau \neq 0$ and $\tau \neq +\infty$. Let $s = \bigoplus_{n \in N} \gamma^n \delta^{\mathcal{D}_s(n)}$ be (ν, τ)-periodic with transient of lenght N. We have:

$$\bigoplus_{n \geq N} \gamma^n \delta^{\mathcal{D}_s(n)} = \left(\bigoplus_{n=N}^{n=N+\nu-1} \gamma^n \delta^{\mathcal{D}_s(n)} \right) \otimes$$

$$\otimes (e \oplus \gamma^\nu \delta^\tau \oplus \gamma^{2\nu} \delta^{2\tau} \oplus \ldots) =$$

$$= \left(\bigoplus_{n=N}^{n=N+\nu-1} \gamma^n \delta^{\mathcal{D}_s(n)} \right)(\gamma^\nu \delta^\tau)^*$$

Taking

$$p = \bigoplus_{n=0}^{N-1} \gamma^n \delta^{\mathcal{D}_s(n)} \quad \text{and} \quad q = \bigoplus_{n=N}^{n=N+\nu-1} \gamma^n \delta^{\mathcal{D}_s(n)}$$

We obviously get a representation like (15).

Conversely, assume we have $\mathcal{D}_s = \max(\mathcal{D}_p, \mathcal{D}_{q(\gamma^n\delta^t)^*})$ and $\mathcal{D}_{q(\gamma^n\delta^t)^*} = \mathcal{D}_q \star \mathcal{D}_{(\gamma^n\delta^t)^*}$. The conclusion results from observing that:

Lemma 4.6
-(i) *The max of two (ν, τ)-periodic functions is (ν, τ)-periodic.*
-(ii) *If d has finite support and d' is (ν, τ)-periodic, then the max-convolution $d \star d'$ is (ν, τ)-periodic.*

Proof of the Lemma: (i) is obvious. Since

$$d \star d'(n) = \max_{u \in \text{Supp}(d)} [d(u) + d'(n - u)] \quad (16)$$

and d has finite support, (16) expresses $d \star d'$ as the max of a finite number of functions $n \mapsto d(u) + d'(n - u)$ which have periodicity (ν, τ). By applying several times the point (i) of the Lemma, we get that $d \star d'$ is (ν, τ)-periodic. This concludes the proof of lemma 4.6 and of proposition 4.5. ∎

Considering $(\gamma\delta)^* = (e \oplus \gamma\delta)(\gamma^2\delta^2)^*$, we see that periodic representations are not unique. However, $(\gamma\delta)^*$ obviously seems to be simpler than $(e \oplus \gamma\delta)(\gamma^2\delta^2)^*$. We explain now which minimality is involved here, in order to obtain a canonical form of periodic series.

Definition 4.7 (Proper representations)

$s = p \oplus q(\gamma^\nu\delta^\tau)^*$ *is proper if: (i)* $\deg p < \text{val } q$ *and (ii)* $\deg q - \text{val } q < (n, t)$.

The meaning of these two conditions is clear: (i) guarantees there is no overlapping between the monomials of p and q, similarly, (ii) guarantees the monomials of $q, q\gamma^\nu\delta^\tau, q\gamma^{2\nu}\delta^{2\tau}, \ldots$ are all different. In the case of proper representations, p can be interpreted as a *transient*, and q as a *pattern* which is moved by the successive translations: $e, \gamma^\nu\delta^\tau, \gamma^{2\nu}\delta^{2\tau}$, etc.

Definition 4.8 (Reduced representation)

Given two proper periodic representations $\mathfrak{P}$: $s = p \oplus q(\gamma^\nu\delta^\tau)^*$ *and* $\mathfrak{P}'$: $s = p' \oplus q'(\gamma^{\nu'}\delta^{\tau'})^*$ *of series* s, *we say that* $\mathfrak{P}$ *is simpler than* $\mathfrak{P}'$ *(denoted by* $\mathfrak{P} \preceq \mathfrak{P}'$*) if* $(\nu, \tau) \leq (\nu', \tau')$ *and* $\deg(p) \leq \deg(p')$.

The $\preceq$ relation is obviously reflexive and transitive. It is also antisymmetric, for if $\deg p$ and (ν, τ) are known, p and q are necessarily equal to:

$$p = \bigoplus_{k=0}^{\deg_\gamma(p)} \gamma^k \delta^{D_s(k)} \qquad q := \bigoplus_{k=\deg_\gamma(p)+1}^{\deg_\gamma(p)+\nu} \gamma^k \delta^{D_s(k)}$$

Indeed, a proper representation is well determined by (N, ν, τ) with $N = \deg_\gamma(p)$ and we shall write

$$\mathfrak{P} = (N, \nu, \tau) \qquad (17)$$

Theorem 4.9 *A periodic series* s *admits a simplest (i.e.* $\preceq$*-minimal) periodic proper representation, called the canonical form of* s.

Sketch of proof If $s = p \oplus q(\gamma^\nu\delta^\tau)^*$ and $s = p' \oplus q'(\gamma^{\nu'}\delta^{\tau'})^*$ are two periodic representations, we have to show that there exists a proper representation $s = (p' \wedge p) \oplus q''(\gamma^{\gcd(\nu,\nu')}\delta^{\gcd(\tau,\tau')})^*$, i.e. that greatest lower bounds exist for the $\preceq$

relation. This can be easily seen by reasoning in terms of dater functions, and proves the uniqueness. Since the set of periodicities, together with $\preceq$ is artinian (by (17), we can identify it to a sub-ordered set of $((\mathbb{N}\cup\{+\infty\})^3, \leq)$, there exists a simplest proper representation. ∎

Remark 4.10 The Theorem 4.9 admits an algorithmic translation, which is a bit involved but presents no conceptual difficulties.

Theorem 4.11 (Main Theorem) *The rational series are precisely the periodic series.*

Proof of theorem 4.11: In proposition 4.5, we have shown that polynomials are periodic, i.e.:

$$\text{MinMax}^+\langle\gamma,\delta\rangle \subset \mathcal{P} \qquad (18)$$

Since $\mathcal{R}$ is the smallest rationally closed subdioid satisfying (18), we only have two check that $\mathcal{P}$ is rationally closed to get $\mathcal{R} \subset \mathcal{P}$. As the other inclusion is obvious, this will conclude the proof.

The next section is devoted to showing that the set of the periodic series $\mathcal{P}$ is stable for $\oplus$, $\otimes$, and $*$, by means of formulae and algorithms. This is the material needed for the proof of theorem 4.11, and it is fundamental in order to make effective rational computations in $\text{MinMax}\langle\langle\gamma,\delta\rangle\rangle$.

5 Rational properties of periodic series

5.1 Sum of periodic series

Let us introduce a new operation over the set of periodicities. The $\sqcup$ of periodicities if the commutative operation defined by

$$\gamma^\nu\delta^\tau \sqcup \gamma^{\nu'}\delta^{\tau'} = \gamma^\nu\delta^\tau \qquad \text{if } \frac{\tau'}{\nu'} < \frac{\tau}{\nu}$$
$$\gamma^\nu\delta^\tau \sqcup \gamma^{\nu'}\delta^{\tau'} = \gamma^{\text{lcm}(\nu,\nu')}\delta^{\text{lcm}(\tau,\tau')} \qquad \text{if } \frac{\tau}{\nu} = \frac{\tau'}{\nu'}$$
$$\gamma^\nu\delta^\tau \sqcup \delta^{+\infty} = \delta^{+\infty}$$
$$\gamma^\nu\delta^\tau \sqcup \gamma = \gamma^\nu\delta^\tau$$

$\sqcup$ is clearly associative and idempotent.

We first consider the sum of two simple periodic series:

Proposition 5.1 *Let* $s = \gamma^n\delta^t(\gamma^\nu\delta^\tau)^*$ *and* $s' = \gamma^{n'}\delta^{t'}(\gamma^{\nu'}\delta^{\tau'})^*$, *then* $s \oplus s'$ *is* $\gamma^\nu\delta^\tau \sqcup \gamma^{\nu'}\delta^{\tau'}$-*periodic. Moreover, if* $\frac{\tau}{\nu} \neq \frac{\tau'}{\nu'}$, *then* $\gamma^\nu\delta^\tau \sqcup \gamma^{\nu'}\delta^{\tau'}$ *is the minimal periodicity.*

Proof -If $\frac{\tau}{\nu} = \frac{\tau'}{\nu'}$: we write $\text{lcm}(\nu, \nu') = k\nu = k'\nu'$, and develop:

$$(\gamma^\nu \delta^\tau)^* =$$
$$(e \oplus \gamma^\nu \delta^\tau \oplus \ldots \oplus \gamma^{(k-1)\nu} \delta^{(k-1)\tau})(\gamma^{k\nu} \delta^{k\tau})^*. \quad (19)$$

Because $k\tau = \lambda(s)k\nu = \lambda(s)k'\nu' = k'\tau'$ and $\gcd(k, k') = 1$, we get $k\tau = \text{lcm}(\tau, \tau')$. Thus (19) is a $\gamma^{\text{lcm}(\nu,\nu')} \delta^{\text{lcm}(\tau,\tau')}$-periodic representation of s. Since a similar representation also holds for s', we obviously obtain the required form for $s \oplus s'$.

-case $\frac{\tau}{\nu} > \frac{\tau'}{\nu'}$. First, we need a technical result:

Lemma 5.2 *Assume* $\frac{\tau}{\nu} > \frac{\tau'}{\nu'}$, *and* n, n', t, t' *are arbitrary integers. Then there exists* $K \geq 0$ *such that*

$$\gamma^{n'} \delta^{t'} \gamma^{K\nu'} \delta^{K\tau'} (\gamma^{\nu'} \delta^{\tau'})^* \leq \gamma^n \delta^t (\gamma^\nu \delta^\tau)^* \quad (20)$$

Applying the Lemma to $s \oplus s'$, we obtain the formula:

$$\gamma^n \delta^t (\gamma^\nu \delta^\tau)^* \oplus \gamma^{n'} \delta^{t'} (\gamma^{\nu'} \delta^{\tau'})^* =$$
$$\gamma^n \delta^t (\gamma^\nu \delta^\tau)^* \oplus$$
$$\oplus \gamma^{n'} \delta^{t'} \oplus \ldots \gamma^{n'+(K-1)\nu'} \delta^{t'+(K-1)\tau'}$$

which shows that $s \oplus s'$ has periodicity $\gamma^\nu \delta^\tau$.

Sketch of proof of Lemma 5.2. The simplest proof consists in observing that $\lim_{n \to +\infty} \mathcal{D}_s(n) - \mathcal{D}_{s'}(n) = +\infty$ (for the slope of $\mathcal{D}_s$ is greater than the one of $\mathcal{D}_{s'}$). Let p_0 such that for all $n \geq p_0$, $\mathcal{D}_{s'}(n) \leq \mathcal{D}_s(n)$: it should be clear that the monomials of s' of valuation in γ greater than p_0 are dominated by s, which yields (20). ∎

Proposition 5.3 *The sum of a two periodic series* $s = p \oplus q \varpi^*$ *and* $s' = p' \oplus q' \varpi'^*$ *is periodic with periodicity* $\varpi \sqcup \varpi'$.

Proof We show that $s \oplus s'$ has a periodic representation of the form:

$$p'' \oplus q''(\varpi \sqcup \varpi')^* \quad (21)$$

Proposition 5.1 gives the result when q and q' are monomials. Applying again 5.1 and using the associativity and idempotence of $\sqcup$, it should be clear that if we add terms like $\gamma^{n_i} \delta^{t_i} \varpi^*$ or $\gamma^{n_j} \delta^{t_j} (\varpi')^*$ to (21), we still obtain an expression of the same form. ∎

5.2 Product of periodic series

We now introduce another law which will play for product the role $\sqcup$ plays for sum. The $\sqcap$ of periodicities is the commutative operation defined by:

$$\gamma^\nu \delta^\tau \sqcap \gamma^{\nu'} \delta^{\tau'} = \gamma^\nu \delta^\tau \qquad \text{if } \frac{\tau'}{\nu'} < \frac{\tau}{\nu}$$
$$\gamma^\nu \delta^\tau \sqcap \gamma^{\nu'} \delta^{\tau'} = \gamma^{\gcd(\nu,\nu')} \delta^{\gcd(\tau,\tau')} \quad \text{if } \frac{\tau}{\nu} = \frac{\tau'}{\nu'}$$
$$\gamma^\nu \delta^\tau \sqcap \delta^{+\infty} = \delta^{+\infty}$$
$$\gamma^\nu \delta^\tau \sqcap \gamma = \gamma^\nu \delta^\tau$$

The $\sqcap$ is also associative and idempotent. The set of periodicities equipped with $\sqcup$ and $\sqcap$ is *not* a lattice, because the absorption property $\varpi \sqcup (\varpi \sqcap \varpi') = \varpi$ does not hold (consider $\varpi = \gamma\delta$ and $\varpi' = \gamma\delta^2$). However, we do have a weaker property which will be sufficient for our purpose:

$$(\varpi \sqcup \varpi') \sqcup (\varpi \sqcap \varpi') = (\varpi \sqcup \varpi') \quad (22)$$

Proposition 5.4 *Let* $s = (\gamma^\nu \delta^\tau)^*$ *and* $s' = (\gamma^{\nu'} \delta^{\tau'})^*$. *Then* ss' *is* $\gamma^\nu \delta^\tau \sqcap \gamma^{\nu'} \delta^{\tau'}$*-periodic, and* $\gamma^\nu \delta^\tau \sqcap \gamma^{\nu'} \delta^{\tau'}$ *is the minimal periodicity.*

Proof
(i) Case $\frac{\tau}{\nu} = \frac{\tau'}{\nu'}$. We have

$$(\gamma^\nu \delta^\tau)^* (\gamma^{\nu'} \delta^{\tau'})^* = \bigoplus_{i,j \geq 0} \gamma^{i\nu + j\nu'} \delta^{i\tau + j\tau'} \quad (23)$$

obviously, the only possible values for $i\nu + j\nu'$ are multiples of $\gcd(\nu, \nu')$. A well known result of the theory of Linear Diophantine equations states there exists a least $n = k \gcd(\nu, \nu')$ such that all the multiples of $\gcd(\nu, \nu')$ greater or equal to n can be expressed as $i\nu + j\nu'$ for some $i, j \geq 0$ (this is explained with more details in the next section). This n is called the *conductor* of (ν, ν'), denoted by $\text{cond}(\nu, \nu')$. Let us show that:

$$\begin{cases} i_0\nu + j_0\nu' = \text{cond}(\nu, \nu') \\ i_1\nu + j_1\nu' = \text{cond}(\nu, \nu') + \gcd(\nu, \nu') \\ i_2\nu + j_2\nu' = \text{cond}(\nu, \nu') + 2\gcd(\nu, \nu') \ldots \end{cases}$$
$$(24)$$

implies

$$\begin{cases} i_0\tau + j_0\tau' = \text{cond}(\tau, \tau') \\ i_1\tau + j_1\tau' = \text{cond}(\tau, \tau') + \gcd(\tau, \tau') \\ i_2\tau + j_2\tau' = \text{cond}(\tau, \tau') + 2\gcd(\tau, \tau') \ldots \end{cases}$$
$$(25)$$

In fact, multiplying (24) by $\lambda = \frac{\tau}{\nu}$, we have $i_0\tau + j_0\tau' = \lambda\mathrm{cond}(\nu,\nu')$, $i_1\tau + j_1\tau' = \lambda\mathrm{cond}(\nu,\nu') + \lambda\gcd(\nu,\nu') = \lambda\mathrm{cond}(\nu,\nu') + \gcd(\tau,\tau')$, and so on. This proves that $\mathrm{cond}(\tau,\tau') \le \lambda\mathrm{cond}(\nu,\nu')$. Applying this result to $\nu = \lambda^{-1}(\lambda\nu)$, $\nu' = \lambda^{-1}(\lambda\nu')$, we obtain the second inequality: $\mathrm{cond}(\nu,\nu') \le \lambda^{-1}\mathrm{cond}(\lambda\nu,\lambda\nu')$. This proves $\lambda\mathrm{cond}(\nu,\nu') = \mathrm{cond}(\tau,\tau')$ and shows that (24) implies (25). Then (23) obviously rewrites:

$$ss' = \left(\bigoplus_{i\nu + j\nu' \le \mathrm{cond}(\nu,\nu')-1} \gamma^{i\nu+j\nu'}\delta^{i\tau+j\tau'} \right) \oplus$$
$$\oplus \gamma^{\mathrm{cond}(\nu,\nu')}\delta^{\mathrm{cond}(\tau,\tau')}(\gamma^{\gcd(\nu,\nu')}\delta^{\gcd(\tau,\tau')})^*$$

which yields a $\gamma^\nu\delta^\tau \sqcap \gamma^{\nu'}\delta^{\tau'}$-periodic representation of ss' an concludes Case (i).

(ii) Case $\frac{\tau}{\nu} > \frac{\tau'}{\nu'}$. We use again the technical lemma 5.2: there exists $k \ge 0$ such that

$$\gamma^{k\nu'}\delta^{k\tau'}(\gamma^{\nu'}\delta^{\tau'})^* \le \gamma^n\delta^t(\gamma^\nu\delta^\tau)^*. \qquad (26)$$

and multiplying this identity by $(\gamma^\nu\delta^\tau)^*$, we obtain $\gamma^{k\nu'}\delta^{k\tau'}(\gamma^{\nu'}\delta^{\tau'})^*(\gamma^\nu\delta^\tau)^* \le (\gamma^\nu\delta^\tau)^*(\gamma^\nu\delta^\tau)^* = (\gamma^\nu\delta^\tau)^*$. Then the identity

$$(\gamma^\nu\delta^\tau)^*(\gamma^{\nu'}\delta^{\tau'})^* =$$
$$= (\gamma^\nu\delta^\tau)^*[e \oplus \gamma^{\nu'}\delta^{\tau'} \oplus \gamma^{2\nu'}\delta^{2\tau'} \oplus \dots$$
$$\oplus \gamma^{(k-1)\nu'}\delta^{(k-1)\tau'} \oplus \gamma^{k\nu'}\delta^{k\tau'}(\gamma^{\nu'}\delta^{\tau'})^*.] \quad (27)$$

rewrites

$$ss' = (\gamma^\nu\delta^\tau)^*[e \oplus \gamma^{\nu'}\delta^{\tau'} \oplus \gamma^{2\nu'}\delta^{2\tau'} \oplus \dots$$
$$\oplus \gamma^{(k-1)\nu'}\delta^{(k-1)\tau'}] \quad (28)$$

This concludes the proof of the proposition. ∎

Proposition 5.5 *The product of two periodic series $s = p\oplus q\varpi^*$ and $s' = p'\oplus q'\varpi'^*$ is periodic with periodicity $\varpi \sqcup \varpi'$.*

Proof $ss' = pp' \oplus pq'\varpi'^* \oplus p'q\varpi^* \oplus qq'\varpi^*\varpi'^*$. Applying proposition 5.4, we obtain $qq'\varpi^*\varpi'^* = p'' \oplus q''(\varpi \sqcap \varpi')^*$. Then the proposition 5.3 shows that ss' has periodicity $(\varpi \sqcup \varpi')\sqcup(\varpi \sqcap \varpi') = (\varpi \sqcup \varpi')$. This concludes the proof. ∎

5.3 Star of periodic series

Let $s = p\oplus q\varpi^*$. The following formula reduces computing the star of s to computing the star of polynomials:

$$(p \oplus q\varpi^*)^* = p^*(e \oplus q(q \oplus \varpi)^*) \qquad (29)$$

We obtain the star of a polynomial p of complexity n (p is the sum of n monomials $\bigoplus_{i=1}^n m_i$) by a recursive application of the rule

$$p^* = m_n^*\left(\bigoplus_{i=1}^{n-1} m_i\right)^* \qquad (30)$$

which reduces the problem to a polynomial of complexity $n - 1$. This provides periodic forms for p^* and s^*, and shows that $\mathcal{P}$ is stable for the $*$ operation. This concludes the proof of the main Theorem 4.11. ∎

Remark 5.6 If the time takes non integer values, the relation between rationality and periodicity vanishes (consider the sum $(\gamma\delta)^* \oplus (\gamma^{\sqrt{2}}\delta^{\sqrt{2}})^*$). This difficulty is far from being an artificial problem. In fact, it is the limit of some pathology which already occurs with integer time. When the slopes of series are very close, the size of the transient becomes too important and the periodic representation is practically impossible to use. In these cases, a truncation has to be made. It should also be possible to use more conventional representations of rationals, like $s = \bigoplus_i a_i b_i^*$, where the a_i, b_i are monomials. The following example illustrates these difficulties:

Example 5.7 $(\gamma^{20}\delta)^* \oplus \delta(\gamma^{21}\delta)^* = \delta \oplus \gamma^{21}\delta^2 \oplus \gamma^{42}\delta^3 \oplus \gamma^{63}\delta^4 \oplus \gamma^{84}\delta^5 \oplus \gamma^{105}\delta^6 \oplus \gamma^{126}\delta^7 \oplus \gamma^{147}\delta^8 \oplus \gamma^{168}\delta^9 \oplus \gamma^{189}\delta^{10} \oplus \gamma^{210}\delta^{11} \oplus \gamma^{231}\delta^{12} \oplus \gamma^{252}\delta^{13} \oplus \gamma^{273}\delta^{14} \oplus \gamma^{294}\delta^{15} \oplus \gamma^{315}\delta^{16} \oplus \gamma^{336}\delta^{17} \oplus \gamma^{357}\delta^{18} \oplus \gamma^{378}\delta^{19} \oplus \gamma^{399}\delta^{20} \oplus \gamma^{420}\delta^{21}(\gamma^{20}\delta)^*.$

6 Star operation and diophantine linear equations

We have already seen that the product of periodic series reduces to computing the conductor of a simple Diophantine Linear equation. We now consider the problem of computing the star of a polynomial a, when all the monomials have the *same* slope. This is a generalization of

Proposition 5.4, which corresponds to a polynomial with complexity 2: $a = \gamma^{\nu}\delta^{\tau} \oplus \gamma^{\nu'}\delta^{\tau'}$. In a similar way, we shall show that the distribution of the monomials of a^* is related with the distribution of solutions of a more general linear diophantine equation, the theory of which affords a more efficient algorithm than the one given in Section 5.3.

Let

$$a = \gamma^{\nu_1}\delta^{\tau_1} \oplus \gamma^{\nu_2}\delta^{\tau_2} \oplus \cdots \oplus \gamma^{\nu_l}\delta^{\tau_l}$$

with $\lambda = \frac{\tau_1}{\nu_1} = \ldots = \frac{\tau_l}{\nu_l}$. For all $i > 0$, each term in a^i is of the form

$$\gamma^{\nu_1 x_1 + \nu_2 x_2 + \cdots + \nu_l x_l} \, \delta^{\tau_1 x_1 + \tau_2 x_2 + \cdots + \tau_l x_l}$$

with $\sum_k x_k = i$. This leads us to consider the following problem:

$$\text{find} \quad (x_1, \ldots, x_l) \in \mathbf{N}^l \text{ such that}$$
$$\nu_1 x_1 + \nu_2 x_2 + \cdots + \nu_l x_l = c \quad (31)$$

It can be shown [1] that if $\gcd(\nu_1, \nu_2, \ldots, \nu_l) = 1$, there exists $K \in \mathbf{N}$ such that for all $c \geq K$, (31) has a solution. The minimal value c_o of K is called the conductor of $(\nu_1, \ldots, \nu_l)$. The question of the determination of c_o is not solved in the general case. It is known as the "Frobenius problem" for linear diophantine equations.

In the case $l = 2$, the conductor for the equation $\nu_1 x_1 + \nu_2 x_2 = c$ is exactly known

$$\text{cond}(\nu_1, \nu_2) = (\nu_1 - 1)(\nu_2 - 1) \quad (32)$$

We now assume that $l > 2$ and $\nu_1 < \nu_2 < \cdots < \nu_l$. A simple upper bound is given by

$$s = (\nu_1 - 1)(\nu_l - 1) \quad (33)$$

A better bound is furnished by

$$t = \nu_2 \frac{d_1}{d_2} + \cdots + \nu_l \frac{d_{l-1}}{d_l} - \sum_i \nu_i \quad (34)$$

where d_i denotes the gcd of $\nu_1, \ldots, \nu_i$.

The existence of these bounds leads to a simple method for computing a^*, which consists in generating all the terms of a^* up to the bound. Then, since the periodicity of a^* is obviously equal to $\gamma^{\nu_1}\delta^{\tau_1} \sqcap \gamma^{\nu_2}\delta^{\tau_2} \sqcap \ldots \sqcap \gamma^{\nu_l}\delta^{\tau_l}$, we obtain

a periodic proper representation of a^*, which can be reduced to minimal form.

Example 6.1 Let us take $a = \gamma^6\delta^6 \oplus \gamma^{10}\delta^{10} \oplus \gamma^{15}\delta^{15}$. Since 6,10,15 are coprime the periodicity of a^* must be $(1,1)$. Indeed, we have:
$a^* = e \oplus \gamma^6\delta^6 \oplus \gamma^{10}\delta^{10} \oplus \gamma^{12}\delta^{12} \oplus \gamma^{16}\delta^{16} \oplus \gamma^{18}\delta^{18} \oplus \gamma^{20}\delta^{20} \oplus \gamma^{22}\delta^{22} \oplus \gamma^{24}\delta^{24} \oplus \gamma^{26}\delta^{26} \oplus \gamma^{28}\delta^{28} \oplus \gamma^{30}\delta^{30}(\gamma\delta)^*$.

Remark 6.2 The formal series point of view should be compared with that presented in [2]. Instead of using formal series, one can introduce the matrix equation in the $(\max, +)$ algebra $x_n = Ax_{n-1} \oplus Bu_n$, where A and B are matrices whith entries in $\mathbf{R} \cup \{-\infty\}$. Then, the study of the periodic behaviour of the *autonomous* system $x_n = Ax_{n-1}$ leads to consider the sequence $\{A^n\}_n$. The approach developed in [2] extends some of the Frobenius' results on the cyclicity of nonnegative matrices to the $(\max, +)$ algebra. Indeed, to the authors' knowledge, the "Frobenius problem" for Linear Diophantine Equations comes from this context, and it is not surprising that this problem still plays a central role in the study of formal series.

7 Illustrating examples

A preliminary implementation has been realized as a set of MAPLE macros. In a further stage, it is intended to support symbolic computation in dioids, which can be used to evaluate and optimize the periodic slope (this is the resource optimization problem, see [4]). Another version devoted to the non symbolic problem is being developed in FORTRAN. The following examples have been dealt with using the MAPLE implementation.

A simple application Let us consider again the Timed Event Graph of Figure 1. It is modeled by the following system:

$$x = \begin{bmatrix} \gamma\delta^5 & \gamma^3\delta^3 \\ \gamma^2\delta & \gamma\delta \oplus \gamma^2\delta^2 \end{bmatrix} x \oplus \begin{bmatrix} e \\ \varepsilon \end{bmatrix} u =$$
$$= Ax \oplus Bu,$$
$$y = \begin{bmatrix} \varepsilon & e \end{bmatrix} x = Cx$$

We obtain the transfer matrix:

$$H = CA^*B = \gamma^2\delta(\gamma\delta^5)^*$$

which indeed is very simple! We can interpret H in terms of impulse response: if an infinite quantity of tokens becomes available at date 0, then the output is simply the dater function associated with H, i.e. two tokens exit at date 1, then, the periodic behavior is reached, and 1 token exits every 5 unit of times. In particular, from the input-output point of view, the Timed Event Graph of Figure 1 is equivalent to the Timed Event Graph of Figure 3.

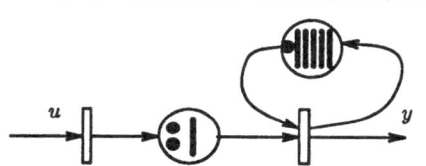

Figure 3: Reduced Timed Event Graph

Interpretation in terms of Timed Event Graphs We classicaly define the weight of a path as the product of the operators associated with the elementary arcs. Then a well known result states that the entry (i, j) of A^* represents the sum of the weights of all paths $j \mapsto i$ (see [5]). From the absorption properties of slopes, it should be clear that the periodicity (ν_i, τ_i) of $(A^*)_{i,i}$ has maximal ratio $\frac{\nu_i}{\tau_i}$ among all the weights of circuits $i \mapsto i$. When the event graph is strongly connected, the maximal ratio $\frac{\nu_i}{\tau_i}$ is the same for all the transitions. In this case, the set of the arcs of circuits which realize this maximal ratio (*critical circuits*) is called the *critical subgraph*, and the periodic throughput associated with any transition i is simply characterized as the slope of the critical circuits, while the gcd and lcm which appear in the computations are related to the structure of the critical subgraph.

Example 7.1 In the case of the event graph of Figure 1, the critical subgraph is reduced to the arc with weigth with $\gamma\delta^5$. We have:

$$A^* = \left[\begin{array}{cc} (\gamma\delta^5)^* & \gamma^3\delta^3(\gamma\delta^5)^* \\ \gamma^2\delta(\gamma\delta^5)^* & \% \end{array} \right]$$

with $\% = e \oplus \gamma\delta \oplus \gamma^2\delta^2 \oplus \gamma^3\delta^3 \oplus \gamma^4\delta^4 \oplus \gamma^5\delta^5 \oplus \gamma^6\delta^9(\gamma\delta^5)^*$. Since the graph is strongly connected, all the entries have the same periodic slope.

Application to a flowshop We consider the simple flowshop with 3 parts P_1, P_2, P_3 and 3 machines M1, M2, M3, shown in Figure 4. We neglect the transportation times between machines. The schedule is the same for all parts : M1 $\rightarrow$ M2 $\rightarrow$ M3 $\rightarrow$ M1. The processing times are given by the following table:

	P1	P2	P3
M1	9	2	1
M2	2	7	2
M3	1	10	1

This means part P1 must be processed at least 9 units of time on machine M1, etc. Then part P1 becomes *instantaneously* available for the next machine M2, and machine M1 is also available for the next job. This leads to draw the Timed Event Graph of Figure 4. The vertical circuits correspond to the circulations of parts, and the the horizontal circuits correspond to machines. The position of tokens in the places is related to a specific initial state of the system. A more detailed account of modeling flowshop and jobshops using dioids can be found in [2].

We have introduced 3 inputs (u_1 in x_1, u_2 in x_2, u_3 in x_3). The output transitions are x_7, x_8, x_9. Then, the following matrix representation can be writen:

$$A = \left[\begin{array}{ccccccccc} \varepsilon & \varepsilon & \gamma\delta & \varepsilon & \varepsilon & \varepsilon & \gamma\delta & \varepsilon & \varepsilon \\ \delta^9 & \varepsilon & \varepsilon & \varepsilon & \varepsilon & \varepsilon & \varepsilon & \gamma\delta^{10} & \varepsilon \\ \varepsilon & \delta^2 & \varepsilon & \varepsilon & \varepsilon & \varepsilon & \varepsilon & \varepsilon & \gamma \\ \delta^9 & \varepsilon & \varepsilon & \varepsilon & \varepsilon & \gamma\delta^2 & \varepsilon & \varepsilon & \varepsilon \\ \varepsilon & \delta^2 & \varepsilon & \delta^2 & \varepsilon & \varepsilon & \varepsilon & \varepsilon & \varepsilon \\ \varepsilon & \varepsilon & \delta & \varepsilon & \delta^7 & \varepsilon & \varepsilon & \varepsilon & \varepsilon \\ \varepsilon & \varepsilon & \varepsilon & \delta^2 & \varepsilon & \varepsilon & \varepsilon & \varepsilon & \gamma\delta \\ \varepsilon & \varepsilon & \varepsilon & \varepsilon & \delta^7 & \varepsilon & \delta & \varepsilon & \varepsilon \\ \varepsilon & \varepsilon & \varepsilon & \varepsilon & \varepsilon & \delta^2 & \varepsilon & \delta^{10} & \varepsilon \end{array} \right]$$

$$B = \begin{bmatrix} e & \varepsilon & \varepsilon \\ \varepsilon & e & \varepsilon \\ \varepsilon & \varepsilon & e \\ \varepsilon & \varepsilon & \varepsilon \\ \varepsilon & \varepsilon & \varepsilon \\ \varepsilon & \varepsilon & \varepsilon \\ \varepsilon & \varepsilon & \varepsilon \\ \varepsilon & \varepsilon & \varepsilon \\ \varepsilon & \varepsilon & \varepsilon \end{bmatrix}$$

$$C = \begin{bmatrix} \varepsilon & \varepsilon & \varepsilon & \varepsilon & \varepsilon & \varepsilon & \delta & \varepsilon & \varepsilon \\ \varepsilon & \varepsilon & \varepsilon & \varepsilon & \varepsilon & \varepsilon & \varepsilon & \delta^{10} & \varepsilon \\ \varepsilon & \varepsilon & \varepsilon & \varepsilon & \varepsilon & \varepsilon & \varepsilon & \varepsilon & \delta \end{bmatrix}$$

We obtain the transfer matrix:

$$H = \begin{bmatrix} \delta^{12} \oplus \gamma \delta^{30} (\gamma \delta^{19})^* & \gamma \delta^{21} (\gamma \delta^{19})^* & \gamma \delta^{13} \oplus \gamma^2 \delta^{31} (\gamma \delta^{19})^* \\ \delta^{28} (\gamma \delta^{19})^* & \delta^{19} (\gamma \delta^{19})^* & \gamma \delta^{29} (\gamma \delta^{19})^* \\ \delta^{29} (\gamma \delta^{19})^* & \delta^{20} (\gamma \delta^{19})^* & \delta^4 \oplus \gamma \delta^{30} (\gamma \delta^{19})^* \end{bmatrix}$$

Because the event graph is strongly connected, the periodic slope is the same for all the entries of the transfer matrix. This periodicity corresponds to the weigth $\gamma \delta^{19}$ of the *unique* critical circuit (the vertical circuit of part P2).

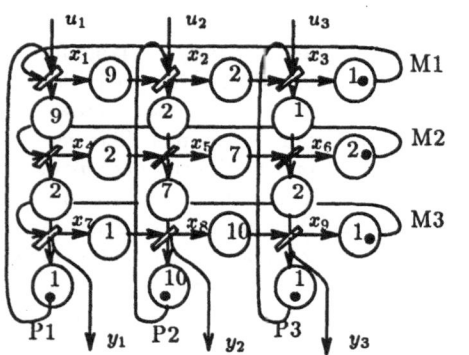

Figure 4: 3-machines,3-parts flowshop

References

[1] A. Brauer. On a problem of partitions. *Am. J. Math.*, 64:299–312, 1942.

[2] G. Cohen, D. Dubois, J.P. Quadrat, and M. Viot. Analyse du comportement périodique des systèmes de production par la théorie des dioïdes. INRIA Report 191, Le Chesnay, France, 1983.

[3] G. Cohen, P. Moller, J.P. Quadrat, and M. Viot. Algebraic tools for the performance evaluation of discrete event systems. *IEEE Proceedings: Special issue on Discrete Event Systems*, 77(1), January 1989.

[4] S. Gaubert. An algebraic method for optimizing resources in timed event graphs. In A.Bensoussan and J.L. Lions, editors, *Proceedings of the 9th International Conference on Analysis and Optimization of Systems*, Antibes, June 1990, number 144 in Lecture Notes in Control and Information Sciences. Springer, 1990.

[5] M. Gondran and M. Minoux. *Graphes et algorithmes.* Eyrolles, Paris, 1979.

[6] B.A. Carré and R.C. Backhouse. Regular algebra applied to path finding problems. *J. of the Inst. of Maths and Appl.*, 15:161–186, 1975.

INTRINSIC DIFFERENTIAL ALGEBRA

J.F. POMMARET

Centre de Mathématiques (CERMA)

Ecole Nationale des Ponts et Chaussées (ENPC)

La Courtine, 93167 Noisy-le Grand Cedex, France

ABSTRACT: Up to now, differential algebra and the corresponding elimination theory for systems of algebraic partial differential equations (PDE) heavily depend on the coordinate system, that is on the choice of the independent variables. Such a fact is particularly clear through the definitions (characteristic sets, ...) and the procedures (differential euclidean algorithm, ...). The purpose of this paper is to make a first step towards an intrinsic approach by using and dualizing the formal theory of systems of PDE introduced in 1960-1975 by D.C. Spencer and coworkers. At the same time, we shall produce and comment a certain number of nontrivial typical examples coming from mathematical physics in order to illustrate the applications of the previous methods to partial differential control theory.

KEY WORDS: Differential algebra, partial differential equations, control theory, computer algebra.

Thanks to the work of M. Janet (7) and J.F. Ritt (13), algebraic concepts have been applied to PDE as early as in 1930 and the two last chapters of Ritt's book constitute a very fine exposition of Janet's work though he is not quoted explicitly. As a matter of fact, the approach of Janet is quite an improvement of the approach of E. Cartan through exterior calculus and, in particular, led Janet in first place to the solution of the Riemannian embedding problem as a difficult problem of differential elimination (any riemannian surface of dimension n can be embedded, at least locally, in an euclidean space of dimension $n(n+1)/2$).

The methods of Ritt have been brought to modern standards and largely improved by E.R. Kolchin (8) or J.M. Thomas (17) after 1945. Meanwhile, elimination theory got benefit from this achievement with Seidenberg (15) and others.

Very recently, according to fast progress in electronics, computer algebra became in fashion and algorithms have been implemented by S. Diop (3), F. Schwarz (14), Wu Wen Tsun (18) and others.

Meanwhile, during the last five years, the importance of differential algebra for ordinary differential control theory has been pointed out by M. Fliess (5) and we have extended these techniques to partial differential control theory, namely to input/output relations defined by systems of PDE in (10,11).

However, a careful examination of all the references, quickly shows that:

1) *The hard steps have a differential geometric origin.* In particular, it is clear that most of the algebraic manipulations come from the differential background, that is, are originally concerned with the formal theory of PDE and only adapted to algebra. The main reason for such a situation is that, whenever one differentiates a PDE, the top order derivatives appear in a linear way and can therefore be eliminated through pure linear algebra.

2) *The link between differential algebra and the formal theory of PDE has been missed by both Spencer and Kolchin.* This fact is rather asthonishing when one knows that their most active periods were almost coinciding. Indeed, a simple examination reveals that the *jet coordinates* of the formal theory are just the *differential indeterminates* of differential algebra, that the *projective limits* used in the differential geometric approach just correspond to the *inductive limits* used in the differential algebraic approach,... . However, the techniques of algebraic geometry have never been used by Spencer, while the δ-cohomology, though crucial in the formal theory, has never been used by Kolchin. Such a link has been exhibited for the first time in our book (9) as a necessary tool towards the solution of the Galois theory for systems of PDE or *differential Galois theory*.

3) *Most of the algorithms of differential algebra are not intrinsic*, that is, depend on the coordinate system. This is particularly clear for the ordering of monomials or the choice of the differential polynomials of a characteristic set and their use by Janet, Ritt, Kolchin and others. As a byproduct, the sub (non-differential)-ideals of a differential ideal made by all polynomials of a given order do not seem to play a role, though such a concept, which dualizes *formal integrability*, should be quite important and is indeed. It follows from (9) that key results of differential algebra like Ritt's basis theorem can be proved without any reference to characteristic sets.

Our purpose, in this paper, is to sketch the general mechanism and to illustrate it by means of a few examples coming from control theory for systems of PDE or *partial differential control theory* (10,11).

Let X be a manifold of dimension n with local coordinates $(x^1, ..., x^n)$ and $\mathcal{E}$ be a fibered manifold over X with fiber dimension m and local coordinates (x^i, y^k). The q-jet bundle of $\mathcal{E}$ is a fibered manifold $J_q(\mathcal{E})$ with local coordinates (x^i, y_μ^k) where $\mu = (\mu_1, ..., \mu_n)$ is a multi-index and $0 \leq |\mu| = \mu_1 + ... + \mu_n \leq q$ or simply (x, y_q). For $\mu = 0$ we shall use to set $y_0^k = y^k$ and we define $\mu + 1_i = (\mu_1, ..., \mu_{i-1}, \mu_i + 1, \mu_{i+1}, ..., \mu_n)$.

The modern way to deal with a nonlinear system of PDE is to use jet coordinates instead of derivatives in order to define a fibered submanifold $\mathcal{R}_q \subset J_q(\mathcal{E})$ by a system of local equations $\Phi^\tau(x, y_q) = 0$. The prolongations $\rho_r(\mathcal{R}_q) = J_r(\mathcal{R}_q) \cap J_{q+r}(\mathcal{E})$ or simply $\mathcal{R}_{q+r}$ will be obtained by substituting derivatives instead of jet coordinates, differentiating r times in the usual way and substituting again jet coordinates. The projections $\pi_{q+r}^{q+r+s} : J_{q+r+s}(\mathcal{E}) \longrightarrow J_{q+r}(\mathcal{E})$ induce maps $\pi_{q+r}^{q+r+s} : \mathcal{R}_{q+r+s} \longrightarrow \mathcal{R}_{q+r}$ which are not in general surjective and we may introduce the sets $\mathcal{R}_{q+r}^{(s)} = \pi_{q+r}^{q+r+s}(\mathcal{R}_{q+r+s}) \subseteq \mathcal{R}_{q+r}$ which may not be fibered manifolds for any $r, s \geq 0$.

Now the symbol M_q of $\mathcal{R}_q$ is defined by the linear equations:

$$\frac{\partial \Phi^\tau}{\partial y_\mu^k}(x, y_q)v_\mu^k = 0 \qquad\qquad |\mu| = q$$

We let the reader check that the symbol M_{q+r} of $\mathcal{R}_{q+r}$ is defined by the linear equations:

$$\frac{\partial \Phi^\tau}{\partial y_\mu^k}(x, y_q)v_{\mu+\nu}^k = 0 \qquad\qquad |\mu| = q, |\mu| = r$$

and only depends on M_q.

If we introduce the cotangent bundle $T^* = T^*(X)$ with corresponding tensor, exterior and symmetric products respectively denoted by $\otimes, \wedge, S$, we may define the Spencer map:

$$\delta : \bigwedge^s T^* \otimes M_{q+r+1} \longrightarrow \bigwedge^{s+1} T^* \otimes M_{q+r}$$

by the local formula $(\delta v)_\mu^k = dx^i \wedge v_{\mu+1_i}^k$ on families of forms. One has $((\delta \circ \delta)v)_\mu^k = dx^i \wedge dx^j \wedge v_{\mu+1_i+1_j}^k \equiv 0$ and thus $\delta \circ \delta = 0$. The cohomology at $\bigwedge^s T^* \otimes M_{q+r}$ of the corresponding sequences is denoted by $H_{q+r}^s(M_q)$ as it only depends on M_q which is said to be s-*acyclic* if $H_{q+r}^1 = ... = H_{q+r}^s = 0, \forall r \geq 0$ or *involutive* if it is n-acyclic. Involutivity can be checked by means of a finite algorithm.

The two key theorems from which all results can be obtained are the following ones (6,10,16):

THEOREM I: If M_q is 2-acyclic and M_{q+1} is a vector bundle (constant rank) over $\mathcal{R}_q$, then M_{q+r} is a vector bundle over $\mathcal{R}_q, \forall r \geq 1$.

THEOREM II: If $\mathcal{R}_q^{(1)}$ is a fibered manifold and M_q is 2-acyclic, then $(\mathcal{R}_q^{(1)})_{+r} = \mathcal{R}_{q+r}^{(1)}, \forall r \geq 0$.

A system is said to be *formally integrable* if $\mathcal{R}_{q+r}^{(1)} = \mathcal{R}_{q+r}$ is a fibered manifold $\forall r \geq 0$ and *involutive* if it is formally integrable with an involutive symbol. Accordingly, the simplest corollary is the following *criterion for formal integrability* which is crucial for any application:

CRITERION: If M_q is 2-acyclic and $\mathcal{R}_q^{(1)} = \mathcal{R}_q$, then $\mathcal{R}_q$ is formally integrable.

Using the fact that the prolongation of any symbol becomes involutive for high enough order, the preceding results shows that any formal result on systems of PDE only depends on three possible intrinsic manipulations:
1) **Prolongation** (involutivity $\subset$ 2-acyclicity)
2) **Rank condition** (theorem I)
3) **Projection** (theorem II)

Hence it is possible to bring differential problems back to linear algebra and non-differential problems. The main idea is now to prove that differential algebra also depends on these results by *duality*.

Indeed, let K be a differential field with commuting derivations $\partial_1, ..., \partial_n$ and $y = (y^1, ..., y^m)$ be a family of differential indeterminates over K. We may introduce the ring $K\{y\} = lim_{q \to \infty} K[y_q]$ of *differential polynomials* as a differential ring with the new derivations $d_i \mid K = \partial_i, d_i y^k_\mu = y^k_{\mu+1_i}$. By quotient, we may introduce the differential field $K < y > = Q(K\{y\})$ of *differential rational functions*. If $\mathfrak{a} \subset K\{y\}$ is a differential ideal, that is an ideal stable by the d_i, we may define $\mathfrak{a}_q = \mathfrak{a} \cap K[y_q]$ and its various prolongations $\rho_r(\mathfrak{a}_q) = \{d_\nu P \mid P \in \mathfrak{a}_q, 0 \leq \mid \nu \mid \leq r\}$ with $\rho_r(\mathfrak{a}_q) \subset \mathfrak{a}_{q+r}$ in general. Theorem 1 may be kept though x has no longer a meaning and theorem 2 just amounts to study $\mathfrak{a}$ from the only knowledge of $\mathfrak{a}_q$. In particular, we obtain the following *criterion for prime differential ideals*(9):
DUAL CRITERION: If $\mathfrak{p}_q$ is a prime ideal, $\mathfrak{p}_{q+1} = \rho_1(\mathfrak{p}_q)$ is also a prime ideal with $\mathfrak{p}_{q+1} \cap K[y_q] = \mathfrak{p}_q$ and M_q is 2-acyclic with M_{q+1} a vector bundle, then $\mathfrak{p} = \rho_\infty(\mathfrak{p}_q)$ is a prime differential ideal with $\mathfrak{p}_{q+r} = \rho_r(\mathfrak{p}_q), \forall r \geq 0$ and we may introduce the *differential extension* $L = Q(K\{y\}/\mathfrak{p})$.

Hence, again, formal integrability is the key ingredient for studying differential ideals and *inductive limits dualize projective limits*. Accordingly, any formal result on differential ideals only depends on three possible intrinsic manipulations:
1) **Prolongation**
2) **Rank condition**
3) **Intersection**

The situation is thus absolutely similar for (algebraic) differential elimination and partial differential control theory, namely the study of input/output relations defined by systems of (algebraic) PDE.

Indeed, by definition, in both cases we have a mixed system of PDE for two sets of indeterminates respectively called *inputs* and *outputs*, such a situation being exactly that of a *differential correspondence*. Hence, any formal result in partial differential control theory will follow from the study of formal integrability for input alone (*output resolvent system*) or for output alone (*input resolvent system*) or for input *and* output together. As a byproduct, *inequalities arising in differential elimination through products of separant may now be produced in an intrinsic way through rank conditions*.

EXAMPLE(Burgers): Let $(t, x) = (time, position)$ and consider the Cole-Hopf differential correspondence:

$$y + 2z_x = 0, 2y_x - y^2 + 4z_t = 0$$

with one input y and one output z. The corresponding resolvent systems are resprctively the Navier-Stokes equation $y_t + yy_x - y_{xx} = 0$ and $z_t - z_x^2 - z_{xx} = 0$ which becomes the heat equation $\theta_t - \theta_{xx} = 0$ if one sets $z = log\theta$. The control system is involutive in z whenever y satisfies the input resolvent system but not involutive in y ,even if z satifies the output resolvent system, though it is involutive in (y, z).

In the algebraic framework, even if the control system determines a prime differential ideal (the reader may check this property on all the examples presented by proving that the residue ring can be generated by purely transcendental elements), the study of formal integrability may be quite tricky but the first equality appearing (case III in example II below) will allow to distinguish the irreducible generic resolvent system to be the "most generic" one. Also, one notices that the *differential transcendence degree*,that is the maximum number of unknowns that can be given completely arbitrary, *cannot* be determined without bringing the system to its formally integrable presentation.

Before looking at two difficult examples, we treat by our approach the control examples given in (3), noticing however at first that 2-acyclicity or involutivity are trivially satisfied by the symbols when $n = 1$.

EXAMPLE: Let us consider the following system $\mathcal{R}_1$ of two ODE:

$$y_x - y = 0, z - y^2 - y = 0$$

This system is easily seen to be a fibered manifold but is not formally integrable because it is a first order system of ODE containing one zero order ODE. Hence $\mathcal{R}_1^{(1)}$ is obtained by addind $z_x - 2y^2 - y = 0$ and is therefore involutive. As polynomials in y, the two zero order ODE must be compatible and the resultant leads to the resolvent system:

$$z_x^2 - (4z + 1)z_x + 4z^2 + z = 0$$

Hence, z_x can be computed from z whenever $2z_x - 4z - 1 \neq 0$ or $z + \frac{1}{4} \neq 0$ and y can be computed from z whenever $y + \frac{1}{2} \neq 0$, the two conditions being equivalent.

EXAMPLE: Let us consider the following control system $\mathcal{R}_1$ of ODE:

$$y_x - uy^2 - u^2y = 0, z - y^2 = 0$$

with one input u, one state or latent variable y and one output z. Similarly, the involutive system $\mathcal{R}_1^{(1)}$ is obtained by adding $z_x - 2uy^3 - 2u^2y^2 = 0$ and the resultant in y, now depending on the input u, gives the resolvent system:

$$z_x^2 - 4u^2zz_x - 4u^2z^3 + 4u^4z^2 = 0$$

Hence, z_x can be computed from (u, z) whenever $z_x - 2u^2z \neq 0$ or $z \neq 0$ and y can be computed from (u, z) whenever $y \neq 0$, the two conditions being equivalent.

Herafter, we shall illustrate the preceding results by means of two delicate examples among the best we know. Both are also typical examples of partial differential control theory.

EXAMPLE I: (*Riemannian geometry*)
Whenever one has a *metric* $\omega \in S_2T^*$ with $det\omega \neq 0$, then one can define the *Christoffel symbols* $\gamma = (\gamma_{ij}^k = \gamma_{ji}^k)$ by the local formula:

$$\gamma_{ij}^k = \frac{1}{2}\omega^{kr}(\partial_i\omega_{rj} + \partial_j\omega_{ir} - \partial_r\omega_{ij})$$

while using the inverse matrix (ω^{kl}) of $\omega = (\omega_{ij})$.
Introducing the *Killing equations* as the subvector bundle $R_1 \subset J_1(T)$ defined by the equations:

$$\omega_{rj}\xi_i^r + \omega_{ir}\xi_j^r + \xi^r\partial_r\omega_{ij} = 0$$

it is easily seen that γ is a *connection*, that is a section of $T^* \otimes R_1$ projecting onto the identity $id_T \in T^* \otimes T$ and that there is an isomorphism $(\omega, \gamma) \sim (\omega, \partial\omega)$. Now one can introduce the *Riemann tensor* of ω to be:

$$\rho_{lij}^k = \partial_i\gamma_{lj}^k - \partial_j\gamma_{li}^k + \gamma_{lj}^r\gamma_{ri}^k - \gamma_{li}^r\gamma_{rj}^k$$

The converse problem may be important for mathematical physics, namely, given a 4-tensor ρ , does it come from a metric ω ?
Of course, certain necessary conditions must be satisfied by ρ, for example:

$$\rho_{lij}^k + \rho_{ijl}^k + \rho_{jli}^k = 0$$
$$\rho_{rij}^r = 0(\Rightarrow \rho_{ij} = \rho_{ijr}^r = \rho_{jir}^r = \rho_{ji})$$

However, one can exhibit counterexamples (12) and the problem is still open because most of the people studying it in mathematical physics seem to ignore almost completely the differential algebraic framework (15) while the use of computer algebra for studying such problems is quite recent (1). Indeed, one has to study the formal integrability of a second order system of nonlinear PDE for ω that depends on ρ. However, one may obtain easily the following zero order PDE:

$$\omega_{rj}\rho_{kij}^r + \omega_{kr}\rho_{lij}^r = 0$$

which must *not at all* be considered as a kind of "symmetry" because ω is unknown (12).In this framework, even the *Bianchi identities*:

$$\nabla_r\rho_{lij}^k + \nabla_i\rho_{ljr}^k + \nabla_j\rho_{lri}^k = 0$$

where ∇ is the covariant derivative with respect to γ, must be considered as first order PDE. As a byproduct, one is led to a very intricate problem of formal integrability with ranks depending on ρ and, *most probably even in the generic situation*, its derivatives. A similar problem could also be asked for knowing γ from ρ and/or ω from γ. The solution of such problems must give rise to "trees" depending on various inequalities but we do not believe that the knowledge of such trees should be quite important for mathematical physics, contrary to what many people claim. Indeed, we may consider the typical problem in gauge theory of knowing wether a *field* is coming from a *potential* whenever the *field equations* are satisfied. More precisely, if $\mathcal{G}$ is a Lie algebra with structure constants $c^\tau_{\rho\sigma}$ describing the bracket [], one can introduce the so-called Maurer-Cartan equations from 1-forms $A = (A^\tau_i) \in T^* \otimes \mathcal{G}$ to 2-forms $F = (F^\tau_{ij}) \in \wedge^2 T^* \otimes \mathcal{G}$ through the formulas:

$$\partial_i A^\tau_j - \partial_j A^\tau_i + c^\tau_{\rho\sigma} A^\rho_i A^\sigma_j = F^\tau_{ij}$$

or simply: $dA + \frac{1}{2}[A, A] = F$ with standard notations. Again, counterexamples for finding A whenever F is given are known (12) and the Bianchi identities:

$$\partial_i F^\tau_{jk} + \partial_j F^\tau_{ki} + \partial_k F^\tau_{ij} + c^\tau_{\rho\sigma}(A^\rho_i F^\sigma_{jk} + A^\rho_j F^\sigma_{ki} + A^\rho_k F^\sigma_{ij}) = 0$$

or simply: $dF + [A, F] = 0$ *cannot be considered as field equations* because they contain the potential. Again, we started from first order PDE for A depending on F and obtain zero order PDE, this fact leading to a delicate problem of formal integrability. We have explained in (10) why the above comment can be avoided by bringing new mathematics where the field equations are the Maurer-Cartan equations and why, as a direct consequence, the differential geometry of gauge theory must be revisited, but this is out of the scope of this paper.

EXAMPLE II: (*Inverse problem of the calculus of variations*)
One of the best examples of differential elimination theory solved according to the work of Janet and containing a tree of inequalities has been provided in 1940 by J. Douglas (4). For simplicity, we quote it on 3-dimensional space with cartesian coordinates (x, y, z) but it could be handled (with much more difficulty) in higher dimensions.

Let us consider the problem of finding an extremum of $\int \Phi(x, y, z, y', z') dx$ for curves $y = f(x), z = g(x)$. The corresponding well known Euler-Lagrange equations are:

$$\frac{d}{dx}\left(\frac{\partial \Phi}{\partial y'}\right) - \frac{\partial \Phi}{\partial y} = 0, \frac{d}{dx}\left(\frac{\partial \Phi}{\partial z'}\right) - \frac{\partial \Phi}{\partial z} = 0$$

where the accent indicates a differentiation and the *total* or *formal derivative* with respect to x is:

$$d_x = \frac{d}{dx} = \frac{\partial}{\partial x} + y'\frac{\partial}{\partial y} + z'\frac{\partial}{\partial z} + y''\frac{\partial}{\partial y'} + z''\frac{\partial}{\partial z'}$$

Under the assumption:

$$det\begin{pmatrix} \frac{\partial^2 \Phi}{\partial y'\partial y'} & \frac{\partial^2 \Phi}{\partial y'\partial z'} \\ \frac{\partial^2 \Phi}{\partial z'\partial y'} & \frac{\partial^2 \Phi}{\partial z'\partial z'} \end{pmatrix} \neq 0 \qquad (hessian)$$

and using linear algebra, the previous system can be written:

$$y'' = F(x, y, z, y', z'), z'' = G(x, y, z, y', z')$$

where (F, G) are easily expressed in terms of Φ and its various derivatives. The converse problem of the calculus of variations is, on the contrary, to suppose (F, G) given and to look for Φ. Equivalently, one has to study the system of two second order PDE for Φ:

$$\frac{\partial \Phi}{\partial y} - \frac{\partial^2 \Phi}{\partial x \partial y'} - y'\frac{\partial^2 \Phi}{\partial y \partial y'} - z'\frac{\partial^2 \Phi}{\partial z \partial y'} - F\frac{\partial^2 \Phi}{\partial y' \partial y'} - G\frac{\partial^2 \Phi}{\partial z' \partial y'} = 0$$

$$\frac{\partial \Phi}{\partial z} - \frac{\partial^2 \Phi}{\partial x \partial z'} - y'\frac{\partial^2 \Phi}{\partial y \partial z'} - z'\frac{\partial^2 \Phi}{\partial z \partial z'} - F\frac{\partial^2 \Phi}{\partial y' \partial z'} - G\frac{\partial^2 \Phi}{\partial z' \partial z'} = 0$$

where *the 5 jet coordinates* (x, y, z, y', z') *are now considered as 5 independent variables.* This is by chance a linear system for Φ (contrary to the previous example !) depending on (F, G) under the only assumption made on the hessian determinant. Of course, such a system can be considered as a nonlinear partial differential control system for (F, G, Φ) with inputs (F, G) and output Φ. Apart from the hessian condition, there is no resolvent system for Φ. However, (F, G) cannot be given arbitrary as explicit computations will prove and *the system for Φ is therefore not formally integrable in general.* The inverse problem just amounts to bring it to a formally integrable equivalent form while taking into account the hessian condition, and then to look for the "degree of generality" of the solutions. However, one must notice that the system is invariant under the transformation $\Phi \implies \Phi + d_x \nu(x, y, z)$ and one must only look for solutions other than the total derivative of a function, the so-called trivial solutions. Successive applications of the criterion for formal integrability eventually bring the answer as a set of differential equalities and inequalities for (F, G) leading to a tree of various cases that we exhibit on a separate page. As *the direct computation is awful,* J. Douglas introduced a trick that we shall present for the general situation $\Phi(x, y^k, y_x^k)$ and illustrate with one particular computation. Let us set:

$$-\omega_k = \frac{d}{dx}\left(\frac{\partial \Phi}{\partial y_x^k}\right) - \frac{\partial \Phi}{\partial y^k} \qquad , \qquad y_{xx}^k = F^k(x, y, y_x)$$

and obtain the following identities:

$$\frac{\partial \omega_k}{\partial y_x^l} + \frac{\partial \omega_l}{\partial y_x^k} + 2\frac{d}{dx}\Phi_{kl} + \frac{\partial F^r}{\partial y_x^k}\Phi_{lr} + \frac{\partial F^r}{\partial y_x^l}\Phi_{kr} \equiv 0$$

$$\frac{d}{dx}\left(\frac{\partial \omega_k}{\partial y_x^l} - \frac{\partial \omega_l}{\partial y_x^k}\right) - 2\left(\frac{\partial \omega_k}{\partial y^l} - \frac{\partial \omega_l}{\partial y^k}\right) + \frac{1}{2}\frac{\partial F^r}{\partial y_x^k}\left(\frac{\partial \omega_l}{\partial y_x^r} + \frac{\partial \omega_r}{\partial y_x^l}\right) - \frac{1}{2}\frac{\partial F^r}{\partial y_x^l}\left(\frac{\partial \omega_k}{\partial y_x^r} + \frac{\partial \omega_r}{\partial y_x^k}\right) + A_l^r\Phi_{kr} - A_k^r\Phi_{lr} \equiv 0$$

with $A_k^r = \frac{d}{dx}\frac{\partial F^r}{\partial y_x^k} - 2\frac{\partial F^r}{\partial y^k} - \frac{1}{2}\frac{\partial F^s}{\partial y_x^k}\frac{\partial F^r}{\partial y_x^s}$ and $\Phi_{kl} = \partial^2 \phi/\partial y_x^k \partial y_x^l$.
Setting $\omega_k = 0$, we are led to a first order system of PDE for the $\Phi_{kl} = \Phi_{lk}$ where we have to notice that:

$$\partial \Phi_{kl}/\partial y_x^m = \partial \Phi_{km}/\partial y_x^l$$

Such a system is surely not formally integrable because *it contains zero order equations* for the Φ_{kl} but the idea is to prove that solving this *first order* system for the Φ_{kl} amounts to solve the previous *second order* system for Φ (see paragraph 5 of reference 4).

Coming back to the previous notations, let us introduce:

$$A = d_x F_{z'} - 2F_z - \frac{1}{2}F_{z'}(F_{y'} + G_{z'})$$

$$B = -d_x F_{y'} + d_x G_{z'} + 2(F_y - G_z) + \frac{1}{2}(F_{y'} - G_{z'})(F_{y'} + G_{z'})$$

$$C = -d_x G_{y'} + 2G_y + \frac{1}{2}G_{y'}(F_{y'} + G_{z'})$$

then:

$$A_1 = d_x A - F_{y'} A - \frac{1}{2}F_{z'} B$$

$$B_1 = d_x B - G_{y'} A - \frac{1}{2}(F_{y'} + G_{z'})B - F_{z'} C$$

$$C_1 = d_x C - \frac{1}{2}G_{y'} B - G_{z'} C$$

while (A_2, B_2, C_2) are derived from (A_1, B_1, C_1) by recursion.

The first branches of the tree are determined by the conditions:
Case I:

$$\mid ABC \mid = 0 \Leftrightarrow A = 0, B = 0, C = 0$$

Case II:

$$\begin{vmatrix} A & B & C \\ A_1 & B_1 & C_1 \end{vmatrix} = 0, |A, B, C| \neq 0$$

Case III:

$$\begin{vmatrix} A & B & C \\ A_1 & B_1 & C_1 \\ A_2 & B_2 & C_2 \end{vmatrix} = 0, \begin{vmatrix} A & B & C \\ A_1 & B_1 & C_1 \end{vmatrix} \neq 0$$

Case IV:

$$\begin{vmatrix} A & B & C \\ A_1 & B_1 & C_1 \\ A_2 & B_2 & C_2 \end{vmatrix} \neq 0$$

The other cases will not be described and we refer the reader to the quoted reference.

However, in order to help the reader, we just give hints for the case $F = y^2 + z^2, G = 0$ Indeed, differentiating the two Euler-Lagrange PDE, the first with respect to z', the second with respect to y' and substracting, we get:

$$\frac{\partial^2 \Phi}{\partial y \partial z'} - \frac{\partial^2 \Phi}{\partial z \partial y'} = 0$$

Taking this result into account and differentiating again the two PDE with respect to z and y respectively, then substracting, we get:

$$y\frac{\partial^2 \Phi}{\partial y' \partial z'} - z\frac{\partial^2 \Phi}{\partial y' \partial y'} = 0$$

Using now the fact that $d_x \Phi_{kl} = 0$, we also get:

$$y'\frac{\partial^2 \Phi}{\partial y' \partial z'} - z'\frac{\partial^2 \Phi}{\partial y' \partial y'} = 0$$

leading to:

$$\frac{\partial^2 \Phi}{\partial y' \partial z'} = 0, \frac{\partial^2 \Phi}{\partial y' \partial y'} = 0$$

and a contradiction with the hessian conditioin.

If we look at this problem in order to construct differential extensions with 5 derivations for $K = Q(x, y, z, y', z')$ while considering (Φ, F, G) as 3 indeterminates, we discover that (F, G) may be expressed by means of Φ and its derivatives, provided that the hessian condition is fulfilled. As Φ alone is purely transcendental, the differential transcendence degree of the total differential extension is 1 and therefore (F, G) *must* be differentially algebraically dependent. Looking through the various cases, we discover that the case III is the "most generic" one allowing for effective solutions. Accordingly, the condition of vanishing determinant, which is specific to this case, gives a necessary differential condition that must be fulfilled by (F, G) and determines the corresponding differential extension of K. Once again, we verify that this result should be awful to look for directly, proving the importance for differential algebra of the techniques coming from the formal theory of PDE.

The following diagram recapitulates all the results in the form of a tree. Doted lines should be very technical to detail while "NO" means that no solution exists other than the trivial one.

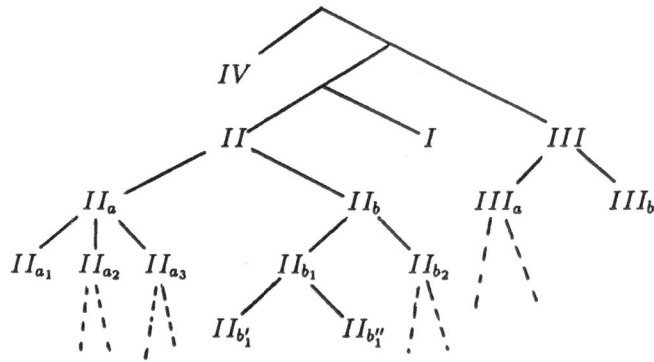

$$I \qquad y'' = 0, z'' = 0$$
$$II_{a_1} \qquad y'' = F(x,y,y'), z'' = G(x,z,z'), B \neq 0$$
$$II_{a_2} \qquad y'' = (1 + y'^2 + z'^2)/y, z'' = 0$$
$$II_{b_1'} \qquad y'' = z, z'' = 0$$
$$II_{b_1''} \quad NO \quad y'' = F, z'' = 0, A \neq 0, B = 0, F_{y'y'y'} \neq 0$$
$$III_a \qquad y'' = z^2, z'' = y^2 \implies \Phi = y'z' + \tfrac{1}{3}y^3 + \tfrac{1}{3}z^3$$
$$III_b \quad NO \quad y'' = y^2 + z^2, z'' = 0$$
$$IV \quad NO \quad y'' = y^2 + z^2, z'' = y$$

BIBLIOGRAPHY

1) M.A.H. Mac CALLUM, G.C. JOLY:Computer-aided classification of the Ricci tensor in general relativity,Class. Quantum Grav.,7,1990,541-556

2) J. DAVENPORT, Y. SIRET, E. TOURNIER:Calcul formel, Masson, Paris,1986,264p

3) S. DIOP:Elimination in control theory,to appear in Mathematics of Control, Signals and Systems

4) J. DOUGLAS:Solution of the inverse problem of the calculus of variatioins, Trans. Amer. Math. Soc.,50,1941,71-128

5) M. FLIESS:Automatique et corps différentiels, Forum Math.,1,1989

6) H. GOLDSCHMIDT:Integrability criteria for systems of nonlinear partial differential equations, J. Diff. Geometry,1,1969,269-307

7) M. JANET:Leçons sur les systèmes d'équations aux dérivées partielles, Cahiers scientifiques, Fasc. IV, Gauthier-Villars, Paris,1929

8) E.R. KOLCHIN:Differential algebra and algebraic groups,Academic Press, New York, 1973, 446p

9) J.F. POMMARET:Differential Galois theory, Gordon and Breach, New York,1983,760p

10) J.F. POMMARET:Lie pseudogroups and mechanics, Gordon and Breach, New York, 1988, 590p

11) J.F. POMMARET:Géométie différentielle algébrique et théorie du contrôle, C. R. Acad. Sc. Paris,302,I,15,1986,547-550

12) A.D. RENDALL:Insufficiency ofthe Ricci and Bianchi identities for Characterising curvature, J. of Geometry and Physics,6,1989,160

13) J.F. RITT:Differential algebra, Dover,1950,1966

14) F. SCHWARZ:The Riquier-Janet theory and its application to nonlinear evolution equations, Physica,110,1984,243-251

15) A. SEIDENBERG:An elimina tion theory for differential algebra, University of California Publications in Math.,3,1956,31-65

16) D.C. SPENCER:Overdetermined systems of PDE, Bull. Amer. Math. Soc., 75, 1965, 1-114

17) J.M. THOMAS:Systems and roots, W. Byrd Press,1962

18) WU WEN TSUN:Mathematics-Mechanization Research Reprints, Institute of Systems Science, Academia Sinica, 3, 1989, 1-26

Truncated bilinear approximants: Carleman, finite Volterra, Padé-type, geometric and structural automata.

C.HESPEL

INSA, 20 av.Buttes de Coësmes, 35043 RENNES Cedex.

15 Octobre 1990

ABSTRACT

In this paper, two original infinite $I\!\!R$-automata representations of an analytic systems are provided, allowing to compute easily a large class of bilinear approximants, including the so called "Carleman linearization method" as well as the "finite Volterra expansions" and "Padé-type ".

Each of these $I\!\!R$-automata is an infinite valued graph, and any "truncation" of this graph is an explicit bilinear approximation of the original system, of which the state space description can be obtain *directly, without any computation*.

The *"geometric $I\!\!R$-automaton"* is an improvement of the Carleman description, replacing the tensor powers of the state vector by its symmetrical powers. (This valued graph has also been obtained independently by Viennot, using combinatorial technics.) The *"structural $I\!\!R$-automaton"* describes an action on differential operators.

Any choice of an iterated truncation procedure produces a family of bilinear approximants, that can be viewed as a family of *noncommutative Padé-type approximants*. A particular choice, detailed here, produces the "up to order k" expansion, that gives the exact value for the *"k-order" Taylor expansion of the output*.

Introduction

The aim of this paper is to give two new representation tools for non linear analytic dynamical systems by infinite automata. The first method, by "geometric bilinearization", is based on the differential calculus in the state-space. It is an automaton representation of the Carleman "linearization method", given in a "non redundant form".

The second method, by "structural bilinearization", is based on computations in the algebra of the differential operators defined by the vector fields of the given system.

These two automata are directly obtained from the system equations, by using the Leibnitz derivation formula. These two infinite automata can be truncated according to several criteria, allowing to compute any well-known bilinear approximation like the Carleman method [2,3,16], the bilinear finite Volterra expansions (when there exist) [2].

To be more exact, any iterated truncation procedure provides a family of noncommutative Padé-type approximants [13].

Moreover, a simple reading of such a truncated bilinear automaton furnishes a bilinear dynamic system approximating the given analytic system. This method was partly presented in [8,9,10], and is illustrated here, for the geometrical and structural approaches, according to an approximation criterion "up to order k". We obtain in that way a bilinear approximation that provides an output Taylor expansion equal to the output Taylor expansion of the given system, up to order k, whatever the input may be.

Let (Σ) be a given nonlinear dynamic analytic system :

$$(\Sigma) \quad \begin{cases} \dot{x}(t) &= A(x) + \sum_{i=1}^{m} u_i(t) B_i(x) \\ y(t) &= h(x) \end{cases}$$

Recall that several methods are used for approximating the input/output behaviour of this system by a linear or bilinear one.

0.1 The Carleman method

The most famous method is the Carleman method [3], of which are knew two versions: the redundant version (called here *version (R)*, see [16]), and the non redundant version (called here version *version (NR)*, [2]). (See also Rotella for a matrix description [15]). It consists in "bilinearizing" the given system, by two successive stages: The first step consists in writing a bilinear differential equation whose dimension is infinite, in the tensor algebra (for the redundant version), or respectively in the symmetrical algebra (for the non redundant version) of the state space. The second step consists in truncating this new system by choosing the tensor powers $x^{\otimes k}$ (resp. the symmetrical power $s^{[k]}$) whose degree is smaller than or equal to $k < N$. We obtain in that way a finite dimensional bilinear system which approximates (Σ).

For a single system, we get the following truncated state equation:

$$\frac{d}{dt} \begin{pmatrix} x^{\otimes 1}(t) \\ \vdots \\ x^{\otimes N}(t) \end{pmatrix} =$$

$$\begin{pmatrix} A_{1,1} & \cdots & A_{1,N} \\ 0 & A_{2,2} & \cdots & A_{2,N} \\ \vdots & \ddots & & \vdots \\ 0 & \cdots & 0 & A_{N,1} \end{pmatrix} \begin{pmatrix} x^{\otimes 1}(t) \\ \vdots \\ x^{\otimes N}(t) \end{pmatrix} + \begin{pmatrix} B_{1,1} & & B_{1,N-1} & 0 \\ B_{2,0} & & B_{2,N-2} & 0 \\ 0 & \ddots & \vdots & \vdots \\ \vdots & \ddots & \ddots & \vdots & 0 \\ 0 & \cdots & 0 & B_{N,0} & 0 \end{pmatrix} \begin{pmatrix} x^{\otimes 1}(t) \\ \vdots \\ x^{\otimes N}(t) \end{pmatrix} u + \begin{pmatrix} B_{10} \\ 0 \\ \vdots \\ 0 \end{pmatrix} u + \cdots$$

$$y(t) = \begin{pmatrix} c(t) & 0 & \cdots & 0 \end{pmatrix} \begin{pmatrix} x^{\otimes 1}(t) \\ \vdots \\ x^{\otimes N}(t) \end{pmatrix},$$

where the state vector $x^{\otimes *}(t)$ has been "truncated" beyond the N^{i-th} tensor power $x^{\otimes N}(t)$. Hence we have set:

$$\begin{aligned} x^{\otimes 1} &= x, \\ \forall i \in I\!\!N^*, \quad x^{\otimes i} &= x^{\otimes(i-1)} \otimes x^{\otimes 1}. \end{aligned}$$

The Carleman representation, in the version (R), is redundant in the following sense: given a basis $(e_i)_{1 \le i \le n}$ of the state space, we have $x = \sum_{i=1}^{n} e_i s_i$. Thus $x \otimes x = \sum_{i,j} (e_i \otimes e_j) x_i x_j$ is a redundant expression, since we see that $x_i x_j$ is at once the coefficient of $e_i \otimes e_j$ and $e_j \otimes e_i$.

The non redundant Carleman representation, used by Brockett [2] in the version (NR), corresponds exactly to computations of Taylor series, and to the calculus in the symmetrical algebra of the state space. We illustrate this representation (NR) and this truncation method with the geometrical $I\!R$-automaton construction: the states of this geometric $I\!R$-automaton are the products $x_1^{i_1} x_2^{i_2} \ldots x_N^{i_N}$ (commutative product, i.e. not redundant products). Then, this automaton is truncated by picking out the states whose degree is smaller than or equal to k, i. e. such that

$$\sum_{j=1}^{N} i_j \le k,$$

and deleting the others states.

0.2 Finite Volterra expansions in an equilibrium point

As showed by Brockett [2], the Carleman method enables to compute exact Volterra kernels as described, if the system is initialized in an equilibrium point, by a bilinear system. An equivalent computation, using enumerative combinatorics, has recently been presented by Viennot [13], and by Lamnabhi-Lagarrigue [12]. Here we rediscover the same result by constructing the "geometrical $I\!R$-automaton": in that case, the autonomous vector field A_0 cannot increase the total degree $i_1 + i_2 + \cdots + i_N = k$. And then, one need only to truncate the geometrical automaton by choosing the states $i_1 + i_2 + \cdots i_N$ whose total degree $\sum_{j=1}^{N} i_j$ is smaller than or equal to k. Thus we obtain a finite $I\!R$-automaton, and then a bilinear approximant.

If the system is not initialized in an equilibrium point, we have need change the approximation test : we have to keep any automaton path containing at most k edges labelled by the non autonomous vector fields of the systems. It is possible, but we may obtain an infinite automaton, and then we may obtain Volterra approximants which are no produced by a bilinear system. (The same test can also be used for the structural $I\!R$-automaton).

0.3 Padé-type approximants

If a differential system has no forcing term, a well known method consists in constructing the Padé-type approximant of the commutative generating series of the given system [1]. For a nonlinear system (Σ), the corresponding method consists in computing a rational approximant of the generating series of (Σ). To be more exact, we define a "Padé-type approximant's family" to be any iterative convergent construction of approximant rational series (in noncommutative variables). (see Leroux-Viennot [12]). We emphasize the generality of the following construction : any iterative method consisting in truncating an $I\!R$-automaton provides a family of non commutative Padé-type approximants of the generating series of (Σ). For any associated bilinear system, a rational expression of its generating series may be directly written, without any calculus, by reading the automaton. (This algorithm has been written in MACSYMA, see [11]).

We expect empirically that the bilinear systems associated to the Padé-type approximants will converge very rapidly. Unfortunately, we are not able to compute generally the input/output functional coded by the rational series.

0.4 General $I\!R$-automaton, and structural $I\!R$-automaton

Given an analytic system (Σ), which is affine in the input variables, the purpose of this paper is to construct two representations of (Σ), by defining two infinite dimensional bilinear $I\!R$-automata, and to truncate these automata in order to compute approximants of (Σ), for order k.

The first construction illustrates the Carleman method. It consists in defining a "geometrical automaton", whose states $\{x_1^{i_1} x_2^{i_2} \ldots x_N^{i_N}\}_{N \geq 0, i_j \geq 0}$ are powers of the state vector components, and whose transitions describe the left action of the system's vector fields.

The second construction consists in defining a "structural automaton", whose states are the partial derivatives $\{D_1^{i_1} D_2^{i_2} \ldots D_N^{i_N}\}_{N \geq 0, i_j \geq 0}$, and whose transitions describe the right action of the system's vector fields.

In the two cases, we obtained so the graph description of an infinite bilinear system producing the same generating series G as the original analytical system (Σ). Then, any finite truncation process gives the graphical description of a finite bilinear system approximating (Σ), that can be directly written without any computation.

In this paper, as an illustration, we detail the truncation process that gives an approximating bilinear system whose generating series coincides with the generating series of (Σ) up to order k. And consequently, the ouput Taylor expansion of the analytical system (Σ) and the one of the approximating bilinear system have the same Taylor expansion up to order k.

1 Notations

We are interested in studying the dynamic system (Σ):

$$(\sigma) \begin{cases} \dot{x}(t) &= A_0(x) + \sum_{i=1}^{m} u_i(t) A_i(x) \\ y(t) &= h(x) \end{cases}$$

where

- x belongs to a finite dimensional $I\!R$-analytic manifold Q, ·

- vector fields $A_0, A_1, \cdots, A_m$ and output function h are defined and analytic in a neighbourhood of the initial state $x(0)$,

$$A_l = \sum_{j=1}^{m} \theta_l^j(x_1, \cdots, x_n) \frac{\partial}{\partial x_j} \qquad for\ 0 \leq l \leq m,$$

- $u_1, \cdots, u_n$ are piecewise continuous.

As shown by M. Fliess ([5]), the input/output behaviour of an analytical system can be coded by the following non commutative formal power series s, called "generating series" of the system:

$$s = h_{|_{x(0)}} + \sum_{\nu \geq 0} \sum_{j_1,\cdots,j_n=0}^{m} A_{j-1}A_{j-2}\cdots A_{j-\nu}h_{|_{x(0)}} z_{j-\nu}\cdots z_{j-2}z_{j-1}$$
$$with \quad Z = \{z_0, z_1, \cdots, z_m\}$$

Our problem consists in building rational series g approximating s up to a given order k, i.e.:

$$ord(s - g) > k.$$

We recall (cf. Schützenberger theorem [17]), that a power series $g \in \mathbb{R} \ll Z \gg$ is said to be *rational, (or regular)* if and only if it can be "recognized" (by a finite $\mathbb{R}$-automaton), that is there exists a $\mathbb{R}$-automaton $\mathcal{A}$ over Z:

$$\mathcal{A} = (Q, q_0, l, \tau), \qquad where$$

- Q is a finite dimensional $\mathbb{R}$-vector space,

- q_0 is the initial state,

- $\tau(z_i)$ is a linear map: $Q \longrightarrow Q$,

- the "observation" l is a linear map: $Q \longrightarrow \mathbb{R}$,

satisfying:

$$g = \sum_{z_{i_1} z_{i_2} \cdots z_{i_n} \in Z^*} l\tau(z_{i_1})\tau(z_{i_2})\cdots \tau(z_{i_n})q_0 \, z_{i_1} z_{i_2} \cdots z_{i_n}.$$

It is convenient to use a more compact notation. We note any finite product $z_{i_1} z_{i_2} \cdots z_{i_n}$, where z_{i_j} are letters in the "alphabet" Z, as a "word", denoted by a single symbol w, belonging to the free monoid Z^* over Z.

Another characterization of recognizable series has been given by Schützenberger, and by M. Fliess [5]: A power series g is regnizable if and only if the rank of it Hankel matrix

$$H(g) = [< g|uv >]_{u,v \in Z^* \times Z^*}$$

is finite. These reminders show that the notions of *bilinear system behaviour*, of *rational generating series*, of *finite Hankel rank series* and of finite $\mathbb{R}$-automaton *recognizable series*, are strictly equivalent. Then , automaton theory is an adequate tool to find rational approximants [9]

To illustrate the connection between these notions, let us consider the following example:

Example 1:

Let $Z = \{z_1, z_2\}$. The following series is recognizable:

$$g = \sum_{w \in Z^*} |w|_{z_1} w.$$

Its Hankel matrix $H(g)$ is given by:

$$H(g) \quad \begin{pmatrix} & C_\epsilon & C_{z_1} & C_{z_2} & C_{z_1^2} & C_{z_1 z_2} & \cdots \\ \hline L_\epsilon & 0 & 1 & 0 & 2 & 1 & \cdots \\ L_{z_1} & 1 & 2 & 1 & 3 & 2 & \cdots \\ L_{z_2} & 0 & 1 & 0 & 2 & 1 & \cdots \\ L_{z_1^2} & 2 & 3 & 2 & 4 & 3 & \cdots \\ L_{z_1 z_2} & 1 & 2 & 1 & 3 & 2 & \cdots \\ \vdots & \vdots & \vdots & \vdots & \vdots & \vdots & \end{pmatrix}$$

Note that $rk(g) = 2$ and that (C_ϵ, C_{z_1}) form a basis of the space of column vectors:

$$(R) = \begin{cases} C_w = C_\epsilon & if \ |w|_{z_1} = 0 \\ C_w = |w|_{z_1}.C_{z_1} - (|w|_{z_1} - 1).C_\epsilon & otherwise. \end{cases}$$

Hence the automaton transitions are given by:

$$\begin{cases} z_1.C_\epsilon = & C_{z_1} \\ z_1.C_{z_1} = & C_{z_1^2} = 2.C_{z_1} - C_\epsilon \\ z_2.C_\epsilon = & C_\epsilon \\ z_2.C_{z_1} = & C_{z_2 z_1} = C_{z_1}. \end{cases}$$

By setting $\mu(z_1) = \begin{pmatrix} 0 & -1 \\ 1 & 2 \end{pmatrix}$ and $\mu(z_2) = \begin{pmatrix} 1 & 0 \\ 0 & 1 \end{pmatrix}$, and $\mu(uv) = \mu(u)\mu(v)$ (for any $u, v \in X^*$. We find again the expression:

$$g = \sum_{w \in Z^*} (0 \ 1) \, \mu(w) \begin{pmatrix} 1 \\ 0 \end{pmatrix}$$

associated with the finite $I\!R$-automaton whose valued graph is:

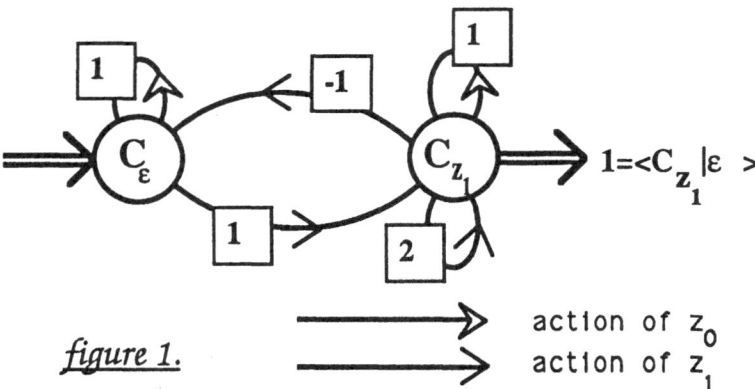

figure 1.

$\longrightarrow$ action of z_0

$\longrightarrow$ action of z_1

To illustrate the connection between the generating series of a dynamic system and the associated $I\!R$-automaton, let us consider the following example :

Exemple 2:

Consider the nonlinear differential equation relating the current excitation $i(t)$ and the voltage $v(t)$ across a capacitor:

$$\dot{v} = -k_1 v - k_1 v^2 + i(t),$$

i.e.
$$\begin{cases} \dot{x} &= A_0(x) + A_1(x)i(t) \\ v(t) &= x(t) \end{cases}$$

where
$$\begin{cases} A_0 &= (-k_1 x - k_2 x^2)\frac{\partial}{\partial x} \\ A_1 &= \frac{\partial}{\partial x} \end{cases}$$

By setting

$$a = -k_1 x - k_2 x^2, \qquad b = -k_1 - 2k_2 x, \qquad c = -2k_2,$$

and by computing the Hankel matrix $H(s)$, we note that the family $(C_{z_1^n})_{b \in N}$ forms a basis of the column vector space. Moreover

$$C_{z_0 z_1^n} w = aC_{z_1^{n+1}} + \frac{n(n-1)}{2}cC_{z_1^{n-1}}$$

The graph associated to this infinite $I\!R$-automaton is :

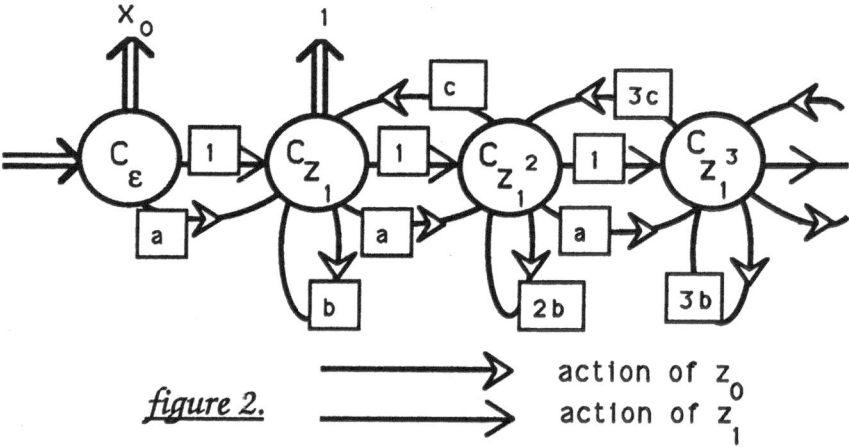

figure 2.

$\longrightarrow$ action of z_0

$\longrightarrow$ action of z_1

2 Geometrical $I\!R$-automaton

This automaton illustrates the Carleman method. Nevertheless, the construction is given in the state space of the symmetrical algebra, instead of the tensor algebra.

Given a dynamic system

$$(S) \qquad \begin{cases} \dot{x} = A_0(x) + \sum_{i=1}^{} u_i(t)A_i(x) \\ \qquad\qquad y(t) = h(x) \end{cases}$$

We can associate with it a "geometrical" $I\!\!R$-automaton $G = (E, V, \tau, l)$, where:

- $E =$ the $I\!\!R$-vector space generated by the family $(x_1^{i_1} \cdots x_n^{i_n})_{i_1, \cdots, i_n \geq 0}$

- $V =$ the initial vector associated to the observation

$$h = \sum_{n \geq 0} \sum_{i_1, \cdots, i_n \geq 0} h_{i_1 \cdots i_n} x_1^{i_1} \cdots x_n^{i_n}$$

- $\tau(z_r)$ is the linear map defined by the Lie derivatives:

$$x_1^{i_1} \cdots x_n^{i_n} \mapsto A_r \circ (x_1^{i_1} \cdots x_n^{i_n})$$

- $l =$ the linear form defined by the row vector whose components are the evaluations of the final states evaluated at $t = 0$:

$$S = (1, x_{1_{|0}}, x_{2_{|0}}, \cdots, x_{n_{|0}}).$$

The series which is recognized by this automaton is:

$$\begin{aligned}
s &= \sum_{w \in Z^*} A_{\tilde{w}} \circ h_{|_{x(0)}} . w \\
&\sum_{p \geq 0} \sum_{\nu_1, \cdots, \nu_p \geq 0} l(\tau(z_{\nu_p}) \cdots \tau(z_{\nu_1} V)) z_{\nu_1} \cdots z_{\nu_p}.
\end{aligned}$$

Here we note $\tilde{w}$ the "mirror image" of $w = z_{\nu_1} z_{\nu_2} \cdots z_{\nu_p}$, that is: $\tilde{w} = z_{\nu_p} z_{\nu_{p-1}} \cdots z_{\nu_1}$.

Exemple 3: Duffing equation.

$$\ddot{y} + a\dot{y} + b^2 y + c y_3 = u(t)$$

By a direct translation, we obtain:

$$(S_1) \quad \begin{cases}
\dot{x} &= A_0(x) + u(t) A_1(x) \\
A_0 &= F \frac{\partial}{\partial x_2} = F D_2 + x_2 D_1 \\
A_1 &= \frac{\partial}{\partial x_2} = D_2 \\
y(t) &= x_1(t),
\end{cases}$$

by setting $F = -a x_2 - b_2 x_1 - c x_1^3$. Then we proceed as follows:

- Let $E =$ the $I\!\!R$ vector space generated by the family $(x_1^i x_2^j)_{i,j \geq 0}$.

- The initial vector $V = (0 \ 1 \ \cdots)^t$.

- By computing the actions of the vector fields A_0 and A_1 on the states $x_1^i x_2^j$, we obtain :

$$\begin{aligned}
A_0 \circ (x_1^i x_2^j) &= -j a x_1^i x_2^j - j b^2 x_1^{i+1} x_2^{j-1} - j c x_1^{i+3} x_2^{j-1} + i x_1^{j-1} x_2^{j+1} \\
A_1 \circ (x_1^i x_2^j) &= j x_1^i x_2^{j-1}.
\end{aligned}$$

This is nothing else as the description of the following "automaton cell":

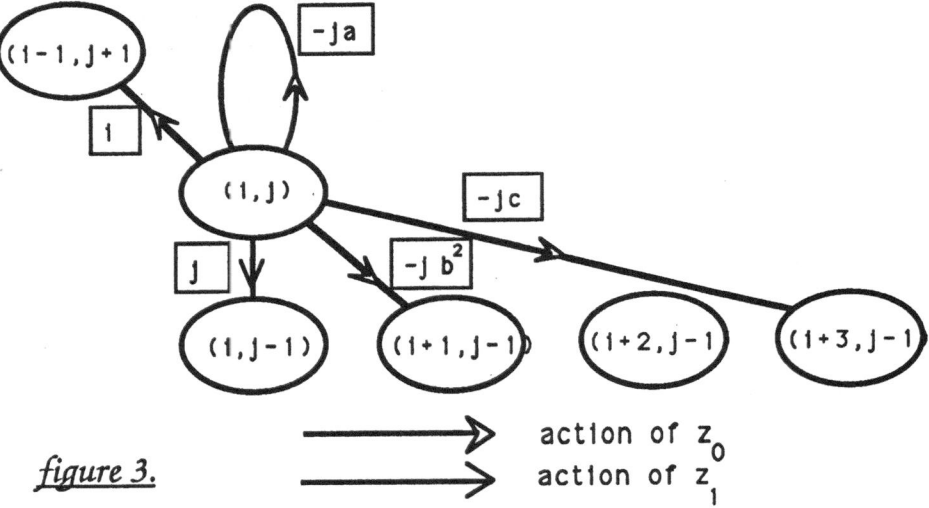

Thus we obtain a "geometrical" $\mathbb{R}$-automaton, described by a labelled graph having an infinite set of edges. *Any finite truncation* of this infinite graph furnishes a finite automaton, describing a *bilinear dynamic system approximating* the given nonlinear analytical system. A "Volterra truncation" is obtain by deleting the states $x_1^i x_2^j$ that satisfy $i + j < k$. In case $k = 3$, we obtain the following automaton, that recognizes exactly the order 3 Volterra approximant (plus higher order terms - a $\mathbb{R}$-automaton recognizing exactly the order 3 Volterra approximant can easily be obtained, see v.g. [7]).

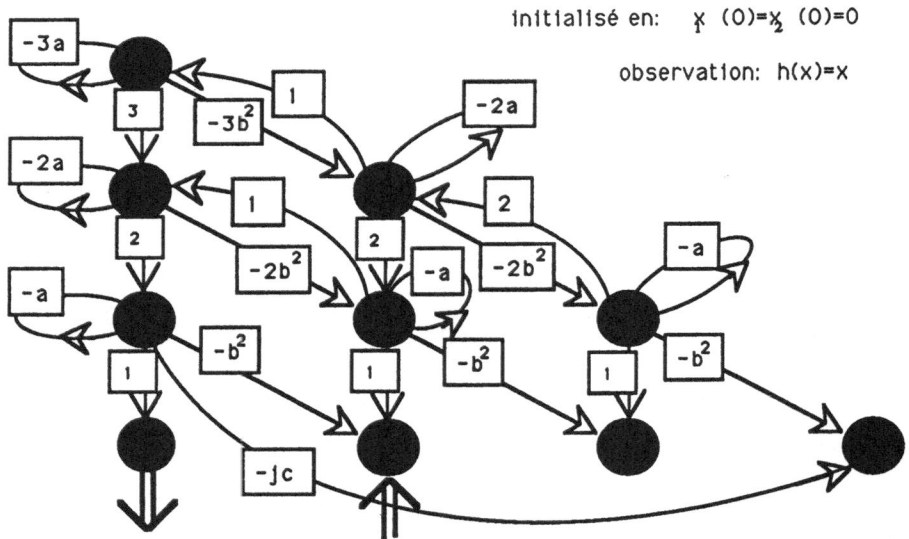

figure 4. Volterra automaton at the order 3.

The valued initial and terminal vertices are given as follows:

$$h = \sum c_{ij} \, x_2^{\,j} x_1^{\,i}$$

$$c_{ij} \Rightarrow \bigcirc\!\!\left(x_1^{\,i} x_2^{\,j} \right)\!\!\Rightarrow x_1^{\,i} x_2^{\,j}\Big|_0$$

figure 5. input and output values.

For the construction of a bilinear system A "Taylor truncation" at the order k is obtained as follows: For the order k, we pick out all the states of the form $x_1^i x_2^j$, which are met along some "successful path" of length $l \, e \, k$. As illustration, we obtain for $k = 3$ the following truncated automaton:

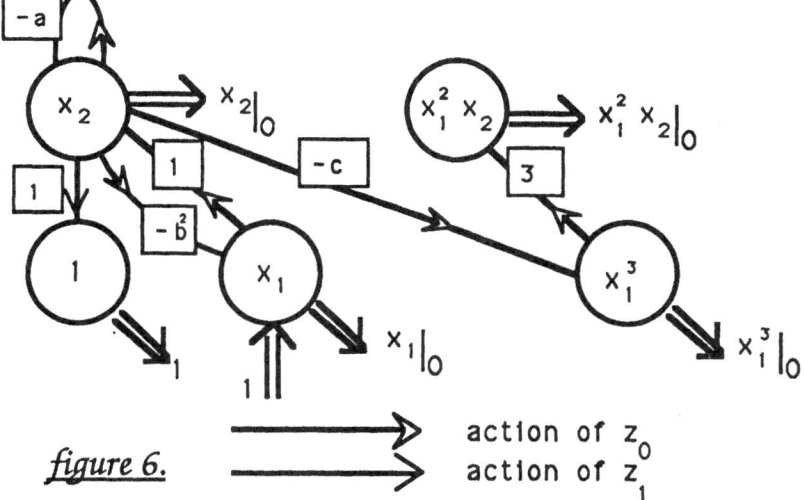

figure 6.

$$\begin{aligned}\Longrightarrow \quad &\text{action of } z_0 \\ \Longrightarrow \quad &\text{action of } z_1\end{aligned}$$

This automaton contains 5 states. It describes the bilinear system (B_1) whose state space dimension is equal to 5. Then (B_1) is a bilinear approximant of order $k = 3$ of the given nonlinear system. Indeed, the state equations of (B_1) are the following:

$$(B_1) \quad \begin{cases} \begin{pmatrix} \dot{q}_1 \\ \dot{q}_2 \\ \dot{q}_3 \\ v\dot{q}_4 \\ \dot{q}_5 \end{pmatrix} = \begin{pmatrix} 0 & 0 & 0 & 0 & 0 \\ 0 & 0 & -b^2 & 0 & 0 \\ 0 & 1 & -a & 0 & 0 \\ 0 & 0 & -c & 0 & 0 \\ 0 & 0 & 0 & 3 & 0 \end{pmatrix} \begin{pmatrix} q_1 \\ q_2 \\ q_3 \\ q_4 \\ q_5 \end{pmatrix} + u(t) \begin{pmatrix} 0 & 0 & 1 & 0 & 0 \\ 0 & 0 & 0 & 0 & 0 \\ 0 & 0 & 0 & 0 & 0 \\ 0 & 0 & 0 & 0 & 0 \\ 0 & 0 & 0 & 0 & 0 \end{pmatrix} \begin{pmatrix} q_1 \\ q_2 \\ q_3 \\ q_4 \\ q_5 \end{pmatrix} \\ \lambda = \begin{pmatrix} 1 & x_{1|_0} & x_{2|_0} & x_{1|_0}^3 & x_1^2 x_{2|_0} \end{pmatrix} \\ y(t) = \lambda q(t) \end{cases}$$

remark

In the *Carleman method* as presented by Rugh [16], a fictitious initial state is added, as we explain now.

Let us define an augmented state vector by adding a new component x_u subject to the condition:

$$x_u(t) = y(t).$$

Then the augmented system is describe as follows:

$$(S) \quad \begin{cases} \dot{x} = A_0(x) + \sum_{i=1} u_i(t) A_i(x) \\ \dot{x}_u = \frac{d}{dt} h(x(t)) \\ \qquad = A_0 \circ h + \sum_{i=1}^{m} u_i(t)(A_i \circ h), \\ output: \ s(t) = x_u(t). \end{cases}$$

The new associated $m+1$ vector fields are:

$$\hat{A}_i = A_i + (A_i \circ h)\frac{\partial}{\partial x_u}.$$

In other words, this method consists in adding a new fictitious initial state to the automaton.

In the example 3 (Duffing equation), the new state equations are deduced from:

$$\begin{cases} A_0 = x_2 D_1 + F D_2 \\ A_1 = D_2 \\ y = h(x_1, x_2) = x_3, \end{cases} \qquad \begin{cases} \dot{x}_1 = x_2 \\ \dot{x}_2 = F + u(t) \\ \dot{x}_3 = \frac{\partial h}{\partial x_1} x_2 + \frac{\partial h}{\partial x_2}(F + u(t)), \end{cases}$$

where $F = -ax_2 - b^2 x_1 - cx_1^3$. We obtain the new augmented vector fields:

$$\begin{aligned} \hat{A}_0 &= x_2 D_1 + F D_2 + (\tfrac{\partial h}{\partial x_1} x_2 + \tfrac{\partial h}{\partial x_2} F) D_3 \\ \hat{A}_1 &= D_2 + \tfrac{\partial h}{\partial x_2} D_3. \end{aligned}$$

The new geometric $I\!R$-automaton holds the new state x_3, in addition to the states $(x_1^i x_2^j)_{i,j \geq 0}$. The actions of the vector fields $\hat{A}_0$ and $\hat{A}_1$ of this new state are given by:

$$\begin{cases} \hat{A}_0 \circ x_3 = \frac{\partial h}{\partial x_1} + \frac{\partial h}{\partial x_2} F \\ \hat{A}_1 \circ x_3 = \frac{\partial h}{\partial x_2} \end{cases} \qquad with \ F = -ax_2 - b^2 x_1 - cx_1^3,$$

and are represented by the following graph:

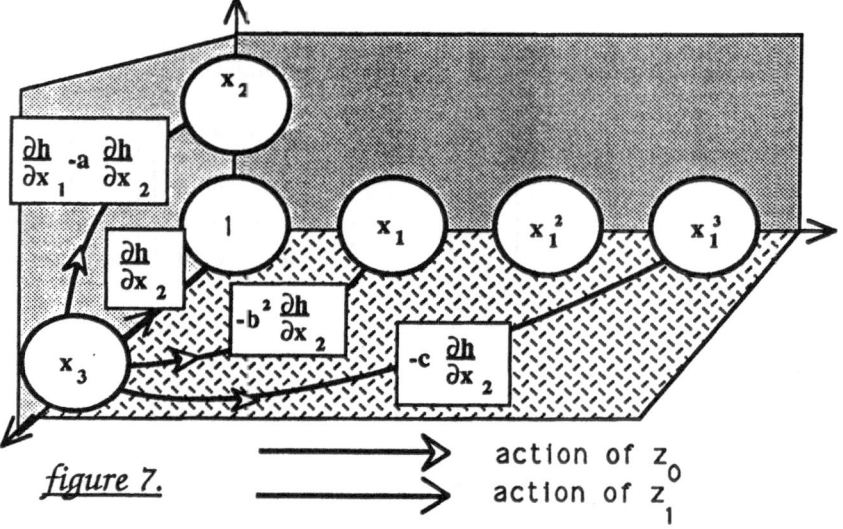

figure 7.

$\longrightarrow$ action of z_0

$\longrightarrow$ action of z_1

3 The structural $I\!R$-automaton

With a given dynamic system

$$(\Sigma) \quad \begin{cases} \dot{x}(t) & = & A_0(x) + \displaystyle\sum_{i=1}^{m} u_i(t) A_i(x) \\ y(t) & = & h(x), \end{cases}$$

we associate a "structural $I\!R$-automaton" S for characterizing the structure of the differential algebra defined by the system's vector fields. Let us note $D_i = \frac{\partial}{\partial x_i}$. In order to present the automaton S, we introduce the multi-indice notation:

$$\begin{aligned} if \quad & \alpha & = & (i_1, i_2, \cdots, i_n), \\ we\ set \quad & D^\alpha & = & D_1^{i_1} D_2^{i_2} \cdots D_n^{i_n}. \end{aligned}$$

Then the structural $I\!R$-automaton S is the 4-uple $(\mathcal{F}, I, \tau, l)$, where

- $\mathcal{F}$ is the $I\!R$-vector space generated by the operators D^α,

- I is the initial state,

- $\tau(z_r)$ is the linear endomorphism of $\mathcal{F}$ describing the right action of A_r on the D^α,

- l is the linear form defined by the row vector whose α^{th} component is $D_\alpha \circ h$.

The action of the vector field $A_r = \displaystyle\sum_{j=1}^{n}(A_r \circ x_j)\frac{\partial}{\partial x_j}$ on D^α is given by the Leibniz formula:

$$D^\alpha A^r = \sum_{j=1}^{n} \sum_{\beta \le \alpha} \binom{\alpha}{\beta} \left(D^\beta \circ A_r \circ x_j \right) D^{\alpha - \beta} D_j$$

where, if $\alpha = (i_1 \cdots i_n)$ and $\beta = (j_1 \cdots j_n)$, we have set:

$$\begin{aligned} & & \beta \le \alpha & \Leftrightarrow & j_1 \le i_1, \quad j_2 \le i_2, \cdots, \quad j_n \le i_n, \\ and\ if \quad \beta \le \alpha: \quad & \binom{\alpha}{\beta} & = & \binom{i_1}{j_1}\binom{i_2}{j_2} \cdots \binom{i_n}{j_n} \end{aligned}$$

In the example 3 (Duffing equation),

- $\mathcal{F}$ is the $I\!R$-vector space generated by $(D_1^i D_2^j)_{i,j \le 0}$

- the linear form l is defined by the row vector

$$(x_1\ 1\ 0\ \cdots\ 0)$$

- the actions of A_0 and A_1 on the states $D_1^i D_2^j$ are given by:

$$D_1^i D_2^j A_1 = D_1^i D_2^{j+1}$$
$$D_1^i D_2^j A_0 = F D_1^i D_2^{j+1}$$
$$+ \binom{i}{1}_{(D_1 F) D_1^{i-1} D_2^{j+1}} + \binom{i}{2}_{(D_1^2 F) D_1^{i-2} D_2^{j+1}} + \binom{i}{3}_{(D_1^3 F) D_1^{i-3} D_2^{j+1}}$$
$$- j a D_1^i D_2^j + x_2 D_1^{i+1} D_2^j + j D_1^{i+1} D_2^{j-1}.$$

Thus we obtain the following "automaton cell":

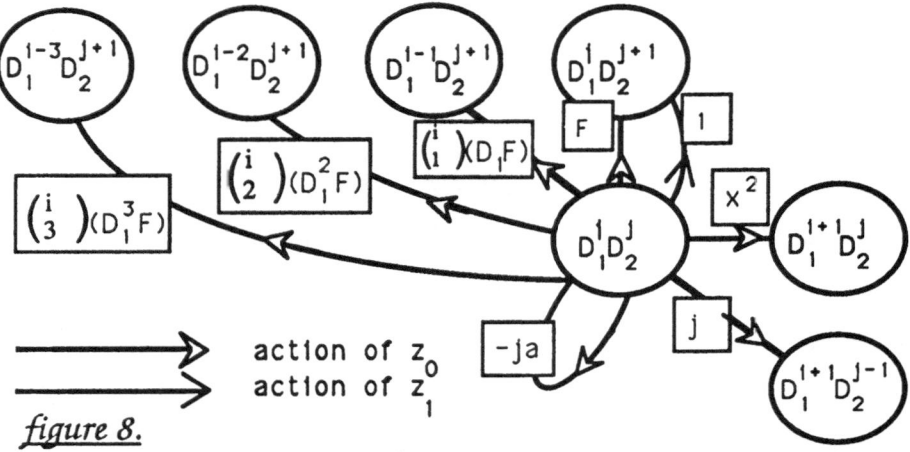

figure 8.

We construct the infinite structural $I\!R$-automaton, and then we construct the truncated automaton by choosing the states $D_1^i D_2^j$ that are met along a successful path and the length of which is smaller than or equal to k.

For $k = 3$, we get the following truncated automaton:

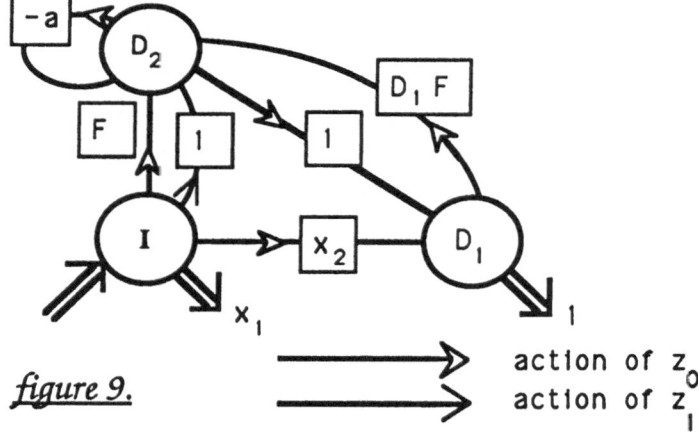

figure 9.

This automaton contains 3 states. It is associated with a bilinear system $(\mathcal{B}_2)$ whose dimension is equal to 3. We deduce the state equation of $(\mathcal{B}_2)$:

$$(\mathcal{B}_2) \quad \begin{cases} \begin{pmatrix} \dot{q}_1 \\ \dot{q}_2 \\ \dot{q}_3 \end{pmatrix} = \begin{pmatrix} 0 & 0 & 0 \\ x_{2|_0} & 0 & 1 \\ F_{|_0} & D_1(F)_{|_0} & a \end{pmatrix} \begin{pmatrix} q_1 \\ q_2 \\ q_3 \end{pmatrix} + \begin{pmatrix} 0 & 0 & 0 \\ 0 & 0 & 0 \\ 1 & 0 & 0 \end{pmatrix} \begin{pmatrix} q_1 \\ q_2 \\ q_3 \end{pmatrix} u(t) \\ y(t) = \lambda q(t) \\ \lambda = \begin{pmatrix} x_{1|_0} & 1 & 0 \end{pmatrix}. \end{cases}$$

4 Conclusion

These two methods allow us to compute a bilinear approximant of the given system (S), at the order k. We emphasize that when we wish to calculate a bilinear approximant, for the order $k+1$, the earlier calculations need not be done again.

These two technics are based on a criterion which is intrinsic and independant of the state representation: the equality of the generating series of the given system and of the approximating bilinear system, up to order k. Because of this remark, these technics defer from the Carleman method.

The geometrical automaton provides a bilinear approximated system, but its dimension is finite if and only if the observation is polynomial.

The structural automaton is a little more difficult to compute . But this method generally provides a bilinear approximated system whose dimension is smaller. But above all, it allows to compute in any case a bilinear approximant system, with an exact representation of the observation.

The algorithm consisting in computing the structural automaton has been implemented in MACSYMA [10].

References

[1] Brezinski C., *Padé-type approximants and general orthogonal polynomials*, INSM, vol. 50, Birkhaüser Verlag, basel, 1980.

[2] Brockett R.W., *Volterra series and Geometric Control Theory*, Automatica, vol.12, p. 167-176, 1976.

[3] Carleman T., *Application de la théorie des équations intégrales singulières aux équations différentielles de la dynamique*, T. Ark. Mat. Astron. Fys. 22B, 1, 1932.

[4] Flajolet P., *Combinatorial aspects of continued fractions*, Discrete Math., n 32, p. 125-161, 1980.

[5] Fliess M., *Fonctionnelles causales non linéaires et indéterminées non commutatives*, Bull. Soc. Math. France, 109, p. 3-40, 1981.

[6] Fliess M., Lamnabhi M., Lamnabhi-Lagarrigue, *An algebraic approach to nonlinear functional expansions,* IEEE Trans. Circuits and Systems, vol. CAS-30, n° 8, p. 554-570, 1983.

[7] Guyon C., Hoang Ngoc Minh, *Séries de Volterra finies, séries rationnelles,* Technical report, LIFL , Lille, 1991.

[8] Hespel C., Jacob G., *Calcul des approximations locales bilinéaires de systèmes analytiques,* RAIRO APII, vol.23, p. 331-349, 1989.

[9] Hespel C., Jacob G., *Rational approximations and automata,* Technical reports LANS n 29, INSA, RENNES, 1990.

[10] Hespel C., Jacob G., *Algorithme d'approximation bilinéaire de systèmes dynamiques par automate structurel,* Technical report LANS n 31, INSA, RENNES, 1991.

[11] Hespel C., *Approximation of nonlinear dynamics systems by rational series,* to appear in "Algebraic and Computing Treatment of Noncommutative Power Series" (ed. G. Jacob and C. Reutenauer), Theoretical Computer Science, 1991. Control Theory, Birkhaüser, 1990.

[12] Lamnabhi-Lagarrigue F., Leroux P., Viennot X.G., *Combinatorial interpretation of Volterra series by bilinear approximations,* to appear in "Analyse des systèmes dynamiques contrôlés", Porgress in Systems and Control Theory, Birkhaüser, 1991.

[13] Leroux P., Viennot X.G., *A combinatorial approach to nonlinear functional expansions: an introduction with example,* to appear in "Algebraic and Computing Treatment of Noncommutative Power Series" (ed. G. Jacob and C. Reuterauer), Theoretical Computer Science, 1991.

[14] Rotella F., *Méthodes algébriques et analytiques pour la simplification et la commande de systèmes bilinéaires à deux dynamiques,* Thèse, Université des Sciences et Techniques de Lille, 1987.

[15] Rotella F., *Generating series for analytic systems,* First european conference on "Algebraic Computing in control", Paris, march 13-15,1991.

[16] Rugh W.J., *Nonlinear system theory,* The John Hopkins university press, Baltimore and London.

[17] Schützenberger M.P., *On the definition of a family of automata,* Information and Control, vol. 4, p. 245-270, 1961.

GENERATING SERIES FOR ANALYTIC SYSTEMS

F.ROTELLA and I.ZAMBETTAKIS

LAII - SDI CNRS I 5980
Institut Industriel du Nord - BP48
59651 Villeneuve d'Ascq - CEDEX
FRANCE

Abstract : *The purpose of this paper is to propose implementable expressions for the determination of the Fliess series of a non linear affine-in-the-control analytic system. The generating series is constructed through an infinite dimensional bilinear realization of the considered non linear system. This point of view overcomes formal derivations of non linear functions which are replaced by algebraic manipulations on matrices.*

1 Introduction

Let us consider the following dynamical system :

$$\dot{x}(t) = g_0(x) + \sum_{i=1}^{m} g_i(x) u_i(t), \tag{1}$$

which can be referred as a non linear affine-in-the-control analytic system. In these equations $g_i(x)$, $i = 0, \ldots, m$, are non linear vector functions of the state vector x of $\mathbb{R}^n$, and when the hypothesis of analycity holds, they can be replaced by their generalized Taylor series expansions [VETTER, 1973]. It leads to the following expressions :

$$
\begin{aligned}
g_0(x) &= \sum_{j=1}^{\infty} G_{0j} x^{[j]}, \\
i = 1, \ldots, m, \quad g_i(x) &= \sum_{j=0}^{\infty} G_{ij} x^{[j]},
\end{aligned}
\tag{2}
$$

where :

- $x^{[i]}$ denotes the i-th power of x in the Kronecker sense, defined as follows :

$$
\begin{aligned}
x^{[0]} &= 1, \quad x^{[1]} = x, \\
\forall i > 0, \quad x^{[i+1]} &= x^{[i]} \otimes x,
\end{aligned}
\tag{3}
$$

where $\otimes$ stands for the Kronecker product [BREWER, 1978].

- the G_{ij} are constant ($n \times n^j$) matrices.

We must note here that if we introduce :

$$u_0(t) \equiv 0, \quad G_{00} = 0, \tag{4}$$

then, the analytic system can be writen more compactly as :

$$\dot{x}(t) = \sum_{i=0}^{m} \sum_{j=0}^{\infty} G_{ij} x^{[j]} u_i(t). \tag{5}$$

The output of this system is firstly supposed to be a linear fonction of the state vector, $y = Cx$, which must not be seen as a restriction but only an hypothesis for simplicity sake. We will see in the last part how the analytic output case can be handled.

1.1 Generating series

It is well known from [FLIESS *et al.*, 1983] that to the system (1) can be associated a generating series encoding the input-output analytic causal functional. In a general setup with a nonlinear output $y = h(x)$, and with the use of the noncommutative symbols $\{\alpha_0, \alpha_1, \ldots, \alpha_m\}$, the generating series $\mathcal{G}$ of the output is given by the Fliess fundamental formula :

$$\mathcal{G} = h \mid_{x^0} + \sum_{\nu \geq 0} \sum_{j_0, j_1, \ldots, j_\nu}^{m} L_{g_{j_0}} L_{g_{j_1}} \cdots L_{g_{j_\nu}} h \mid_{x^0} \alpha_{j_\nu} \cdots \alpha_{j_1} \alpha_{j_0}, \tag{6}$$

where x^0 denotes the initial condition on the state vector, $\mid_{x^0}$ an evaluation in x^0, and $L_g f$ the Lie derivative of f with respect to g defined by :

$$L_g f = \frac{\partial f}{\partial x} g(x), \tag{7}$$

where $\frac{\partial f}{\partial x}$ is the Jacobian matrix of f. Then, in the general case, the determination of the generating series involves successive differentiations of nonlinear functions.

In the following, the generic coefficient of $\mathcal{G}$ will be denoted as $\mathcal{G}_{j_0, \ldots, j_\nu}$.

1.2 Bilinear systems

The previous differentiations are avoided in the particular case of bilinear systems which are defined by a non linear state space equation linear with respect to the state. More precisely, we have :

$$\dot{x}(t) = A_0 x + \sum_{i=1}^{m} A_i x u_i(t) + B u(t), \tag{8}$$

$$y = Cx,$$

where $u(t) = [u_1(t), \ldots, u_m(t)]^T$, and A_i, B, and C are constant matrices.

The use of the symbolic calculus [FLIESS et al., 1983] allows to transforms the differential system (8) in an algebraic one, expressed directly in terms of generating series :

$$\mathcal{X} = x^0 + \sum_{i=0}^{m} A_i \alpha_i \mathcal{X} + B\alpha,$$
$$\mathcal{G} = C\mathcal{X}, \tag{9}$$

where $\mathcal{X}$ is the generating series associated to the state vector, and $\alpha = [\alpha_1, \ldots, \alpha_m]^T$. It leads then to :

$$\mathcal{G} = C(I - \sum_{i=0}^{m} A_i \alpha_i)^{-1}(x^0 + B\alpha). \tag{10}$$

It is obvious that this expression is a compact notation for the true writing of the generating series, which is :

$$\mathcal{G} = C(I + \sum_{\nu \geq 0} \sum_{j_0, j_1, \ldots, j_\nu}^{m} A_{j_\nu} \cdots A_{j_1} A_{j_0} \alpha_{j_\nu} \cdots \alpha_{j_1} \alpha_{j_0})(x^0 + B\alpha). \tag{11}$$

The generic element of this series appears then (from the analogy with linear systems) as the superposition of two terms :

- one which depends on the initial condition :

$$C A_{j_\nu} \cdots A_{j_1} A_{j_0} x^0 \alpha_{j_\nu} \cdots \alpha_{j_1} \alpha_{j_0}; \tag{12}$$

- another purely dependant on the input :

$$C A_{j_\nu} \cdots A_{j_1} A_{j_0} B \alpha_{j_\nu} \cdots \alpha_{j_1} \alpha_{j_0} \alpha. \tag{13}$$

This points outs the first interest of bilinear systems which is to allow to obtain a generating series only by matric manipulations, this specific feature will be used in the following. The second advantage is that we can approach, with an arbitrary accuracy, a non linear system by a bilinear one [FLIESS, NORMAND-CYROT, 1981; SUSSMAN, 1976]. Several technics for approximation have been proposed, the first one [HESPEL, JACOB, 1986] is based on the equality of the first coefficients of the generating series of the non linear system and the approximating bilinear one. As our goal is to determine the generating series of the non linear system, this method cannot be used. Another method, which has been systematised in [ROTELLA et al., 1986], consists in the use of a dilated state vector composed of the power of the state of the non linear model. As this technic will be a central point of this paper, the next part will recall it.

2 Bilinear approximations

To construct a bilinear approximation of a nonlinear system, we can use the Carleman linearization method [CARLEMAN, 1932], which has been used frequently in System Theory [BROCKETT, 1972, 1976; RUGH, 1981]. These approaches are based on the use of the non-redondant power of x in the Kronecker sense, which is defined as the power of x in the Kronecker sense (3) where are deleted all excepted one of the repeated components. If we define the infinite dimensional vector :

$$X = \left[x^{[1]T}, x^{[2]T}, \ldots, x^{[k]T}, \ldots \right]^T,$$

(14)

it has been shown in [ROTELLA et al., 1986], that the non linear system (5) is locally equivalent to the bilinear realization :

$$\dot{X} = \mathcal{A}_0 X + \sum_{i=1}^{m} \mathcal{A}_i X u_i(t) + \mathcal{B}u(t),$$
$$x = \mathcal{C}X,$$

(15)

where $\mathcal{A}_i$, $\mathcal{B}$, and $\mathcal{C}$ are infinite dimensional constant matrices. The matrices $\mathcal{A}_i$ are block-upper Hessenberg matrices defined by :

$$i = 0, \ldots, m, \quad \mathcal{A}_i = \begin{bmatrix} A_{11}^{(i)} & A_{12}^{(i)} & \cdots & \cdots & A_{1k}^{(i)} & \cdots \\ A_{21}^{(i)} & A_{22}^{(i)} & \ddots & \ddots & A_{2k}^{(i)} & \ddots \\ 0 & A_{32}^{(i)} & \ddots & \ddots & A_{3k}^{(i)} & \ddots \\ \vdots & 0 & A_{43}^{(i)} & \ddots & \ddots & \ddots \\ \vdots & & \ddots & \ddots & \ddots & \ddots \end{bmatrix},$$

(16)

where $\forall i = 0, \ldots, m$, the blocks $A_{jk}^{(i)}$ are recursively constructed as :

$$k > 1, \quad A_{1k}^{(i)} = G_{ik}, \quad A_{10}^{(i)} = 0,$$

$$j > 1, \quad k \geq j - 1, \quad A_{jk}^{(i)} = A_{j-1,k-1}^{(i)} \otimes I_n + I_{n(j-1)} \otimes G_{i,j-k+1}.$$

(17)

The two others constant matrices have the following simpler expressions :

$$\mathcal{B} = \begin{bmatrix} G_{1,0} & G_{2,0} & \cdots & G_{m,0} \\ 0 & 0 & \cdots & 0 \\ \vdots & \vdots & \vdots & \vdots \end{bmatrix} = \begin{bmatrix} G_0 \\ 0 \\ \vdots \end{bmatrix}, \quad \mathcal{C} = [I_n \quad 0 \quad \ldots].$$

(18)

We must note that, from $G_{00} = 0$, the matrix $\mathcal{A}_0$ is a block-upper triangular matrix :

$$j > 1, \quad A_{j,j-1}^{(0)} = 0,$$

(19)

and the initial condition on (15), is obviously given by :

$$X^0 = \left[x^{0[1]T}, x^{0[2]T}, \ldots, x^{0[k]T}, \ldots \right]^T. \tag{20}$$

A bilinear approximating system of the non linear system can be performed by a simple truncation, at a given order, of the vector X and the infinite dimensional matrices. The accuracy increase with the chosen order.

3 Calculus of the generating series

To calculate the generating series of the non linear system (5) we can apply the formulas obtained in the bilinear case, but now, in the case of the infinite dimensional bilinear realization (15). From the local equivalence, and the unicity of the generating series, it comes directly that in a formal way :

$$\mathcal{X} = \mathcal{C}(\mathcal{I} - \sum_{i=0}^{m} \mathcal{A}_i \alpha_i)^{-1}(X^0 + \mathcal{B}\alpha), \tag{21}$$

where $\mathcal{X}$ is the generating series associated to the state vector x of (5), and $\mathcal{I}$ is the infinite dimensional identity matrix. From the discussion of the previous part, the development of this expression leads to the calculation of two kinds of terms :

$$\begin{aligned} \mathcal{X}^u_{j_\nu,\ldots,j_0} &= \mathcal{C}\mathcal{A}_{j_\nu} \cdots \mathcal{A}_{j_1}\mathcal{A}_{j_0}\mathcal{B}, \quad \mathcal{X}^u = G_0, \\ \mathcal{X}^0_{j_\nu,\ldots,j_0} &= \mathcal{C}\mathcal{A}_{j_\nu} \cdots \mathcal{A}_{j_1}\mathcal{A}_{j_0}X^0, \quad \mathcal{X}^0 = x^0. \end{aligned} \tag{22}$$

In the linear output case, $y = Cx$, the generating series $\mathcal{G}$ will be obtained simply as $\mathcal{G} = C\mathcal{X}$.

3.1 Calculus of the first term

From the structure of $\mathcal{B}$ and $\mathcal{C}$, the matrix :

$$\mathcal{X}^u_{j_\nu,\ldots,j_0} = \mathcal{C}\mathcal{A}_{j_\nu} \cdots \mathcal{A}_{j_1}\mathcal{A}_{j_0}\mathcal{B}, \tag{23}$$

is a finite dimensional matrix, obtained by successive products of block Hessenbergn matrices. It leads directly to the following expressions for the first matrices :

- $i \in \{0, \ldots, m\}$:

$$\mathcal{X}^u_i = A^{(i)}_{11} G_0; \tag{24}$$

- $i, j \in \{0, \ldots, m\}^2$:

$$\mathcal{X}^u_{ij} = \left[A^{(i)}_{11} A^{(j)}_{11} + A^{(i)}_{12} A^{(j)}_{21} \right] G_0; \tag{25}$$

- $i, j, k \in \{0, \ldots, m\}^3$:

$$
\begin{aligned}
\mathcal{X}^u_{ijk} = [&A^{(i)}_{11} A^{(j)}_{11} A^{(k)}_{11} + A^{(i)}_{11} A^{(j)}_{12} A^{(k)}_{21} + \\
&A^{(i)}_{12} A^{(j)}_{21} A^{(k)}_{11} + A^{(i)}_{12} A^{(j)}_{22} A^{(k)}_{21} + \\
&A^{(i)}_{13} A^{(j)}_{32} A^{(k)}_{21}] G_0;
\end{aligned}
\tag{26}
$$

- $i, j, k, l \in \{0, \ldots, m\}^4$:

$$
\begin{aligned}
\mathcal{X}^u_{ijkl} = [A^{(i)}_{11} \ \ (&A^{(j)}_{11} A^{(k)}_{11} A^{(l)}_{11} + A^{(j)}_{11} A^{(k)}_{12} A^{(l)}_{21} + \\
&A^{(j)}_{12} A^{(k)}_{21} A^{(l)}_{11} + A^{(j)}_{12} A^{(k)}_{22} A^{(l)}_{21} + \\
&A^{(j)}_{13} A^{(k)}_{32} A^{(l)}_{21}) + \\
A^{(i)}_{12} \ \ (&A^{(j)}_{21} A^{(k)}_{11} A^{(l)}_{11} + A^{(j)}_{21} A^{(k)}_{12} A^{(l)}_{21} + \\
&A^{(j)}_{22} A^{(k)}_{21} A^{(l)}_{11} + A^{(j)}_{22} A^{(k)}_{22} A^{(l)}_{21} + \\
&A^{(j)}_{23} A^{(k)}_{32} A^{(l)}_{21}) + \\
A^{(i)}_{13} \ \ (&A^{(j)}_{32} A^{(k)}_{21} A^{(l)}_{11} + A^{(j)}_{32} A^{(k)}_{22} A^{(l)}_{21} + \\
&A^{(j)}_{33} A^{(k)}_{32} A^{(l)}_{21}) + \\
A^{(i)}_{14} \ \ (&A^{(j)}_{43} A^{(k)}_{32} A^{(l)}_{21})] G_0;
\end{aligned}
\tag{27}
$$

- etc...

3.2 Calculus of the second term

We must note here that only the previous term corresponding to a nul initial condition can be exactly calculated, indeed the second one can be only known from a truncation of X^0. In order to calculate $\mathcal{X}^0_{j_\nu, \ldots, j_0}$, we will decompose it as :

$$
\mathcal{X}^0_{j_\nu, \ldots, j_0} = \sum_{\mu=1}^{\infty} \mathcal{X}^{0,\mu}_{j_\nu, \ldots, j_0},
\tag{28}
$$

where :

$$
\mathcal{X}^{0,\mu}_{j_\nu, \ldots, j_0} = \mathcal{C} \mathcal{A}_{j_\nu} \cdots \mathcal{A}_{j_1} \mathcal{A}_{|,} X^{0,\mu},
\tag{29}
$$

where $X^{0,\mu}$ is X^0 with all $x^{0[k]}$ equal to 0 except $x^{0[\mu]}$. It is obvious that we obtain then :

- $i \in \{0, \ldots, m\}$:

$$
\mathcal{X}^{0,\mu}_i = A^{(i)}_{1,\mu} x^{0[\mu]};
\tag{30}
$$

- $i, j \in \{0, \ldots, m\}^2$:

$$\mathcal{X}_{i,j}^{0,\mu} = \left[A_{11}^{(i)} A_{1,\mu}^{(j)} + A_{12}^{(i)} A_{2,\mu}^{(j)} + \cdots + A_{1,\mu+1}^{(i)} A_{\mu+1,\mu}^{(j)} \right] x^{0[\mu]}; \qquad (31)$$

- $i, j, k \in \{0, \ldots, m\}^3$:

$$
\begin{aligned}
\mathcal{X}_{i,j,k}^{0,\mu} =\ & [A_{11}^{(i)}(A_{11}^{(j)} A_{1,\mu}^{(k)} + A_{12}^{(j)} A_{2,\mu}^{(k)} + \cdots + A_{1,\mu+1}^{(j)} A_{\mu+1,\mu}^{(k)}) + \\
& A_{12}^{(i)}(A_{21}^{(j)} A_{1,\mu}^{(k)} + A_{22}^{(j)} A_{2,\mu}^{(k)} + \cdots + A_{2,\mu+1}^{(j)} A_{\mu+1,\mu}^{(k)}) + \\
& A_{13}^{(i)}(A_{32}^{(j)} A_{2,\mu}^{(k)} + A_{33}^{(j)} A_{3,\mu}^{(k)} + \cdots + A_{3,\mu+1}^{(j)} A_{\mu+1,\mu}^{(k)}) + \\
& A_{14}^{(i)}(A_{43}^{(j)} A_{3,\mu}^{(k)} + \cdots + A_{4,\mu+1}^{(j)} A_{\mu+1,\mu}^{(k)}) + \\
& \vdots \\
& A_{1,\mu+1}^{(i)}(A_{\mu+1,\mu}^{(j)} A_{\mu,\mu}^{(k)} + A_{\mu+1,\mu+1}^{(j)} A_{\mu+1,\mu}^{(k)}) + \\
& A_{1,\mu+2}^{(i)} A_{\mu+2,\mu+1}^{(j)} A_{\mu+1,\mu}^{(k)}] x^{0[\mu]};
\end{aligned}
\qquad (32)
$$

- etc...

We must remark that for $\mu = 1$, $\mathcal{X}_{j_\nu,\ldots,j_0}^{0,1}$ can be obtained from $\mathcal{X}_{j_\nu,\ldots,j_0}^{u}$ by replacing G_0 by x^0.

4   Non linear output case

Let us consider now the more general case where to the state space equation (5), is associated an analytic output function given by :

$$y = \sum_{i=1}^{\infty} H_i x^{[i]}, \qquad (33)$$

where the H_i are constant matrices. In order to calculate the generating series of this new system, two points of view can be proposed. The first one consists in writing :

$$\mathcal{G} = \sum_{i=1}^{\infty} H_i \mathcal{X}^i, \qquad (34)$$

where $\mathcal{X}^i$ is the generating series of $x^{[i]}$. This last one can be calculated from $\mathcal{X}$ with the use of the shuffle product of generating series [ROTELLA, 1990], but this method leads to complicated algebraic manipulations when i is large. In our context it is preferable to calculate the generating series from a bilinear realization which is given by :

$$\dot{X} = \mathcal{A}_0 X + \sum_{i=1}^{m} \mathcal{A}_i X u_i(t) + \mathcal{B} u(t), \tag{35}$$
$$y = \mathcal{H} X,$$

where :

$$\mathcal{H} = [\, H_1 \quad H_2 \quad \ldots \quad H_k \quad \ldots \,]. \tag{36}$$

The generating series is then of the form :

$$\mathcal{G} = \mathcal{H}(\mathcal{I} - \sum_{i=0}^{m} \mathcal{A}_i \alpha_i)^{-1}(X^0 + \mathcal{B}\alpha). \tag{37}$$

To have a shorter explanation, we can see how is calculated $\mathcal{G}$ with $X^0 = 0$. It then comes out the following first terms for $\mathcal{G}$:

$$
\begin{aligned}
\mathcal{G} = \quad & H_1 G_0 \alpha + \sum_{i=0}^{m}[H_1 A_{11}^{(i)} + H_2 A_{21}^{(i)}]G_0 \alpha_i \alpha \\
& + \sum_{i,j=0}^{m} [H_1(A_{11}^{(i)} A_{11}^{(j)} + A_{12}^{(i)} A_{21}^{(j)}) + H_2(A_{21}^{(i)} A_{11}^{(j)} + A_{22}^{(i)} A_{21}^{(j)}) + \\
& \quad H_3 A_{32}^{(i)} A_{21}^{(j)}]G_0 \alpha_i \alpha_j \alpha \\
& + \cdots
\end{aligned}
\tag{38}
$$

As we can see, these expressions need not extra calculations with respect to the case of a linear output function, because the terms appearing in $\mathcal{G}$ can be obtained directly from the terms $\mathcal{X}_{j_\nu,\ldots,j_0}^u$ by simply replacing $A_{1,k}^{(j_\nu)}$ by H_k. This important remark is also available for the terms which are purely dependant on the initial condition. Then the calculus of the generating series in the general case of a non linear output function can be performed through the subroutines constructed for the linear output function case.

5 Conclusion

We have seen that, in the analytic case, the generating series of a non linear system can be calculated only by matric manipulations, as in the bilinear case. More precisely, only subroutines to calculate products of Hessenberg matrices are needed. A further application of this work will be to construct an analytic realization of a given series in noncommutative variables, this will be done in another paper.

6 References

BREWER, J. W. , "Kronecker Products and Matrix Calculus in System Theory", *IEEE Trans. Circuits and Systems*, **CAS-25**, 9, 772–781, 1978.

BROCKETT, R.W. , "On the Algebraic Structure of Bilinear Systems", in *Theory and Applications of Variable Structure Systems*, ed. MOHLER R.R. and RUBERTI A., Academic Press, 153–168, 1972.

BROCKETT, R.W. , "Volterra Series and Geometric Control Theory", *Automatica*, **12**, 167–176, 1976.

CARLEMAN, T. , "Application de la théorie des équations intégrales singulières aux équations différentielles de la dynamique", *T. Ark. Mat. Astron. Fys.*, **22**, 1–30, 1932.

FLIESS, M., LAMNABHI, M., LAMNABHI-LAGARRIGUE, F. , "An Algebraic Approach to Nonlinear Expansions",*IEEE Trans. Circuits and Systems*, **CAS-30**, 554–570, 1983.

FLIESS, M., NORMAND-CYROT, D. , "La propriété d'approximation des systèmes réguliers (ou bilnéaires)", *Outils et Modèles Mathématiques pour l'Automatique, l'Analyse de Systèmes et le Traitement du Signal*, Editions du CNRS, 1, 379–384, 1981.

HESPEL, C., JACOB, G. , "Approximations of Nonlinear Systems by Bilinear ones", in *Algebraic and Geometric Methods in Nonlinear Control Theory*, ed. FLIESS, M. and HAZEWINKEL, M., Reidel, 511–520, 1986.

ROTELLA, F. , "Shuffle Product of Generating Series", *Algabraic and Computing Treatment of Noncommutative Power Series*, Ed G. Jacob and C. Reutenauer, Theoret. Comput. Sci., 1990, to appear.

ROTELLA, F., ZAMBETTAKIS, I., DAUPHIN-TANGUY, G., BORNE, P. , "Modelling and Simulation of Non Linear Systems Based on Infinite Bilinear Realizations", Proc. Symp. Simulation of Control Systems, Vienne, 413–416, 1986.

RUGH, W.J. , *Non Linear System Theory, The Volerra-Wiener approach*, John Hopkins University Press, 1981.

SUSSMAN, H., , "Semigroup representations, bilinear approximation of input-output maps and generalized inputs", *Mathematical Systems Theory*, Lecture Notes in Econ. Math. Syst., **131**, 172–191, Springer-Verlag, 1976.

VETTER, W., J., , "Matrix Calculus Operations and Taylor Expansions", *SIAM Review*, 15, 2, 352–369, 1973.

POINCARE NORMAL FORM AND CARLEMAN LINEARIZATION

(SYSTEM OF NON LINEAR ORDINARY DIFFERENTIAL EQUATIONS)

J. DELLA DORA
L.STOLOVITCH

Laboratoire LMC - IMAG
B.P 53 X
38041 - GRENOBLE CEDEX
FRANCE

ABSTRACT

The goal of this paper is to present a generalization of the POINCARE - DULAC theorem about normal forms of vectors fields near an equilibrium point.
We show that a derivation whose terms are formal power series in non commutative variables without constant terms can be transformed by a change of variables into a normal form whose "non - linear " terms are "resonnant".

This research was supported by the french PRC Math-Info and the IMAG project DESIR - II.

1) NOTATIONS

Let $\underline{X}$ be a non empty set, of finite cardinality ; denoted by :

$$\{x_1, x_2....x_n\}$$

its elements . $\underline{X}^*$ shall be the free monoid generated by $\underline{X}$. The composition law on $\underline{X}$ is the concatenation and e shall be the neutral element of $\underline{X}$ (that is the empty word).The unique morphism from $\underline{X}^*$ into the integers whose value is 1 on the letters of $\underline{X}$ shall be called the length .

$$| \ | : X^* \rightarrow IN$$

Let m be an element of the free monoid $\underline{X}^*$. We define the *content* of m as :

$$C(m) = (\sigma_1, \cdots, \sigma_n)$$

where σ_i is the number of occurences of the letter x_i in the word m.

If k is a commutative field (of characteristic 0), $k< \underline{X} >$ shall denote the free associative algebra on k generated by $\underline{X}$; it's the algebra of polynomials in non commutative variables $\underline{X}$. $k<< \underline{X} >>$ denotes the algebra of formal power series in non commutative variables on k generated by $\underline{X}$.

H(i) denotes the homogeneous polynomials (of i - length) elements of $k < \underline{X} >$, and $k_i < \underline{X} >$ stands for the subspace of $k< \underline{X} >$ of polynomials of length less or equal to i.

If f belongs to $k< \underline{X} >$ we have two useful notations :

$$f = \sum_{m \in X^*} < m , f > m \ \text{ where } < m , f > \text{ is the coefficient of f in the word m.}$$

$$f = \sum_{i = 0}^{+\infty} f_i \ \text{ where } f_i \in H(i)$$

J^i denotes the space of i - jets. It is defined by the quotient $k<<\underline{X}>>/\mathcal{M}^{i+1}$, where $\mathcal{M}$ is the maximal ideal of $k<<\underline{X}>>$. J^i as an algebra structure by means of a product defined by obvious troncature.

All these notations are well known in the context of formal power series in non commutative variables, but for our purpose it is better to use tensor product notation. In these notations a word

$$x_{i_1}...x_{i_p} \text{ will be denoted by } x_{i_1} \otimes ... \otimes x_{i_p}.$$

The introduction of this notation gives a very appropriate formalism for our purpose.

2) PROPERTIES OF THE TENSOR PRODUCT

Let us remember properties of tensor products.

First of all, if A is a matrix of order p x q and B another matrix of order r x s then $A \otimes B$ denote the pr x qs matrix :

$$A \otimes B = \begin{pmatrix} a_{11} B & \dots & a_{1q} B \\ a_{p1} B & \dots & a_{pq} B \end{pmatrix}$$

one of the properties of this product is the following:

If A, B, C, D are suitable matrices then:

$$AB \otimes CD = (A \otimes C)(B \otimes D)$$

We shall denote by:

$H(0) = (0)$

$H(1) = k\, x_1 \oplus \dots \oplus k\, x_n$

Then if we consider the following vector:

$$m_1 = \begin{pmatrix} x_1 \\ \\ x_n \end{pmatrix}$$

built with the letters of $\underline{X}$, an element of $H(1)$ can be written in the form:

$$h = a^t\, m_1$$

where a is vector of k^r , and $a^t = (a_1, \dots, a_n)$ is the transpose of a .

With this notation we introduce the following sequence of vectors:

$$m_2 = m_1 \otimes m_1 \;,\; m_3 = m_2 \otimes m_1 \;,\; \dots$$

it is very important to note that, due to the form of the construction:

$$m_3 = m_2 \otimes m_1 = m_1 \otimes m_2$$

It is clear that $H(i)$ is a k vector space whose basis is built with the elements of m_i and so is of dimension n^i .In any case we take as a convention that $m_0 = 1$.

We add two more definitions before introducing the main result.

If q is a positive integer, and m an element of $\underline{X}^*$ such that $|m| = q$, we define

$$v_q(m) \in N$$

as the index of m in the vector

$$(m_1)^{\otimes q} = m_q$$

(i.e the index of m in the ordered basis of terms of length q).

Let Q be an n-dimensonal vector of integer whose norm q is greater than two.

We then define the set :

$$Res(Q) = \left\{ i \in N_q / v_q(m) = i \text{ and } C(m) = Q \text{ for } m \in X^* \right\}$$

(that is, given a m in $\underline{X}^*$ such that C(m)=Q, Res(Q) is the set of indices of all the words obtained by permutation of the letters of m.).

Examples :

Let n = 2, Q = (1, 2). Then, the following elements :

$$x_1 \otimes x_2 \otimes x_2 ,$$
$$x_2 \otimes x_1 \otimes x_2 ,$$
$$x_2 \otimes x_2 \otimes x_1$$

have their content equal to Q, and since our ordered base of elements of length 3 is

$$(x_1 \otimes x_1 \otimes x_1, \ x_1 \otimes x_1 \otimes x_2, \ x_1 \otimes x_2 \otimes x_1, \ x_1 \otimes x_2 \otimes x_2,$$
$$x_2 \otimes x_1 \otimes x_1, \ x_2 \otimes x_1 \otimes x_2, \ x_2 \otimes x_2 \otimes x_1, \ x_2 \otimes x_2 \otimes x_2 \ )$$

then the set Res(Q) is equal to :

$$\{ 4, 6, 7 \}$$

We recall the following well known result

THEOREM I:

> Let A and B be two square matrices, (possibly of different order) .
> If A and B do not have any common eigenvalue then for evey C there exists a unique matrix X solution of the following equation:
>
> $$A X - X B = C$$

3) THE AUTOMORPHISM GROUP OF k<<X>>

We plan to study these automorphisms using the notations of the preceding paragraph. If f belongs to k<< $\underline{X}$ >> we can write it :

$$f = \sum_{i=0}^{+\infty} m_i^t f^i$$

(f^i belongs to k^{n^i}).

If F is an element of the automorphic group, than it is characterised by its action on the letter x_i :

$$F(x_i) = \sum_{j=0}^{+\infty} f_i^j m_j \quad i = 1,\ldots, n$$

Then we have:

$$(F(x_1) ,\ldots,F(x_n)) = \sum_{j=0}^{+\infty} m_j^t \left(f_1^j,\ldots,f_n^j\right)^t = \sum_{j=0}^{+\infty} m_j^t F_{j,1}$$

It is well known that we have the following lemma

Lemma:

> F belongs to Aut (k<< $\underline{X}$ >>) if and only if $F_{0,1} = 0$ and det $(F_{11}) \neq 0$

We can also see the previous formula as the action of F on m_1^t:

$$F \left(m_1^t\right) = \sum_{j=1}^{+\infty} m_j^t F_{j,i}$$

(remember that $F_{j,i}$ is a matrix of order n^j x n)

This notation is very useful ifor calculating the $F\left(m_j^t\right)$, as an example let us calculate:

$$F\left(m_2^t\right) = F\left(m_1^t \otimes m_1^t\right) = \left(\sum_{j=1}^{+\infty} m_j^t F_{j,1}\right) \otimes \left(\sum_{l=1}^{+\infty} m_l^t F_{l,1}\right)$$

$$= \sum_{j=1}^{+\infty} \sum_{l=1}^{+\infty} \left(m_j^t F_{j,1}\right) \otimes \left(m_l^t F_{l,1}\right)$$

we have noticed in the first paragraph that, this expression can be written as:

$$\sum_{j=1}^{+\infty} \sum_{l=1}^{+\infty} \left(m_j^t \otimes m_l^t\right) \left(F_{j,1} \otimes F_{l,1}\right)$$

which is equal to:

$$\sum_{j=1}^{+\infty} \sum_{l=1}^{+\infty} \left(m_{j+l}^t\right) \left(F_{j,1} \otimes F_{l,1}\right)$$

if $r = 1 + j$ then:

$$\sum_{r=1}^{+\infty} (m_r^1) \left(\sum_{j=1}^{r-1} F_{j,1} \otimes F_{r-j,1} \right)$$

Setting :

$$\sum_{r=2}^{+\infty} (m_r^1) \; F_{r,2} \; \text{where} \; F_{r,2} = \left(\sum_{j=1}^{r} F_{j,1} \otimes F_{r-j,1} \right)$$

The generalisation of this kind of calculation is easy, notice that we can also write:

$$F(m_j^1) = F(m_1^1) \otimes \ldots \otimes F(m_1^1)$$

$$= \sum_{i_1}^{+\infty} m_{i_1}^1 \; F_{i_1 \, 1} \otimes \ldots \otimes \sum_{i_j}^{+\infty} m_{i_j}^1 \; F_{i_j \, 1}$$

$$= \sum_{r=j}^{+\infty} m_r^1 \left(\sum_{\substack{j \\ \sum_{s=1} i_s}} F_{i_1 \, 1} \otimes \ldots \otimes F_{i_j \, 1} \right)$$

we have a clear description of the algorithms for the F calculations.

Another way to proceed for these calculations, is to write :

$$\text{For } k \geq 2, \; 1 \leq j \leq k : \; F(m_k^1) = F(m_j^1 \otimes m_{k-j}^1) =$$

$$\sum_{i=k}^{\infty} m_i^1 \left(\sum_{m=1}^{i-1} F_{i-m,k-j} \otimes F_{m,j} \right)$$

Consequently, we have the following formula :

$$\forall \; i \geq k, \; \forall \; 1 \leq j \leq k, \quad F_{i,k} = \sum_{m=1}^{i-1} F_{i-m,k-j} \otimes F_{m,j} \qquad (1_j)$$

Matricial representation of F

We can represent F by an infinite (formal) matrix T (F), which represents the action of F on the m_i basis. The previous formula describe the inner structure of this matrix. In fact we have the following lemma :

lemma:

T (F) is a lower block triangular matrix. On its diagonal we have the $n_j \times n_j$ matrix F_{jj} .

Notice that :

$$F_{jj} = F_{11} \otimes \ldots \otimes F_{11} = (F_{11})^{\otimes j}$$

is the tensorial j-power.

We know that:

$$\det\left((F_{11})^{\otimes j}\right) = \det(F_{11})^{\sum_{i=1}^{j} n^i}$$

and so all the diagonal blocks of T (F) are regular.

The representation of T (F) is :

$$T(F) = \begin{pmatrix} F_{11} & 0 & \ldots \\ F_{21} & F_{22} & \ldots \\ \ldots & \ldots & \ldots \end{pmatrix}$$

4) THE ALGEBRA DER (k<< $\underline{X}$ >>)

A derivation D of k<< X >> is defined by its action on the letter x_i with the following formula:

$$D(x_i) = \sum_{j=1}^{+\infty} d_i^j \, m_j$$

and so we can associate an infinite matrice T (D) representing the action of D on the m_i basis. For this we introduce the same notation as before:

$$D\left(m_i^l\right) = \sum_{j=1}^{+\infty} m_j^l \, D_{j\,1}$$

Let's calculate for example $D\left(m_2^l\right)$ which is the derivation of $m_1^l \otimes m_1^l$. The extension of D to the action on a tensor product is defined by :

$$D\left(m_1^l \otimes m_1^l\right) = D\left(m_1^l\right) \otimes m_1^l + m_1^l \otimes D\left(m_1^l\right)$$

which is equal to:

$$= \left(\sum_{j=1}^{+\infty} \left(m_j^l \, D_{j\,1}\right) \otimes m_1^l \right) + \left(\sum_{l=1}^{+\infty} m_1^l \otimes \left(m_l^l \, D_{l\,1}\right) \right)$$

In order to transform these expressions we use the following important tool:

$$\left(m_j^l \, D_{j\,1}\right) \otimes m_1^l = \left(m_j^l \, D_{j\,1}\right) \otimes m_1^l \, I_n$$
$$= \left(m_j^l \otimes m_1^l\right) \left(D_{j1} \otimes I_n\right)$$
$$= m_j^l \left(D_{j1} \otimes I_n\right)$$

So we obtain:

$$D\left(m_2^l\right) = \sum_{j=1}^{+\infty} m_{j+1}^l \left(D_{j\,1} \otimes I_n + I_n \otimes D_{j\,1}\right)$$

$$= \sum_{j=2}^{+\infty} m_j^l \, D_{j\,2}$$

where we introduce $D_{j+1,\,2} = D_{j,\,1} \otimes I_n + I_n \otimes D_{j,\,1}$

More generally, we have the following result :

$$D(m_k^l) = \sum_{i=k}^{\infty} m_i^l \, D_{i,\,k}$$

By writing that $D(m_k^l) = D(m_{k-j}^l \otimes m_j^l)$, we obtain the following formula :

$$\forall \; i \geq k, \; \forall \; 1 \leq j \leq k, \quad D_{i,\,k} = D_{i+j-k,\,j} \otimes I_{n^{k-j}} + I_{n^j} \otimes D_{i-j,\,k-j} \qquad (2_j)$$

The structure of $T(D)$ as a block lower triangular matrix is clear.

Moreover, we have the following

Lemma:

> If v is an eigenvector of the matrix A, associated to the eigenvalue l and if w is another eigenvector associated to m then $v \otimes w$ is an eigenvector of $A \oplus A$ associated to the eigenvalue $l + m$.

Since $D_{k+1, k+1} = D_{1,1} \otimes I_{n^k} + I_n \otimes D_{k,k}$, it can be easily seen that :

$$D_{k, k} = D_{1, 1} \oplus \cdots \oplus D_{1, 1} = D_{1, 1}^{\oplus k} \quad \text{where } \oplus \text{ is the Krönecker sum.}$$

So, if $\lambda_1, \cdots, \lambda_n$ denote the eigenvalues of $D_{1, 1}$, then those of $D_{k, k}$ are $\langle \lambda, K \rangle$

$$\text{where } K^t = (k_1, \cdots, k_n) \text{ with } |K| \equiv \sum_{i=1}^n k_i = k \text{ and } \lambda^t = (\lambda_1, \cdots, \lambda_n)$$

5) POINCARE NORMAL FORM OF A DERIVATION

The problem and the method we want to speak about, have their roots in the classical and seminal work of Poincaré [1]. Since those old days, this method has been studied and explained several times .The path we plan to follow has its origins in a classical work of Carleman [2] (The Carleman linearization process). Bellman [3] and others have made severals suggestions in this direction and papers by Steeb and Wilhelm [4] show that a lot of research is being done on this subject.

In order to build Computer Algebra programs for a general study of non linear differential equations, we have the feeling that it's important to gather all these scatered results, to give a general framework and of course to try to add new results.

The aim of this study is to answer the following question:

D being a derivation of k<< X >>, is it possible to find f belonging to Aut (k<< X >>), such that it's action on D defined by:

$$f*D \equiv f^{-1} o D o f$$

is as simple as possible.

This is the classical : **normal form problem** .

As we shall see this problem has (in the formal sense of the word) a solution that we call the POINCARE-DULAC normal form.

Our approach to the problem use intensively the infinite dimensional matrix formalism, a great number of other formalisms can be used (inductive limit in Lie Algebra, formal vector fields,...). But it appears that our formalism fits well the algorithmic point of view.

The idea is to see the previous problem as a JORDAN normal form problem, that is we shall build f so that $T (f^{-1} o D o f)$ is on JORDAN form.

But as we can write $T (f^{-1} o D o f) = T (f)^{-1} . T (D) . T (f)$ we have to build $T (f)$ so that the result is in desired form .

So, we have the following

Theorem: (Poincaré-Dulac normal form).

Let D be a derivation of k<< $\underline{X}$ >> whose value at 0 is 0.(i.e singular at 0)
Let D_{11} be the linear part of D and $T(A_{11})$ the Jordan's Form of $T(D_{11})$
Let $\lambda_1, \cdots, \lambda_n$ be the eigenvalues of $T(D_{11})$

Then, there exists an automorphism f of k<< $\underline{X}$ >> such that its action on the derivation D is :

$$f*D = A_{11} + \sum_{i=1}^{n} (\sum_{\substack{|Q| \geq 2, Q \in N^n \\ C(m)=Q, m \in H(|Q|) \\ (Q,\lambda)=\lambda_i, \alpha_m \in K}} \alpha_m \, m) \partial_i$$

That is , the matrix representation $A = T(f*D)$ of $f*D$ satisfies the following proprieties :

1) For any integer $q \geq 2$, A satisfies to :

For any $Q \in N^n$ such that $|Q| = q$, for any integer s , $1 \leq s \leq n$ then :

(P_q) :

$$\text{if} \left(Q, \lambda \right) \neq \lambda_s \text{ then } \forall j \notin \text{Res}(Q), \ a_{j,s}^{(q,1)} = 0$$

where $A = (A_{i,j})$, $A_{i,j} \in M_{n^i, n^j}(k)$ and $A_{i,j} = \left(a_{k,l}^{(i,j)} \right)_{1 \leq i \leq n^i, \ 1 \leq j \leq n^j}$

2) A_{11} is a Jordan Form

We shall prove, by induction on $k \geq 1$, that there is an automorphism f_k of J^k such that the representation A_k of f_k*D_k satisfies to $(P_i)_{1 \leq i \leq k}$ (if $k \geq 2$).
Its linear part has a Jordan canonical form. The theorem II will then be proved since $k<<X>>$ is the projective limit of the J^i's.

* $\underline{k = 1}$

Since the field k is closed, then is always possible to transform a matrix into its Jordan Form .Consequently, we can write :

$$A_{1,1} = F_{1,1}^{1} D_{1,1} F_{1,1}$$

where $A_{1,1}$ is a Jordan Form of $D_{1,1}$, $F_{1,1}$ being the transformation matrix.

* <u>We assume that the proposition is true at order k-1</u>

So, we have $F_{k-1} A_{k-1} = D_{k-1} F_{k-1}$.

Let

$$F_k = \begin{pmatrix} & & F_{k-1} & & & & 0 \\ & & & & & & \vdots \\ & & & & & & 0 \\ & F_{k,1} & \cdots & F_{k,k-1} & & F_{k,k} \end{pmatrix}$$

be a matrix representation of an automorphism on J^k those restriction to J^{k-1} is given by induction.

We define A_k and D_k in the same way.

We want to find the $A_{k,j}$'s and the $F_{k,j}$'s in order to solve the equation :
$$(3): F_k A_k = D_k F_k$$
where F_k (resp. A_k) is the matrix representation of the restriction to J^k
of an automorphism (resp. derivation). Moreover, A_k has to satisfy to $(P_i)_{2 \leq i \leq k}$
In fact, by induction, A_k already satisfies (P_i) $_{2 \leq i \leq k-1}$
Since A_k and D_k are derivations, the sub-matrices $A_{k,j}$ and $D_{k,j}$, $2 \leq j \leq k$, are known.
It is the same for F_k : the $F_{k,j}$'s are known, by induction, if $2 \leq j \leq k$.

The matricial equation (3) leads to the following equations:

$$(E_j): \sum_{l=j}^{k} D_{k,l} F_{l,j} = \sum_{l=j}^{k} F_{k,l} A_{l,j} \ , \ 1 \leq j \leq k$$

The problem is split into two steps :

1st step : Determination of the automorphism and its action on the derivation D

So, by setting $S_k = F^1_{k,k} F_{k,1}$, the equation (E_1) to solve becomes :

$$\boxed{(4): A_{k,k} S_k - S_k A_{1,1} = A_{k,1} + \alpha_k}$$

where $\alpha_k = F^1_{k,k} (\sum_{l=2}^{k-1} F_{k,l} A_{l,1} - \sum_{l=1}^{k-1} D_{k,l} F_{l,1})$

and, where S_k and $A_{k,1}$ are the unknown.
The equation (4) is the fundamental one.
We have to solve it in such a way that A_k satisfies (P_k).

Resolution of equation (4) :

1st case : There is no resonnance of order k

Then, $Sp(A_{1,1}) \cap Sp(A_{k,k}) = \emptyset$ since the eigenvalues of $A_{k,k}$ are the (Q, λ) with $|Q| = q$.
So, by theorem I, for any matrix $A_{k,1}$, there is an unique solution S_k of the equation (2).
In particular, for $A_{k,1} = 0$, there is a unique solution of the equation.
Obviously, the matrix A_k satisfies the condition (P_k).

2nd case : There are resonnances of order k

In this case, we can't say anything, a priori, about the existence of solutions of (2).
We have to look more precisely for what happens.

By developping the generic term of (2) and since the matrices $A_{1,1}$ and $A_{k,k}$ are lower triangular , we have :

$\forall i \in \text{Res}(Q), \forall j \in N_n,$

$$((Q, \lambda) - \lambda_j) s_{i,j}^{(k)} = a_{i,j}^{(k,1)} + (\sum_{p=j+1}^{n} s_{i,p}^{(k)} a_{p,j}^{(k,k)}) - (\sum_{p=1}^{i-1} a_{i,p}^{(k,k)} s_{p,j}^{(k)}) + \alpha_{i,j}^{(k)}$$

where $S_k = (s_{i,j}^{(k)})$ and $\alpha_k = (\alpha_{i,j}^{(k)})$

We do a top-down right-left computation of the $s_{i,j}^{(k)}$'s, so that the $a_{i,j}^{(k,1)}$'s satisfies (P_k). Starting at colum n and at row 1, we have :

$$((Q, \lambda) - \lambda_n) s_{1,n}^{(k)} = a_{1,n}^{(k,1)} + \alpha_{1,n}^{(k)} , \text{ where } Q \in N^n \text{ is such that } 1 \in \text{Res}(Q).$$

if $(Q, \lambda) \neq \lambda_n$, then we can take $a_{1,n}^{(k,1)} = 0$; we then compute $s_{1,n}^{(k)}$.

if this is not the case, then we set $a_{1,n}^{(k,1)} = -\alpha_{1,n}^{(k)}$; $s_{1,n}^{(k)}$ can be chosen arbitrarily.

Suppose we have computed all the $s_{i,j}^{(k)}$'s until $i = i_0$ and $j = j_0$.
So, we have :

$$((Q, \lambda) - \lambda_{j_0}) s_{i_0,j_0}^{(k)} = a_{i_0,j_0}^{(k,1)} + (\sum_{p=j_0+1}^{n} s_{i_0,p}^{(k)} a_{p,j_0}^{(k,k)}) - (\sum_{p=1}^{i_0-1} a_{i_0,p}^{(k,k)} s_{p,j_0}^{(k)}) + \alpha_{i_0,j_0}^{(k)}$$

where $i_0 \in \text{Res}(Q)$.

So, if $(Q, \lambda) \neq \lambda_{j_0}$, then we can set $a_{i_0,j_0}^{(k,1)} = 0$ and then compute $s_{i_0 j_0}^{(k)}$.
If this is not the case, then we have :

$$a_{i_0,j_0}^{(k,1)} = -(\sum_{p=j_0+1}^{n} s_{i_0,p}^{(k)} a_{p,j_0}^{(k,k)}) + (\sum_{p=1}^{i_0-1} a_{i_0,p}^{(k,k)} s_{p,j_0}^{(k)}) - \alpha_{i_0,j_0}^{(k)}$$

and $s_{i_0,j_0}^{(k)}$ can be chosen arbitaly.

Once the matrix $A_{k,1}$ is computed in this way, the matrix A_k satisfies (P_k).
This ends the induction.

2nd step : check up on the final n-1 equations

This is the verification of the identities defined by the n-1 equations (E_i), $2 \leq i \leq k$.
It is an interesting induction because we can understand the recursive construction of the matrices $F_{n,i}$ (for $i > 1$ and $i < n$).
We want to check up on the following equations :

$$(E_j) : \sum_{l=j}^{k} F_{k, l} A_{l, j} = \sum_{l=j}^{k} D_{k, l} F_{l, j} \quad , \quad 2 \leq j \leq k$$

We shall compute all the terms $F_{k, l} A_{l, j}$.

Using the formula (1_{j-1}), we have :

$$F_{k, l} = \sum_{m=1}^{k-1} F_{k-m, l-j+1} \otimes F_{m, j-1}$$

and, by the formula (1_l), we have :

$$F_{k, l} = \sum_{m=1}^{k-1} F_{k-m, l} \otimes F_{m, l-1}$$

but

$$A_{k, j} = A_{k-j+1, 1} \otimes I_{n^{j-1}} + I_n \otimes A_{k-1, j-1}$$

Consequently, we have :

$$F_{k, l} A_{l, j} = \sum_{m=1}^{k-1} \left(F_{k-m, l-j+1} \otimes F_{m, j-1} \right) \left(A_{l-j+1, 1} \otimes I_{n^{j-1}} \right) + \sum_{m=1}^{k-1} \left(F_{k-m, l} \otimes F_{m, l-1} \right) \left(I_n \otimes A_{l-1, j-1} \right)$$

and then, by the tensor product's properties, we have :

$$\sum_{l=j}^{k} F_{k, l} A_{l, j} = \sum_{m=1}^{k-1} \left(\left(\sum_{l=j}^{k} F_{k-m, l-j+1} A_{l-j+1, 1} \right) \otimes F_{m, j-1} \right) + \sum_{m=1}^{k-1} \left(F_{k-m, l} \otimes \left(\sum_{l=j}^{k} F_{m, l-1} A_{l-1, j-1} \right) \right)$$

We apply, now, the induction :

$$\sum_{l=j}^{k} F_{k-m, l-j+1} A_{l-j+1, 1} = \sum_{l=1}^{k-j+1} F_{k-m, l} A_{l, 1}$$

$$= \sum_{l=1}^{k-m} D_{k-m, l} F_{l, 1} \quad , \text{if } m \geq j-1$$

and :

$$\sum_{l=j}^{k} F_{m,l-1}A_{l-1, j-1} = \sum_{l=j-1}^{k-1} F_{m,l}A_{l, j-1}$$

$$= \sum_{l=j-1}^{m} D_{m,l}F_{l, j-1} \text{ , if } m \geq j-1$$

$$= 0 \quad \text{unless.}$$

Using the fact that the matrix are lower triangular, and the properties of the tensor product, we have :

$$\sum_{l=j}^{k} F_{k, l}A_{l, j} = \sum_{m=j-1}^{n-1} \sum_{l=1}^{n-m} \left(D_{n-m, 1} \otimes I_{n^m}\right)\left(F_{1,1} \otimes F_{m,j-1}\right) +$$

$$\sum_{m=j-1}^{n-1} \sum_{l=j-1}^{nm} \left(I_{n^{n-m}} \otimes D_{m, 1}\right)\left(F_{n-m,1} \otimes F_{l,j-1}\right)$$

We have, now, to put terms together correctly, in order to rebuild the sum. But, we have :

$$\sum_{m=j-1}^{n-1} \sum_{l=j-1}^{nm} \left(I_{n^{n-m}} \otimes D_{m, 1}\right)\left(F_{n-m,1} \otimes F_{l,j-1}\right) =$$

$$\sum_{m=j-1}^{n-1} \left(I_{n^{n-m}} \otimes D_{m, m}\right)\left(F_{n-m,1} \otimes F_{m,j-1}\right) +$$

$$\sum_{m=j-1}^{n-2} \left(I_{n^{n-m-1}} \otimes D_{m+1, m}\right)\left(F_{n-m-1,1} \otimes F_{m,j-1}\right) +$$

$$\sum_{m=j-1}^{n-3} \left(I_{n^{n-m-2}} \otimes D_{m+2, m}\right)\left(F_{n-m-2,1} \otimes F_{m,j-1}\right) +$$

$$\vdots$$

$$\sum_{m=j-1}^{j-1} \left(I_{n^{j-m}} \otimes D_{m+(n-1)-(j-1), m}\right)\left(F_{j-m,1} \otimes F_{m,j-1}\right)$$

Moreover we have :

$$\sum_{m=j-1}^{n-1} \sum_{l=1}^{n-m} \left(D_{n-m, 1} \otimes I_{n^m}\right)\left(F_{1,1} \otimes F_{m,j-1}\right) =$$

$$\sum_{m=j-1}^{n-1} \left(D_{n-m, n-m} \otimes I_{n^m}\right)\left(F_{n-m,1} \otimes F_{m,j-1}\right) +$$

$$\sum_{m=j-1}^{n-2} \left(D_{n-m,\,n-m-1} \otimes I_{n^m}\right)\left(F_{n-m-1,1} \otimes F_{m,j-1}\right) +$$

$$\sum_{m=j-1}^{n-3} \left(D_{n-m,\,n-m-2} \otimes I_{n^m}\right)\left(F_{n-m-2,1} \otimes F_{m,j-1}\right) +$$

$$\vdots$$

$$\sum_{m=j-1}^{j-1} \left(D_{n-m,\,j-m} \otimes I_{n^m}\right)\left(F_{j-m,1} \otimes F_{m,j-1}\right)$$

But, since we have the formula :

$$\forall\ i \geq k,\ \forall\ 1 \leq j \leq k, \quad D_{i,\,k} = D_{i+j-k,\,j} \otimes I_{n^{k-j}} + I_{n^j} \otimes D_{i-j,\,k-j} \qquad (2_j)$$

we can easely rebuild the sum :

$$\sum_{l=j}^{k} F_{k,\,l} A_{l,\,j} =$$

$$\sum_{m=j-1}^{n-1} D_{n,\,n}\left(F_{n-m,1} \otimes F_{m,j-1}\right) +$$

$$\sum_{m=j-1}^{n-2} D_{n,\,n-1}\left(F_{n-m-1,1} \otimes F_{m,j-1}\right) +$$

$$\sum_{m=j-1}^{n-3} D_{n,\,n-2}\left(F_{n-m-2,1} \otimes F_{m,j-1}\right) +$$

$$\vdots$$

$$\sum_{m=j-1}^{j-1} D_{n,\,n-j}\left(F_{j-m,1} \otimes F_{m,j-1}\right)$$

But, by the formula (1_{j-1}), we have

$$\forall\ i \geq k,\ \forall 1 \leq j \leq k, \quad F_{i,\,k} = \sum_{m=1}^{i-1} F_{i-m,\,k-j} \otimes F_{m,j} \qquad (1_j)$$

So, it finally becomes:

$$(E_j): \quad \sum_{l=j}^{k} F_{k,\,l} A_{l,\,j} = \sum_{l=j}^{k} D_{k,\,l} F_{l,\,j}\ ,\ 2 \leq j \leq k$$

EXEMPLE:

Let D be the derivation defined by :

$$\left(2\,x_1 + x_1 \otimes x_2 + x_2 \otimes x_1 + x_2 \otimes x_2\right)\partial_1 + \left(x_2 + x_1 \otimes x_1\right)\partial_2$$

The resonnance equation is : $2\,m_1 + m_2 = 2$ or $2\,m_1 + m_2 = 1$ the only solution is given by $2 \cdot 0 + 2 \cdot 1 = 2$.
So that the only resonnant monomial is given by :

$$x_2 \otimes x_2 \,\partial_1$$

Let us start the computation for the 2 - jets :

$$F_2\,A_2\,. = D_2F_2$$

$$
\begin{pmatrix}
1 & 0 & 0 & 0 & 0 & 0 \\
0 & 1 & 0 & 0 & 0 & 0 \\
F_{11} & F_{12} & 1 & 0 & 0 & 0 \\
F_{21} & F_{22} & 0 & 1 & 0 & 0 \\
F_{31} & F_{32} & 0 & 0 & 1 & 0 \\
F_{41} & F_{42} & 0 & 0 & 0 & 1
\end{pmatrix}
\begin{pmatrix}
2 & 0 & 0 & 0 & 0 & 0 \\
0 & 1 & 0 & 0 & 0 & 0 \\
a_{11} & a_{12} & 1 & 0 & 0 & 0 \\
a_{21} & a_{22} & 0 & 1 & 0 & 0 \\
a_{31} & a_{32} & 0 & 0 & 1 & 0 \\
a_{41} & a_{42} & 0 & 0 & 0 & 1
\end{pmatrix}
=
$$

$$
\begin{pmatrix}
2 & 0 & 0 & 0 & 0 & 0 \\
0 & 1 & 0 & 0 & 0 & 0 \\
0 & 1 & 4 & 0 & 0 & 0 \\
1 & 0 & 0 & 3 & 0 & 0 \\
1 & 0 & 0 & 0 & 3 & 0 \\
1 & 0 & 0 & 0 & 0 & 2
\end{pmatrix}
\begin{pmatrix}
1 & 0 & 0 & 0 & 0 & 0 \\
0 & 1 & 0 & 0 & 0 & 0 \\
F_{11} & F_{12} & 1 & 0 & 0 & 0 \\
F_{21} & F_{22} & 0 & 1 & 0 & 0 \\
F_{31} & F_{32} & 0 & 0 & 1 & 0 \\
F_{41} & F_{42} & 0 & 0 & 0 & 1
\end{pmatrix}
$$

We have:

$$F_{12} + a_{12} = 1 + 4\,F_{12}$$
$$F_{22} + a_{22} = 3\,F_{22}$$
$$F_{32} + a_{32} = 3\,F_{32}$$
$$F_{42} + a_{42} = 2\,F_{42}$$

So

$$a_{i2} = 0 \; i = 1, ..,4$$
$$F_{12} = -\frac{1}{3}, F_{i2} = 0 \; i = 2, 3, 4$$

For the first colum, we have

$$2\,F_{11} + a_{11} = 4\,F_{11}$$
$$2\,F_{21} + a_{21} = 1 + 3\,F_{21}$$
$$2\,F_{31} + a_{31} = 1 + 3\,F_{31}$$
$$2\,F_{31} + a_{31} = 1 + 2\,F_{41}$$

So :

$$a_{i\,1} = 0 \quad i = 1, 2, 3$$
$$F_{1\,1} = 0, \; F_{2\,1} = F_{3\,1} = -1$$
$$a_{4\,1} = 1$$

Consequently, by setting

$$y_1 = x_1 - x_1 \otimes x_2 - x_2 \otimes x_1 + \beta x_2 \otimes x_2$$
$$y_2 = x_2 - \frac{1}{3} x_1 \otimes x_1$$

β belonging to the complex field

D is transform into :

$$D' = \left(2\, y_1 + y_2 \otimes y_2 + \dots\right) \partial_1 + \left(y_2 + \dots\right) \partial_2$$

More formally, the action of the automorphism F defined by $F(1)$, $F(x_1) = y_1$, $F(x_2) = y_2$, on the derivation D is D'.

6) REFERENCES

[1] H. Poincaré: Thèse (1879). Oeuvres I, Gauthier-Villars, Paris, 1928, 69 - 129;

[2] T. Carleman, Ark. Math. Astron. Fys. 22B, 1 (1932)

[3] R. Bellman and J.M. Richardson Quart. Appl. Math. 20 (1963), 333-339

[4] W.-H. Steeb and F. Wilhelm Journal of Mathematical Analysis and Applications 77, 601-611 (1980)

Control Algorithms for Chaotic Systems

Elizabeth Bradley

Department of Electrical Engineering and Computer Science

Massachusetts Institute of Technology

Abstract

The techniques presented in this paper actively exploit chaotic behavior to accomplish otherwise-impossible control tasks. The state space is mapped by numerical integration at different system parameter values and trajectory segments from several of these maps are automatically combined into a path between the desired system states. A fine-grained search and high computational accuracy are required to locate appropriate trajectory segments, piece them together and cause the system to follow the composite path. The sensitivity of a chaotic system's state-space topology to the parameters of its equations and of its trajectories to the initial conditions make this approach rewarding in spite of its computational demands. Boundaries of basins of attraction can be breached using these techniques, vastly altering both global and local convergence properties. Strange attractor bridges can be found that connect previously unreachable points. Examples of both are shown.

1 Introduction

This paper presents control techniques that can be applied to chaotic systems. The unique attributes of chaos are exploited by these algorithms in order to accomplish control tasks that could not otherwise be realized. Extensive simulation and global reasoning about state-space features are used to identify nonlinear trajectory segments that can be pieced together into a path between the specified states. The driving concept behind this approach to control of nonlinear and chaotic systems is to combine fast computers with deep knowledge of nonlinear dynamics to improve performance in systems whose performance is rich but whose analysis is mathematically and computationally demanding.

In a nonlinear system, the distance between neighboring state-space trajectories grows exponentially with time, so small trajectory perturbations can have serious global repercussions. The Voyager missions used Jupiter as a slingshot for just this reason: near the point of closest approach, a small change in angle, via a short rocket burn, drastically changed the spacecraft's overall path in a fashion simply unobtainable in a linear system. Small errors in that course correction can, however, have equally dramatic effects. This leverage is the power of and, paradoxically, the difficulty with nonlinearity.

The system's state space is explored for different parameter values. Since nonlinear systems are exquisitely sensitive to these parameters, a small range of parameter variation can give the control algorithm a large range of behaviors to exploit. Taking advantage of this range and understanding these behaviors, the automatic path-finding algorithm selects a set of trajectory segments from the maps and combines them to form a path through the state space between the desired system states. The controller causes the system to follow this path by monitoring the system state and switching parameter values when the segment junctions.are reached. Striking results are achieved with this technique: a very small control action, delivered precisely at the right time and place, can accurately direct the system to a distant point on the state space. In one of the examples in this paper, a small control action briefly pushes a system in a counterintuitive direction in order to reach a path that travels directly to the goal state. In another example, an equally small change moves a particular state from the basin of attraction of one fixed point to the basin of another.

These techniques can be applied to any system — linear or nonlinear — but chaotic systems have several properties that make them especially useful in control. Trajectories in such systems cover a subset of the state space densely, visiting arbitrarily small neighborhoods of every point in that subset. This denseness has obvious implications for reachability: chaotic attractors can be used as bridges between otherwise-unconnected points. Furthermore, such attractors contain an infinite number of unstable periodic orbits that can be located and stabilized.

The rest of this paper is organized as follows. Section 2 briefly reviews pertinent aspects of nonlinear dynamics theory. Section 3 outlines the control algorithms that are used here to find and follow paths through state space. Section 4 illustrates the algorithm with several numerical examples and section 5 summarizes the work, its implications and its connections to previous research.

2 Theory

The equations for an n-dimensional nonlinear system can be written:

$$\frac{d\vec{x}}{dt} = F(\vec{x}, k_1, \ldots, k_p, t) \tag{1}$$

where $\vec{x}$ is an n-vector whose elements are the system's state variables and F is a nonlinear function of the state $\vec{x}$ and the parameters k_i. Necessary conditions for chaos are, in addition to the nonlinearity of F, that $n \geq 3$ and $F(\vec{x}, k_1, \ldots, k_p, t)$ be non-integrable[9]. The state-space trajectories of a dissipative chaotic system[1] separate exponentially over time and yet remain on a bounded fractal subset of the state space, called a chaotic or strange attractor[15], within which are embedded an infinite number of unstable periodic orbits. The distance between nearby trajectories grows as $O(e^{\lambda t})$, where λ is the largest positive Lyapunov exponent[6]. On a surface of section through the attractor, the unstable periodic orbits appear as unstable fixed points at the intersections of the stable and unstable manifolds of the system (1) above.

The direction and magnitude of the vector field described by equations (1) depend strongly on the equations' parameters. Locally, this sensitivity can be defined as

$$V_{k_i}(\vec{x}, k_1, \ldots, k_p, t) = \frac{\partial F(\vec{x}, k_1, \ldots, k_p, t)}{\partial k_i} \tag{2}$$

Changes in parameter values (Δk_i) also affect the large-scale features of the state-space plot, causing fixed points to split and give birth to other fixed points, limit cycles or chaotic attractors. These topological changes are known as bifurcations. Between bifurcations, the Δk_i can also cause dramatic changes in the position, shape and size of existing attractors.

Both small- and large-scale changes can be exploited by control algorithms if the responsible parameters are accessible. For example, V_{k_i} can be used to identify regions of state and parameter space where small Δk_i have locally large effects. Changes in attractor size or position can make a target state reachable from different areas of state space. Slow, roundabout paths and fast, direct paths between two points are often separated only by a small parameter difference.

If a stable fixed point can be found near the target state for some parameter value, the control problem is more or less solved, provided that the initial state is in its basin of attraction. However, while the stable fixed points of a chaotic system do move about

---

[1] Hamiltonian or non-dissipative systems do not have attractors, as their equations preserve state-space volumes.

the state space as the parameters are varied, they may only wander over a small region before bifurcating into more complex attractors. In general, it is unlikely that any choice of parameter value would place a fixed point near the target state, unless the path of the fixed point were fortuitous. Moreover, convergence to such points is often slow.

If no suitable fixed points exist, stabilization of the system state at a particular point is, in the classic sense, impossible. An alternative control objective is a steady-state orbit that returns to the target point every m cycles. The unstable periodic orbits embedded within a chaotic attractor can be located using the method of Gunaratne et al[8]. Points on a section that return to their own small neighborhoods after m piercings are assumed to be very close to m-cycles; averages of tight bunches of such points are taken to be good approximations to unstable fixed points.

These unstable periodic orbits can be stabilized using the control scheme developed by Ott et al[13], wherein the system's dependence upon the parameter is linearized about the fixed point on the $n-1$-dimensional surface of section. This works where the linearization is a good approximation: in the $n-1$-dimensional "control parallelogram" around the point, whose size is determined by the control parameter's range, its effects on the orbit and the orbit's unperturbed stability properties. A small change in k causes a $k = k_0$ periodic orbit to return, after m cycles, not to its original coordinates $\vec{\mathcal{P}}_0$, but to some nearby point $\vec{\mathcal{P}}$. The vector $\vec{g}_k$ measures this effect:

$$\vec{g}_k \equiv \frac{\partial \vec{\mathcal{P}}_0}{\partial k}|_{k_0} \approx \frac{1}{k - k_0}(\vec{\mathcal{P}} - \vec{\mathcal{P}}_0) \tag{3}$$

The stability properties are determined by integrating the variations

$$\frac{d\delta_{ij}}{dt} = \sum_k \frac{\partial f^j}{\partial x_k}\delta_{ik}$$

around the orbit. f^j is the jth component of the system equations (1) and the δ_{ik} are variations around $\vec{\mathcal{P}}_c$. The unstable eigenvector $\hat{e}_u$ and eigenvalue λ_u of this matrix, together with the admissible variation of the parameter k^* around k_0 and the vector $\vec{g}_k$, determine P_{c_k}, the size of the control parallelogram, according to

$$P_{c_k} = k^*|(1 - \lambda_u^{-1})\vec{g}_k \cdot \hat{e}_u| \tag{4}$$

Details about the derivation of these formulae are given in [13].

Since the control parallelogram surrounds a point that is embedded within a chaotic attractor, all trajectories will eventually enter the controller's domain, be driven to the orbit, and, in the absence of noise, remain there indefinitely. The denseness of these orbits makes this technique very practical if a chaotic attractor overlapping the target state exists; nevertheless, target acquisition is a problem. The delay before any particular trajectory wanders into the parallelogram is unpredictable, although it does depend stochastically on the ratio of the areas of the parallelogram and of the entire attractor.

3 The Control Algorithm

The following algorithm can be used to find a path between two state-space points **A** and **B** to within a tolerance T. The system is stabilized either at the target state or upon a nearby periodic orbit. Some restrictions on **B** do apply; this issue is discussed at the end of the section. This algorithm applies to linear, nonlinear and chaotic systems, but only in the latter can it exploit the dense unstable periodic orbits that are embedded within chaotic attractors. For expositional clarity, this presentation assumes that the system has a single parameter k; more parameters would simply increase the size of the search space and the number of indices needed to keep track of it.

1. Map the state space for different parameter values:

   - Pick an initial range (Δk) and interval $[k_{low}, k_{high}]$ for k.
   - Construct state-space portraits at each $k_{low} + n\Delta k$ for $(k_{low} + n\Delta k) \in [k_{low}, k_{high}]$, using **A** as one of the initial conditions.
   - Construct portraits with a smaller Δk in ranges where successive plots exhibit large differences (e.g., large V_{k_i}, stable fixed points near **B**, bifurcations or changes in areas of chaotic zones.)

2. Establish the goal state:

   - Examine the collection of state-space maps and locate a stable fixed point that lies within T of **B**.
   - Failing that, identify parameter values that create chaotic attractors overlapping **B**. Choose a surface of section S through **B**. Locate an unstable periodic orbit within T of **B**. Determine the size P_{c0} of the parallelogram of control around the point where that orbit pierces S.

3. Choose an initial grid size ϵ and find the best (e.g., fastest, shortest euclidean distance) path between the grid squares containing **A** and **B**. This segment can be

a portion of any trajectory on any of the state-space maps constructed in step 1. It is designated S^0, starts at S^0_{init} and ends at S^0_{final} with $k = k_0$.

4. Reduce ϵ and find the best path between the grid squares containing $\mathbf{B}$ and S^0_{final}. Iterate on successively smaller scales until $\epsilon \leq P_{c0}$. Record the k_i value and the starting and ending states S^i_{init} and S^i_{final} of each segment.

5. Similarly, find the best path between S_0 and $\mathbf{A}$. The final ϵ depends not on P_{c0} but on the largest positive Lyapunov exponent of the attractor of which S^0 is a section.

The system state can be caused to evolve along a trajectory consisting of a series of path segments $\{S^0, \ldots, S^l\}$ via the following set of control actions. Because of the recursive, longest-first nature of the path-finding algorithm, the segments are not followed in the order in which they are found, so the list must first be sorted into the proper order $\{S'^0, \ldots, S'^l\}$. Beginning at $\mathbf{A}$, the parameter is set to k'_0 to initiate the first segment S'^0 and the state is monitored until $\vec{x} = S'^0_{final}$. The parameter is then changed to k'_1, rerouting the system onto S'^1. Clearly, it is vital that parameter switches take place much faster than the system's time scales. This procedure is repeated through all segments in the path. After the final switch, k is set to the value, determined in step 2, that creates the desired fixed point or unstable periodic orbit; in the case of the latter, the linearized control scheme of [13] is then activated.

This algorithm finds globally good paths that have locally bad segments (e.g., driving east to an airport to catch a westward flight) — the sort of path that purely-local control schemes miss. Even paths between regions of state space that are apparently not connected can be found; the control parameters add dimensions to the space that can open conduits between those regions. The examples presented in the next section illustrate both of these cases.

This particular version of the path-finding algorithm, simplified for presentation, does not apply to nonautonomous systems, problems where the final state is unspecified, or problems that require a specific path to be followed. However, adapting the algorithm to fit these cases requires only simple modifications. Time dependence simply adds a dimension to the problem. Where the state can acceptably settle anywhere in a given range, control problems could be solved with a broadened fixed-point search in step 2. Matching a specified path would require a different criterion for path choice in step 3 — not the fastest or most direct segment, but the best match to the specified path.

Several caveats accompany this method:

- **B** may not fall near an unstable periodic orbit or a fixed point for any value of k, in which case this algorithm will fail. The non-zero fractal dimension of a chaotic attractor and the denseness with which trajectories cover it make the former less likely.

- This method applies to systems of any dimension, but the periodic orbit stabilization method of [13] requires that there be at least as many accessible parameters as there are unstable eigenvalues of the orbit.

- If state variables are not directly accessible, information about the system state must be synthesized from outputs and other accessible signals. Systems in which this is not possible cannot be controlled using this approach.

- Slight timing or parameter value errors (e.g., quantization error) can be magnified exponentially, particularly if they occur at the beginning of a long segment or in an area of large $\frac{\partial F(\vec{x})}{\partial x}$. The property elucidated in the Beta Shadowing Lemma[2] keeps these errors from being truly disastrous, but they still place a fundamental upper bound on realizable path length.

Constructing and examining state-space portraits is time-consuming. This motivates the attempt in step 1 of the algorithm to restrict attention to the useful ones — new maps only being considered useful if their portraits differ from the existing ones, regardless of the k-interval between plots. The region of state-space considered is also restricted: it must be somewhat larger than the bounding box of **A** and **B** in order to allow for locally counterintuitive moves. The size of this region is determined heuristically and can be varied if the algorithm fails to find a path on its first pass. There are many obvious points in this procedure where computation can be traded for accuracy: a smaller Δk and a larger range $[k_{low}, k_{high}]$, location and study of different unstable periodic points to find the one that is closest to the target state or whose control parallelogram is largest, variational analysis around each segment to check whether, for example, the $k = 49.9$ path is better than the $k = 50$ path, etc.

Preliminary versions of these path-finding and control algorithms have been implemented and tested. The programs are written in the Lisp dialect Scheme[14] and run on an HP series-300 workstation. State-space portraits are computed from a given set of

---

[2]Because "with high probability, the sample paths of the problem with external noise follow *some* orbit of the deterministic system closely"[7] and the deterministic orbit lives on a bounded attractor.

system equations using a fourth-order Runge-Kutta adaptive time-step integrator. Trajectories are indexed on a state-space grid and represented as lists of the grid squares that they enter. The grid size ϵ is manipulated as part of the search in step 3 of the path-finding algorithm.

The next section of this paper illustrates these techniques with several numerical examples.

4   An Example

The Lorenz equations[11] are:

$$F(\vec{x}, a, r, b) = \begin{bmatrix} \frac{dx}{dt} \\ \frac{dy}{dt} \\ \frac{dz}{dt} \end{bmatrix} = \begin{bmatrix} a(y - x) \\ rx - y - xz \\ xy - bz \end{bmatrix} \tag{5}$$

These well-known equations approximately describe convection in a sheet of fluid heated from below. The state variable x is proportional to convection intensity; y and z quantify temperature variations. a and r are physical parameters of the fluid — the Prandtl and Rayleigh numbers — and b is an aspect ratio. Numerical integration of these equations from an initial state $\vec{x}_0 = (x_0, y_0, z_0)$ yields a time-parametrized trajectory $\vec{x}(t)$ in state space. F is dissipative and nonlinear. It is also non-integrable for some parameter values; for these values, its trajectories converge to a chaotic attractor. A particular case of this, shown in figure 1, occurs for $a = 16$, $r = 45$ and $b = 4$. As this figure is an $x - z$ projection of a three-dimensional object, the apparent trajectory crossings do not represent uniqueness violations. The parameter values and the initial state are shown at the bottom of the figure. The values in the upper right and lower left corners are normalized axis coordinates. An $x - z$ section[3] of the same attractor at $y = -15$ is shown in figure 2. More details about the structure and properties of Lorenz attractors may be found in [19].

Consider the task of navigating between the two points marked by crosses in figure 3, starting at the rightmost (A) and ending at the leftmost (B.) On the axes of the figure, the coordinates (x, y, z) of these points are (8, 29, 64) and (-24.5, -20, 68). r is used as the control parameter and a and b are fixed. We make no assertions about whether changing this parameter is either physical or practical; this is purely a mathematical example[4].

---

[3]Points appear on this section where the trajectory pierces the $y = -15$ plane with a positive velocity.
[4]Lorenz himself explored the parameter space outside the range ($r \approx 1$) within which the equations

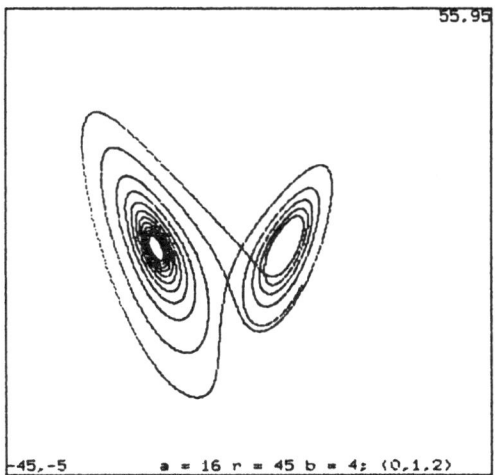

Figure 1: Lorenz Attractor for a = 16, r = 45 and b = 4

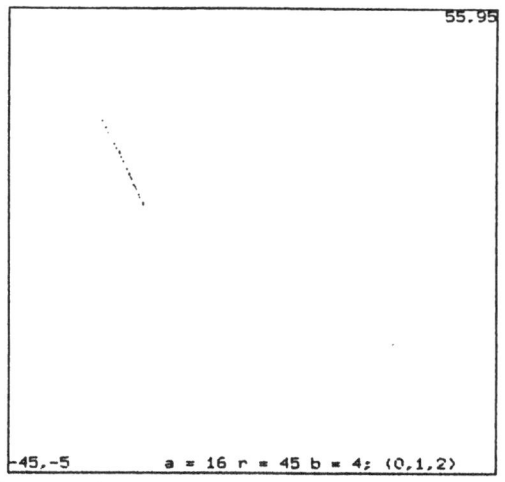

Figure 2: x-z Section of Lorenz Attractor for a = 16, r = 45 and b = 4

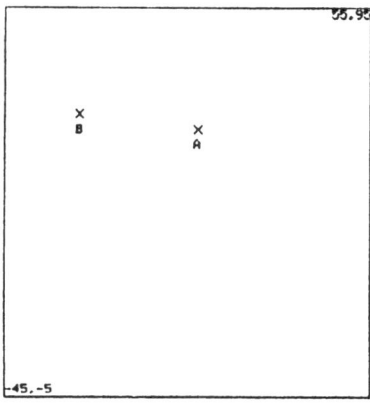

Figure 3: Origin and Destination Points

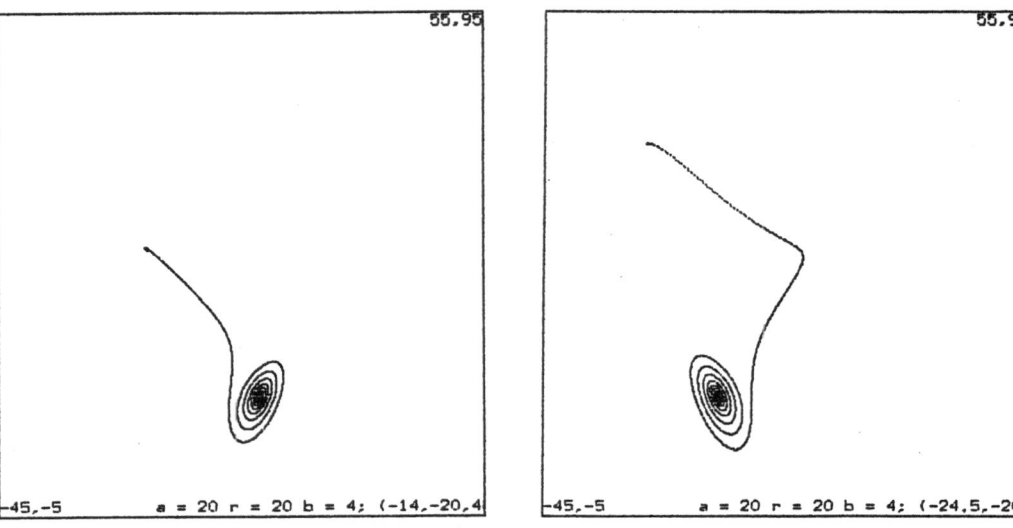

Figure 4: Two Stable Fixed Points

As set out in step 1 of the algorithm in section 3, the state space is mapped with an initial r-step of five and an r-range of [20,60]. Portraits are also constructed on a finer grain ($\Delta r = 1$) near, for example, the bifurcation at $r = 24$ that creates a chaotic attractor and in the range where that attractor expands to cover **B**.

Proceeding to the first part of step 2, the maps are examined for stable fixed points of the system (5) at or near **B**. The Lorenz attractor collapses to a pair of stable fixed points at low r, as demonstrated by the $r = 20$ trajectories in figures 4(a) and 4(b). Note that the two starting points are in different basins of attraction. Both of the fixed points — whose coordinates are (8.72, 8.72, 19.00) and (-8.72, -8.72, 19.00)[5] — move as r is varied, but neither approaches **B** for any value of r.

Since no appropriate fixed points exist, the second part of step 2 indicates that the maps should be examined for nearby chaotic attractors and unstable periodic orbits. Figure 5 shows a trajectory for $r = 30$, just above the bifurcation that changes the fixed points of figure 4 into a chaotic attractor. Neither A nor B happens to lie upon this particular attractor, so it cannot be used as a bridge between them or as a source of nearby unstable periodic orbits. As r is raised further, however, the attractor expands; one of its lobes overlaps **B** when $r = 42$ (figure 6). For $r = 50$, an unstable seven-cycle is found on the $y = -20$ $x - z$ surface of section at $\vec{\mathcal{P}}^7 = $ (-24.673, 68.207). This orbit is found by computing 3000 piercings of the $y = -20$ plane, identifying, sorting into

---

are considered to be an accurate physical model.

[5]The symmetry is not coincidence; see [19] for details.

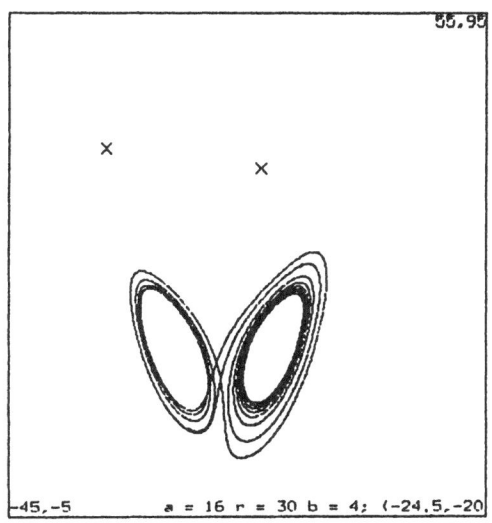

Figure 5: Lorenz Attractor for a = 16, r = 30 and b = 4

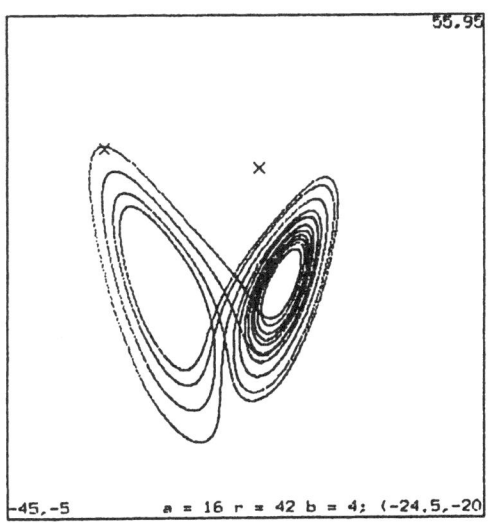

Figure 6: Lorenz Attractor for a = 16, r = 42 and b = 4

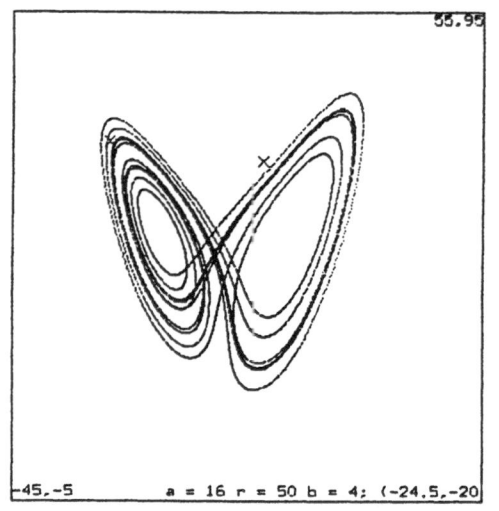 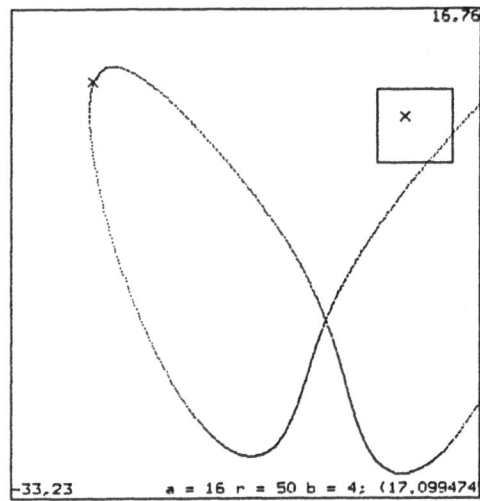

Figure 7: Lorenz Attractor for a = 16, r = 50 and b = 4

bunches and averaging those that return after m cycles, and then choosing the one closest
to **B**. We assume, for the purposes of this problem, that errors smaller than 1% of the
total signal amplitude $\sqrt{x^2 + y^2 + z^2}$ are admissible, so $\vec{\mathcal{P}}^7$ meets the specifications. The
eigenvalues of the variational system integrated around the seven-cycle are $\lambda_u = 0.734$
and $\lambda_s = -502.077$, with the associated eigenvectors $\hat{e}_u = 0.834\hat{x} + 0.552\hat{z}$ and $\hat{e}_s =$
$0.685\hat{x} + 0.728\hat{z}$. A 1/2% change in r causes the point $\vec{\mathcal{P}}^7$ to return not to itself (-24.673,
68.207), but to $\vec{\mathcal{P}}^{7\prime}$ =(-23.929, 66.498), so

$$\vec{g}_r \equiv \frac{\partial \vec{\mathcal{P}}^7}{\partial r}\Big|_{r=50} \approx \frac{1}{\Delta r}(\vec{\mathcal{P}}^{7\prime} - \vec{\mathcal{P}}^7) = 70.8\hat{x} - 170.9\hat{z}$$

The allowed range of variation of r, for a 1% control tolerance, is ±0.0075. Using these
values in equation (4), we obtain $P_{c_r} = 0.096$ to complete step 2.

Using an initial choice of $\epsilon = 20$, the collection of portraits is examined, as described
in step 3, for paths between the grid squares containing **A** and **B**. The $r = 50$ portrait of
figure 7(a) contains one such segment, hereafter designated S^0. An edited version of this
portrait, plotted on a smaller region around the two points and showing only the useful
part of the trajectory, is shown in figure 7(b). S^0 pierces the $y = -20$ plane at (-24.51,
-19.99, 68.05), which is actually within the control parallelogram, so no additional path
segments need be found to connect S^0 and **B**.

To find trajectory segments that connect point **A** to the other end of S^0, step 4 is
iterated in the region outlined by the square in figure 7(b). The nature of the projection
makes the distances deceptive; **A** is actually quite far above the nearest threads of the
$r = 50$ attractor. The segment S^1, found on the $r = 40$ map with $\epsilon = 1$, spans most of the

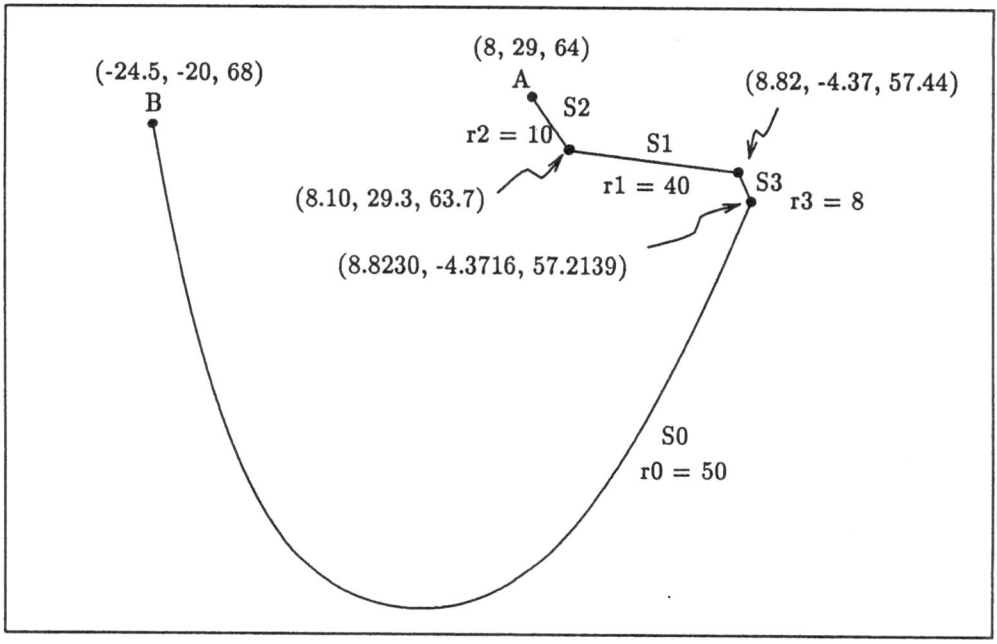

Figure 8: Schematized Segmented Path

distance from **A** to S^0. One more iteration is required on a yet-finer scale ($\epsilon = 0.0001$) to find the two very short segments S^2 and S^3, at $r = 10$ and $r = 8$, needed to connect S^1 to **A** and to S^0. Since S^0 actually enters the control parallelogram around **B**, this process of connecting to it amounts to a solution to the "target acquisition problem" of [13].

A schematized version of the overall path $\{S^2, S^1, S^3, S^0\}$ is shown in figure 8. It is composed of the four segments discussed in the previous two paragraphs. The two longer segments are segments of chaotic attractors; the two shorter ones, enlarged so as to be visible, are sections of transient trajectories ultimately destined for one of the system's fixed points. The former are examples of a "strange attractor bridge", connecting two otherwise-unconnected points. The values for the r_i and the S^i_{final} are shown next to the segments and the transition points where they meet. The actual numerical integration of the path from **A** to **B** along $\{S^3, S^1, S^4, S^0\}$ is shown in figure 9. On this scale, the smaller connecting segments are invisible. Note that S^1 actually moves the system state *directly away from* **B**. This locally counterintuitive move is made in order to reach a globally good path.

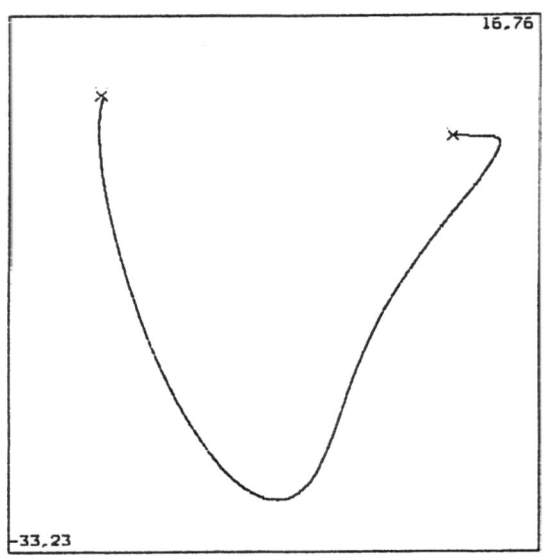

Figure 9: Numerical Integration of Segmented Path

The control program causes the system to follow this segmented path by monitoring the system state and switching the parameter accordingly. The transition that initiates S^0 is the most critical of the three, as nonlinear expansion along its great length can severely magnify any error; this is where the four decimal place accuracy is important. The path length between **A** and **B** is 130.1 normalized distance units, requiring 0.3567 normalized time units to traverse. The contrast to the case *without* active target acquisition (see [13]) is striking: if a trajectory is started at **A** and simply allowed to evolve with $r = 50$, it enters the control parallelogram around **B** after traveling 25223 normalized distance units around the attractor in 104 normalized time units. See figure 10.

If **B** were near one of the system's low-r fixed points, the first part of step 2 would succeed. Though a single segment might then converge to the desired state, use of a segmented path can alter macroscopic quantities like convergence speed and reachability. For example, a trajectory starting from the point (22.4, 30.5, 60) at the value $r = 25$ would normally converge to the left hand fixed point (-9.80, -9.80, 24.00) along the tightly-wound spiral at the bottom left of figure 11. The right-hand path in the figure was found by a single pass of the first two steps of the algorithm with $\epsilon = 5$. This path contains two segments: an $r = 60$ trajectory that travels most of the way from (22.4, 30.5, 60) to the *other* fixed point, which lies at (10.25, 10.31, 26.77), and a short section of the $r = 25$ spiral that surrounds the second fixed point. Not only has this manipulation allowed the trajectory to jump the basin boundary and converge to the opposite fixed point, but it has also bypassed much of the slowly-converging spiral around that point.

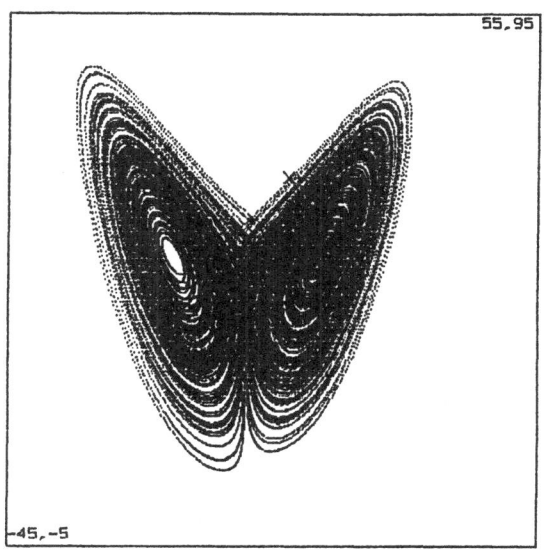

Figure 10: Without Target Acquisition

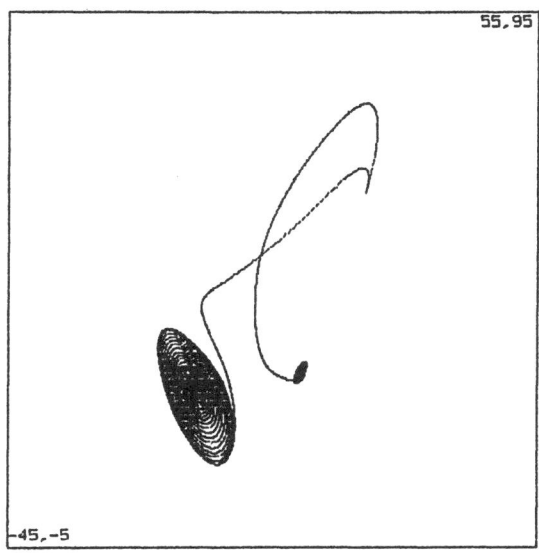

Figure 11: Segmented Path to a Fixed Point·

5   Conclusion

The state-space features of chaotic systems are strongly affected by parameter values. Moreover, the trajectories that make up those features separate exponentially over time. Small changes in parameters or state can thus have large and rapid effects; this leverage is a powerful basis for control schemes. In this paper, we have indicated how fast and accurate computation can be used to synthesize paths through a chaotic system's state space that exploit this leverage to accomplish otherwise-unattainable control tasks. Nonlinear dynamics provides the mathematical tools used to choose values, tolerances, heuristics and limits for the algorithms that select and piece together trajectory segments to create these paths. The Lorenz system is only one of the many systems to which these techniques can be applied. Other examples, with perhaps more practical benefits, are phase-locked loops[4], robot manipulators and spacecraft controllers[18], and single and double pendulums[5]. This brief list is far from complete.

The algorithm is not yet fully automated; for example, the determination of when an attractor "covers" a point was made by eye. Algorithms that produce qualitative descriptions of state space, like the Bifurcation Interpreter[1], KAM[20] or PLR[16] will be involved in the ultimate mechanization of this. Mathematical models are currently used to construct state-space portraits; errors in these models can cause spectacularly bad control decisions. Experimentally exploring and mapping the state space of a physical system [10] would probably be much faster and would also obviate modeling error. Although these techniques have, to date, only been applied to simulated devices, the ultimate aim of this project is to control actual physical systems. Instrumented versions of several chaotic systems — a double pendulum, a driven single pendulum, and several phase-locked loops — have been built. The I/O channel that transmits the state information and control parameter values between controller and system is under construction. These tools will be used to obtain experimental verification of the results presented in this paper.

In the most general terms, the implications of this work are that:

- A broader view and understanding of chaotic state-space features and the effects of parameters upon them is a powerful tool, but its application requires great computational effort.

- Understood and controlled, chaotic behavior can be profitably used to improve a system's design and performance.

This approach can be thought of as a new flavor of adaptive control — one that takes a global viewpoint and eschews almost all linearization[6]. It extends the active use of chaos in control, which presently consists of (1) using knowledge of chaotic zones' boundaries to site an operating point in the middle of the widest part of a system's largest stable zone to maximize noise immunity[17] and (2) the recent work on unstable periodic orbits [13] discussed at various points in this paper. The main difference between this work and [13] lies in the breadth of the aims: we wish, rather than to stabilize a system on a particular type of orbit, to navigate dynamically through *all* of state space with minimal restrictions. This project also extends the ongoing program of research in our group[3] with the overall goal of investigating the use of combined numerical and symbolic methods in scientific and engineering computing.

The complexity of the tasks that are executed by this control program and the accuracy with which it must perform make computation speed a vital issue. The program must compute, store, search through and recognize features in a large number of intricate state-space portraits whose topologies are extremely sensitive to parameter variations. At the same time, many computational approximations are out of bounds because the small errors that they introduce can be amplified exponentially. Physical constraints require that control actions take place about an order of magnitude faster than the actions of the system itself, which further exacerbates the demands on the program's speed. All of this work is worth it: allowing a system to operate in its chaotic regimes opens up new possibilities for better designs. Faster computers[2] or special-purpose hardware might be part of the ultimate solution, together with the understanding and algorithms gained in the course of this research, to attaining novel and effective control of useful systems via knowledge of nonlinear theory and intensive computation.

---

[6] Some authors in the adaptive control literature have hinted in the direction of controlling chaos, but they fall short of active pursuit of the idea, admitting only that "chaotic parameter estimates are not necessarily a bad thing to have"[12].

Acknowledgements

This report describes research done at the Artificial Intelligence Laboratory of the Massachusetts Institute of Technology. Support for the laboratory's artificial intelligence research is provided in part by the Advanced Research Projects Agency of the Department of Defense under Office of Naval Research contracts N00014-85-K-0124 and N00014-86-K-0180. Hal Abelson's encouragement, proofreading and suggestions vastly improved the content and presentation of this paper.

References

[1] H. Abelson. The Bifurcation Interpreter: A step towards the automatic analysis of dynamical systems. *International Journal of Computers and Mathematics with Applications*, 20:13, 1990.

[2] H. Abelson, A. A. Berlin, J. Katzenelson, W. H. McAllister, G. J. Rozas, and G. J. Sussman. The Supercomputer Toolkit and its applications. Technical Report AI Memo 1249, M.I.T. Artificial Intelligence Lab, July 1990.

[3] H. Abelson, M. Eisenberg, M. Halfant, J. Katzenelon, G. J. Sussman, and K. Yip. Intelligence in scientific computing. *Communications of the ACM*, June 1989.

[4] G. Bernstein, M. A. Lieberman, and A. J. Lichtenberg. Nonlinear dynamics of a digital phase-locked loop. *IEEE Transactions on Communications*, 37:1062, 1989.

[5] E. Bradley. *Taming Chaotic Circuits*. PhD thesis, M.I.T., May 1990. Proposal.

[6] R. L. Devaney. *An Introduction to Chaotic Dynamical Systems*. Benjamin/Cummings, Menlo Park, CA, 1986.

[7] J.-P. Eckmann. Roads to turbulence in dissipative dynamical systems. *Reviews of Modern Physics*, 53:643, 1981.

[8] G. H. Gunaratne, P. S. Linsay, and M. J. Vinson. Chaos beyond onset: A comparison of theory and experiment. *Physics Review Letters A*, 63:1, 1989.

[9] R. H. G. Helleman. Self-generated chaotic behavior in nonlinear mechanics. *Fundamental Problems in Statistical Mechanics*, 5:165, 1980.

[10] M. Lee. Summarizing qualitative behavior from measurements of Nonlinear Circuits. Technical Report AI-TR 1125, M.I.T. Artificial Intelligence Lab, May 1989.

[11] E. N. Lorenz. Deterministic nonperiodic flow. *Journal of the Atmospheric Sciences*, 20:13, 1963.

[12] I. M. Y. Mareels and R. R. Bitmead. Nonlinear dynamics in adaptive control: Chaotic and periodic stabilization. *Automatica*, 22:641, 1986.

[13] E. Ott, C. Grebogi, and J. A. Yorke. Controlling chaos. In *Chaos: Proceedings of a Soviet-American Conference*. American Institute of Physics, 1990.

[14] J. Rees and W. Clinger. The revised3 report on the algorithmic language Scheme. *ACM SIGPLAN Notices*, 21:37, 1986.

[15] D. Ruelle. Strange attractors. *The Mathematical Intelligencer*, 2:126, 1980.

[16] E. P. Sacks. Automatic qualitative analysis of ordinary differential equations using piecewise-linear approximations. Technical Report AI-TR 1031, M.I.T. Artificial Intelligence Lab, March 1988.

[17] F. M. Salam and S. Bai. Complicated dynamics of a prototype continuous-time adaptive control system. *IEEE Transactions on Circuits and Systems*, 35:842, 1988.

[18] J.-J. E. Slotine. Putting physics back in control. In J. Descusse, editor, *New Trends in Nonlinear Control Theory*. Springer-Verlag, 1988.

[19] R. F. Williams. The structure of Lorenz attractors. *Publications Mathematiques de l'IHES*, 50:101, 1979.

[20] K. M.-K. Yip. KAM: Automatic planning and interpretation of numerical experiments using geometrical methods. Technical Report AI-TR 1163, M.I.T. Artificial Intelligence Lab, August 1989.

Analytical method of solving the differential equations, describing a developing macroeconomical system that is under the influence of scientific and technical progress

B.L. Kuchin* and L.A. Ovcharov†

Moscow Institute of Oil and Gas
Soviet Union, Moscow 117917, Leninsky Prospect 65.
Tel. 135-71-56. Fax 1358895. Telex 411637 NAFTA SU.

Abstract: In the paper we suggest an algebraic method that may be useful to get an analytical solution for nonlinear differential equations of evolutionary type, that describe the behaviour of a developing macroeconomic system. As an example we analyse the system in two-dimensional space, that describes the influence of scientific and technical progress (STP) on a complicated macroeconomical structure of parallel type; illustrations of application of this method is given in the paper. The results of the solution of macromodel (obtained on a personal computer) is given for an interaction between the oil–gas complex and the adjacent branches of industry. There is a good efficiency and accuracy of the given algebraic method, that has a general character. The method could be used for a wide class of Volterra type systems, that describe not only the economic systems, but also biological and ecological developing systems.

Topic: The algebraic computing tools that are available in the field of control theory.

1 The statement of the problem

We suppose that the complicated macroeconomical structure consists of some simple macroeconomical systems, that are given by four elements $\{\bar{X}_S, Y_S, F_S, \Phi_S\} \in \Omega_S$, where $\bar{X}_S$ – is the vector of input resources that are processed by the S system, Y_S – is the output of the S macrosystem, F_S – is the production function of the S macrosystem, Φ_S – is the operator, that describes the influence of the STP on the S macrosystem. Besides

---

*Prof. of the department of Automated Control Systems,
†Head of the department of Automated Control Systems.

that there is given an operator "A" that describes the structure of interactions between the "simple" macrosystems inside the 'complicated" macroeconomic structures.

For every "simple" macroeconomic system we use the functional–dynamic model, that was described in [1]. As it was shown in [2] in this case the complicated macroeconomic structure could be given by a system of evolutionary type differential equation:

$$\frac{dY_S}{dt} = \alpha_S(t)Y_S + Y_S \sum_{\substack{k=1 \\ k \neq S}}^{r} \gamma_{Sk}(t)Y_k + \beta_S(t)Y_S^2 \tag{1}$$

$$S = 1, r$$

where $\alpha_S, \beta_S, \gamma_{KS}$ are functions that are defined by $\bar{X}_S, F_S, \Phi_S A$.

Let us analyse a two–dimension space :

$$\begin{cases} \frac{dY_1}{dt} = \alpha_1 Y_1 + \gamma_{12} Y_1 Y_2 + \beta_1 Y_1^2 \\ \frac{dY_2}{dt} = \alpha_2 Y_2 + \gamma_{21} Y_1 Y_2 + \beta_2 Y_2^2 \end{cases} \tag{2}$$

This system of differential equations describes quite well the complicated macroeconomical structures, that consist of two parallel macroeconomic systems. By its form it reminds us of the well–known Volterra model of interaction between two biological population in an organic environment. Nowever system (2) is much richer, because the coefficients $\alpha_i, \beta_i, \gamma_{ij}$ are strongly dependent on time. That is why the qualitative analysis of the system (2) with analytical methods causes big difficulties.

In the paper we suggest an analytical method of finding the solution of the system using the results of the works [4, 5, 6].

2 The method of solution

In the first step we transform the two nonlinear differential equations (2) into one nonlinear differential equation of the 2nd order. For that reason we differentiate the first equation of system (2) by time t and get the following equation:

$$\frac{d^2 Y_1}{dt^2} = \frac{d\alpha_1}{dt}Y_1 + \alpha_1 \frac{dY_1}{dt} + 2\beta_1 Y_1 \frac{dY_1}{dt} + \frac{d\beta_1}{dt}Y_1^2 + \gamma_{12}\frac{dY_1}{dt}Y_2$$
$$+ \gamma_{12}\frac{dY_2}{dt}Y_1 + \frac{d\gamma_{12}}{dt}Y_1 Y_2 \tag{3}$$

From the first equation of system (2) we find :

$$Y_2 = \frac{\frac{dY_1}{dt} - \alpha_1 Y_1 - \beta_1 Y_1^2}{\gamma_{12} Y_1} \tag{4}$$

Then we substitute (4) in the second equation of system (2) :

$$\frac{dY_2}{dt} = \frac{Y_1}{\gamma_{12}}\left(\frac{2\alpha_1\beta_1\beta_2}{\gamma_{12}} - \alpha_1\gamma_{21} - \alpha_2\beta_1\right) + \frac{Y_1^2}{\gamma_{12}}\left(\frac{\beta_1^2\beta_2}{\gamma_{12}} - \beta_1\gamma_{21}\right)$$
$$+ \frac{dY_1}{dt}\frac{1}{Y_1\gamma_{12}}\left(\alpha_2 - \frac{2\alpha_1\beta_2}{\gamma_{12}}\right) + \left(\frac{dY_1}{dt}\right)^2\frac{1}{Y_1^2}\frac{\beta_2}{\gamma_{12}^2} \qquad (5)$$
$$+ \frac{dY_1}{dt}\left(\frac{\gamma_{21}}{\gamma_{12}} - \frac{2\beta_1\beta_2}{\gamma_{12}^2}\right) + \frac{\alpha_1}{\gamma_{12}}\left(\frac{\alpha_1\beta_2}{\gamma_{12}} - \alpha_2\right).$$

Substituting (4), (5) in (3), after some algebraic transformations, we get:

$$\ddot{Y}_1 + \frac{\dot{Y}_1^2}{Y_1}\left(-1 - \frac{\beta_2}{\gamma_{12}}\right) + \dot{Y}_1\left(\frac{2\alpha_1\beta_2}{\gamma_{12}} - \alpha_2 - \frac{\dot{\gamma}_{12}}{\gamma_{12}} + Y_1\left(\frac{2\beta_1\beta_2}{\gamma_{12}} - \gamma_{21} - \beta_1\right)\right)$$

$$+ Y_1\left(-\dot{\alpha}_1 - \alpha_1\left(-\frac{\dot{\gamma}_{12}}{\gamma_{12}} + \frac{\beta_2\alpha_1}{\gamma_{12}} - \alpha_2\right)\right)$$

$$+ Y_1^2\left(\alpha_1\gamma_{21} + \alpha_2\beta_1 + \frac{\beta_1\dot{\gamma}_{12}}{\gamma_{12}} - \dot{\beta}_1 - \frac{2\beta_1\beta_2\alpha_1}{\gamma_{12}}\right) + Y_1^3\left(\beta_1\gamma_{21} - \frac{\beta_1^2\beta_2}{\gamma_{12}}\right) = 0 \qquad (6)$$

where $(\dot{\ }) = \frac{d(\ )}{dt}$.

Introducing some new variables we finally get

$$\ddot{Y}_1 + (r-1)\frac{\dot{Y}_1^2}{Y_1} + (pY_1 + q)\dot{Y}_1 + sY_1 + mY_1^2 + nY_1^3 = 0 \qquad (7)$$

where

$$r = -\frac{\beta_2}{\gamma_{12}}$$

$$p = \frac{2\beta_1\beta_2}{\gamma_{12}} - \gamma_{21} - \beta_1$$

$$q = \frac{2\alpha_1\beta_2}{\gamma_{12}} - \alpha_2 - \frac{\dot{\gamma}_{12}}{\gamma_{12}}$$

$$s = -\dot{\alpha}_1 - \alpha_1\left(-\frac{\dot{\gamma}_{12}}{\gamma_{12}} + \frac{\beta_2\alpha_1}{\gamma_{12}} - \alpha_2\right) \qquad (8)$$

$$m = \alpha_1\gamma_{21} + \alpha_2\beta_1 + \frac{\beta_1\dot{\gamma}_{12}}{\gamma_{12}} - \dot{\beta}_1 - \frac{2\alpha_1\beta_1\beta_2}{\gamma_{12}}$$

$$n = \beta_1\gamma_{21} - \frac{\beta_1^2\beta_2}{\gamma_{12}}$$

We can show that using some transformations of variables:

$$Y = \int_1^{Y_1}(pY_1 + q)Y_1^{r-1}dY_1 = \frac{qY_1^r}{r} + \frac{pY_1^{r+1}}{r+1}; \quad r = \begin{cases} 0 \\ -1 \end{cases} \qquad (9)$$

$$T = \int_1^t (q + pY_1)\, dt_1 \tag{10}$$

the nonlinear equation (7) may be transformed to a linear on

$$Y'' + QY' + PY = 0 \tag{11}$$

where $(\dot{}\,)' = \frac{d(\dot{})}{dT}$.

Let us find the derivative of (9) by a new variable (10); then we get:

$$Y' = Y_1^{r-1}\, \dot{Y}_1 \tag{12}$$

The repeated differentiation has the result:

$$Y'' = \frac{Y_1^{r-2}\left[Y_1\, \ddot{Y}_1 + (r-1)Y_1^2\right]}{(pY_1 + q)} \tag{13}$$

Substituting (9), (10), (12), (13) in the equation (11) we get

$$Y_1 + \frac{(r-1)\dot{Y}_1^2}{Y_1} + Q\,(pY_1 + q)\,\dot{Y}_1 + \frac{Pq^2 Y_1}{r} + \frac{Ppq(2r+1)Y_1^r}{r(r+1)} + \frac{Pp^2 Y_1^3}{r+1} = 0. \tag{14}$$

The equation (14) is equivalent to equation (7) if we use some additional restriction for the coefficients

$$\frac{\alpha_1(\beta_1 - \gamma_{21})}{\beta_1} + \frac{\alpha_2(\beta_2 - \gamma_{12})}{\beta_2} = 0. \tag{15}$$

From this equality we find that of six parameters, that describe the process of interaction, only five of them are independent. The sixth parameter must satisfy the restriction (15) and subordinate to the following conditions:

$$Q = 1$$
$$\frac{Pq^2}{r} = s$$
$$\frac{Ppq(2r+1)}{r(r+1)} = m \tag{16}$$
$$\frac{Pp^2}{(r+1)} = n$$

From equation (16) we can write

$$Q = 1,$$
$$P = \frac{sr}{q^2} = \frac{mr(r+1)}{pq(2r+1)} = \frac{n(r+1)}{p^2}. \tag{17}$$

Taking into consideration the formulas that we got above we get the linear differential equation in the following form:

$$Y'' + Y' + PY = 0$$

$$P = \frac{\beta_2(\dot{\alpha}_1\gamma_{12} + \alpha_1(\beta_2\alpha_1 - \dot{\gamma}_{12} - \alpha_2\gamma_{12}))}{(2\alpha_1\beta_2 - \alpha_2\gamma_{12} - \dot{\gamma}_{12})^2}$$

$$= \frac{\beta_2(\beta_2 - \gamma_{12})(\alpha_1\gamma_{21}\gamma_{12} + \alpha_2\beta_1\gamma_{12} + \beta_1\dot{\gamma}_{12} - \dot{\beta}_1\gamma_{12} - 2\alpha_1\beta_1\beta_2)}{(\gamma_{12} - 2\beta_2)(2\beta_1\beta_2 - \gamma_{21}\gamma_{12} - \beta_1\gamma_{12})(2\alpha_1\beta_2 - \alpha_2\gamma_{12} - \dot{\gamma}_{12})} \tag{18}$$

$$= \frac{\beta_1(\gamma_{12}\gamma_{21} - \beta_1\beta_2)(\gamma_{12} - \beta_2)}{(2\beta_1\beta_2 - \gamma_{12}\gamma_{21} - \beta_1\gamma_{12})^2}$$

and

$$2\alpha_1\beta_2 \neq \alpha_2\gamma_{12} + \dot{\gamma}_{12}, \quad \text{that is} \quad q \neq 0;$$

$$2\beta_1\beta_2 \neq \gamma_{21}\gamma_{12} + \beta_1\gamma_{12} \quad \text{that is} \quad p \neq 0;$$

$$\gamma_{12} \neq 2\beta_2$$

The solution of equation (18) has the form [7]

$$Y(T) = K_1 \exp(-S_1 T) + K_2 \exp(-S_2 T) \tag{19}$$

where

$$S_1 = \frac{1}{2}\left[1 + (1 - 4P)^{1/2}\right], \quad S_2 = \frac{1}{2}\left[1 - (1 - 4P)^{1/2}\right].$$

P is found from formula (18).

$$T = \int_1^t (q + p Y_1)\, dt;$$

$$q = \frac{2\alpha_1\beta_2}{\gamma_{12}} - \alpha_2 - \frac{\dot{\gamma}_{12}}{\gamma_{12}}$$

$$P = \frac{2\beta_1\beta_2}{\gamma_{12}} - \gamma_{21} - \beta_1$$

K_1, K_2 are found from the initial conditions.

Using (9), (10), after some transformations, we get the final formula for interaction $Y_j(t)$ in the following form:

$$\frac{q Y_j^r}{r} + \frac{p Y_j^{r+1}}{r+1} - \sum_{i=1}^2 K_i \exp\left[-S_i \int_1^t (q + p Y_j)\, dt_1\right] = 0 \tag{20}$$

$$i = 1, 2; \quad j = 1, 2.$$

3 An example of forecasting the development of a macrosystem that is under the influence of STP

Let us analyse a complicated two–domensional macroeconomic system (the model of interaction between the oil–gas complex and the adjacent branches of industry) which is under the influence of STP, neutral in the sense of Hicks. In this case the parameters $\alpha_1, \alpha_2, \gamma_{12}, \gamma_{21}, \beta_1, \beta_2$ have the form:

$$\alpha_1 = \eta_1 - \frac{F_1}{c_{11}}; \qquad \eta_1 = \sum_i b_{i1} \frac{X'_{i1}}{X_{i1}}; \qquad c_{11} = t\,F_1;$$

$$\alpha_2 = \eta_2 - \frac{F_2}{c_{22}}; \qquad \eta_2 = \sum_i b_{i2} \frac{X'_{i2}}{X_{i2}}; \qquad c_{22} = t\,F_2;$$

$$\gamma_{12} = -\frac{a_{12}}{c_{11}}; \qquad \gamma_{21} = -\frac{a_{21}}{c_{22}};$$

$$\beta_1 = \frac{1}{c_{11}}; \qquad \beta_2 = \frac{1}{c_{22}}.$$

Let us set the production functions as exponential functions of time (that means that we have a degree production functions relatively to the input resources).

$$F_1 = a_{01} e^{\lambda_1 t},$$
$$F_2 = a_{02} e^{\lambda_2 t}.$$

If the parameters are given the variables in the equation (20) would be as follows: $i = 1,2$; $\;j = 1,2$

$$q = \frac{2\left(\frac{1}{t} - \eta_j\right) a_{0j}}{a_{0i}\, a_{ji}\, e^{t\,(\lambda_i - \lambda_j)}} + 2\frac{1}{t} + \lambda_j - \eta_i;$$

$$p = \frac{a_{ij}\, e^{-\lambda_i t}}{a_{0i}\, t} - \frac{2\, e^{-\lambda_i t}}{a_{0i}\, a_{ji}\, t} - \frac{e^{-\lambda_j t}}{a_{0j}\, t};$$

$$r = \frac{a_{0j}}{a_{0i}\, a_{ji}\, e^{t\,(\lambda_i - \lambda_j)}};$$

$$S_1 = \frac{1}{2}\left[1 + \sqrt{1 - 4P}\right];$$

$$S_2 = \frac{1}{2}\left[1 - \sqrt{1 - 4P}\right];$$

$$P = \frac{\left(1 + \frac{a_{ji}\, a_{0i}\, e^{t\,(\lambda_i - \lambda_j)}}{a_{0j}}\right)\left(1 - a_{ji}\, a_{ij}\right)}{\left(2 - a_{ji}\, a_{ij} + \frac{a_{ji}\, a_{0i}}{a_{0j}}\, e^{t\,(\lambda_i - \lambda_j)}\right)^2}$$

Coefficients K_1, K_2 are determined from the initial conditions.

The computation are made for following data in the interval of time $t = 1 \div 10$:

$a_{01} = 310; \qquad a_{02} = 290; \qquad a_{12} = 0.2; \quad a_{21} = 0.3;$

$\lambda_1 = 0.1; \qquad \lambda_2 = 0.1; \qquad \eta_1 = 0.2; \quad \eta_2 = 0.2;$

$Y_1(2) = 301; \quad Y_2(2) = 280;$

Year	The amount of transfered gas Y_1 (in relative units); the fact	The amount of transfered gas Y_1 (in relative units); the results of calculation	The amount of progressive technologies, that is given to the gas industry Y_2 (in relative units); the fact	The amount of progressive technologies, that is given to the gas industry Y_2 (in relative units); the results of calculation
1	1.000000	1.000000	1.000000	1.000000
2	1.115085	1.114752	1.093935	1.092751
3	1.277297	1.248214	1.213978	1.207988
4	1.480365	1.467203	1.351943	1.319413
5	1.722267	1.701219	1.503922	1.482802
6	1.999598	1.986957	1.666890	1.618937
7	2.305022	2.310054	1.838028	1.840771
8	2.625816	2.685022	2.014537	2.072135
9	2.943975	2.990070	2.193511	2.211004
10	3.238437	3.272113	2.371722	2.412395

Table 1: The calculated output of macrosystems

The output in the basic year $Y_1(1) = 270$; $Y_2(1) = 256$.

The calculated output of macrosystems Y_1, Y_2, reduced to the basic year, are given in the Table 1.

It is seen from the Table 1. that the errors are at most 3%.

References

[1] B.L. Kuchin, L.A. Ovcharov.– The functional-dynamic model of the Scientific and Technical Progress, *Transactions of X Allunion conference on automatic control*, V II. Alma-Ata. 1986.

[2] B.L. Kuchin.– The modeling of the Scientific and Technical Progress in complicated macroeconomic structures, *XI Allunion school-seminar "The control of big systems"*, the thesis's of papers. Vilnus. 1988.

[3] J.M. Svirijev.– Nonlinear waves, dissipative structures and catastrophes in ecology, *M. "Nauka"*, 1987.

[4] F.V. Doljansky, V.I. Kljatsky, A.M. Obuschov, M.A. Chusev.– The nonlinear systems of hydrodynamical type, *M. "Nauka"*, 1974.

[5] B.V. Dasarathy.– Dynamics of a class of social interaction systems, *M. Intern. J. Syst. Sci.*, 1974, V 5, N 4, p. 329-333.

[6] B.V. Dasarathy.– On a generalized dynamic model of bi-state social interaction process, *Intem. J. Syst. Sci.,* 1974, V 5, p. 499-506.

[7] I.N. Bronstein, K.A. Semendjaev.– *Reference book on mathematics,* M. "Nauka". 1986.

Practical Computation of Accessibility, Controllability and Nilpotent Approximations

Matthias Kawski[1]

Department of Mathematics

Arizona State University

Tempe, Arizona 85287

1 Introduction.

The recent past has seen a tremendous effort in the theoretical analysis of nonlinear control systems, that is aimed at eventually even surpassing the almost ubiquituous successful implementation of linear control theory into engineering applications of all kinds. While there are huge areas of nonlinear control theory which still need to be discovered, or are still far from completely understood, many concepts have been developed, necessary and sufficient conditions for many properties have been found, and even algorithms for various kinds of computational problems have been developed.

Nonetheless we still have to see the large engineering community to adopt this so far mainly theoretical work. While actual implementation of nonlinear control designs may often be hampered by still incomplete theories, or just a different language (e.g. *Lie brackets* instead of *matrix products*) we have repeatedly seen that the complexity of the necessary computations is perceived (often erroneously) as a main obstacle.

This need not be: Though admittedly, e.g. nonlinear controllability is harder to check than the Kalman rank condition for linear controllability, it is in most instances still a very straightforward task. Indeed, the verification of the known conditions theoretically involves little more than the tools of elementary calculus (and of course knowledge of the literature for finding these conditions).

This is where this article and the accompanying software-programs are intended to throw a bridge. In particular, our programs provide easy-to-use facilities to compute Lie-brackets of vector fields, to check a set of formal Lie brackets for linear dependence relations (*Jacobi relatedness*), and using these, to test any given system (with at this time still polynomial vector fields) for accessibility, small-time local controllability and also return nilpotent approximations that preserve controllability.

This software is intended to serve a dual purpose: Firstly, it shall allow the practical engineer to effortlessly carry out the computations needed for determining above proper-

---

[1]This work was partially supported by NSF grant DMS 90-07547 and an Arizona State Univerity FGIA grant

ties and finding a simple approximating system, and also eventually win him/her over to the nonlinear theory; secondly, it is supposed to be research tool, helpful for addressing as fundamental questions: "Which ones are the *good / bad* brackets?" "What is a suitable basis for a free Lie algebra adapted to the properties of control systems with a nontrivial drift (=uncontrolled) vector field?" (compare the last section of this note). These programs are intended to be expanded to allow e.g. nonpolynomial vector fields (the next step is to allow transcendental functions as well), and e.g. to return asymptotically stabilizing feedback laws or feedback laws that achieve asymptotic tracking.

This paper is organized as follows: First, we give a brief overview of the precise definitions of the properties we want to check, an almost complete collection of the presently available criteria (both necessary and sufficient conditions). In the following section we address the problems that arise when trying to make these conditions into computer algorithms. Specifically, we address problems of decidability, computational complexity, and some of the combinatorial reasonings that went into these programs. In the final section we provide two small examples (one illustrating practical, the other ptheoretical applications), and demonstrate how these are handled by our programs.

2   Terminology.

We consider smooth control system that are affine in the control of the form

$$(1) \quad \dot{x} = f(x) + \sum_{j=1}^{m} u^j g_j(x)$$

with state $x \in \mathbf{R}^n$, f and g_j real analytic vector fields on $\mathbf{R}^n$, and controls that are bounded $|u^j(\cdot)| \leq 1$ measurable functions of time t.

We work in a neighbourhood of the equilibrium point $x_0 = 0$, of the uncontrolled system, i.e. $f(0) = 0$. (However, this is not required in order to compute Lie products, but it is the typical setting for problems centering on controllability.) The reachable set $R(t)$ at time t is the set of all points $\{x(t, u) : u \text{ admissible}\}$ where $t \longrightarrow x(t, u)$ denotes the solution of the initial value problem (1) together with $x(0) = 0$ corresponding to the control $u(\cdot)$.

The system (1) is *accessible* (from $x = 0$) if $R(t)$ has nonempty n-dimensional interior for all $t > 0$. The system (1) is *small-time locally controllable (STLC)* about $x = 0$ if $0 \in \text{int } R(t)$ for all $t > 0$. We say that the system is *small-time locally controllable with small controls* STLC$_\varepsilon$ if the system is still STLC for every $\varepsilon > 0$ when in addition one requires $|u^i(\cdot)| \leq \varepsilon$.

Since the fundamental work of Sussmann in the 70's (compare e.g. [19]) it is understood that the local geometric properties of systems like (1) are encoded in the relations satisfied by the elements of the Lie algebra generated by the vector fields f and g^j. In order to be able to state the conditions most concisely we first introduce the free Lie algebra $L(X, Y) = L(X, Y_1, \ldots, Y_m)$ on $(m + 1)$ *indeterminants* $X, Y_1, \ldots, Y_m$. We denote by $L^{(k,l)}(X, Y)$ the homogeneous subspace spanned by all Lie-products with k factors X and l_j factors Y_j (or l factors Y if $m = 1$). We also use the direct sums $L^{((r))} = \sum_{k+|l|=r} L^{(k,l)}$ and $S^{((r))} = \sum_{|l|=r} L^{(k,l)}$ where $|l| = \sum_{j=1}^m l_j$.

We do not introduce further symbols for the canonical evaluation homomorphism from $L(X, Y)$ into $L(f, g)$, defined on the generators by $(X, Y_1 \ldots, Y_m) \longrightarrow (f, g_1, \ldots, g_m)$. Instead, we simply keep superscripts. e.g. if $X^\pi = [[X, Y_1], Y_3]$ then $f^\pi = [[f, g_1], g_3]$.

A Lie algebra L is nilpotent if its central descending series consists only of finitely many nontrivial subspaces. I.e. if we use $L^{<1>} = L$ and $L^{<i+1>} = [L^{<i>}, L]$, then $L^{<k+1>} = \{0\}$ for some smallest integer k, called the rank the nilpotent Lie algebra L.

If $L(f, g)$ is nilpotent it is a straightforward task to integrate the differential equation (1) explicitly for any given control $u(t)$ – i.e. this involves simple quadratures only. Thus it often desirable to approximate any given system (1) by a system that generates a nilpotent Lie algebra, yet sufficiently well approximates the local geometric properties of the original system (e.g. preserves STLC, optimality of certain controls). Nilpotent approximating systems may be considered as generalizations of linearizations and they play an invaluable role in the proofs of sufficient conditions for STLC (compare e.g. [21]) and the construction of locally asymptotically stabilizing feedback laws [16].

3 Necessary and sufficient conditions.

In this section we give a brief survey of currently known conditions for controllability etc. For a detailed survey with some background discussion see [14].

The system (1) is accessible from $x = 0$ if and only if $\dim L(f, g)(\_, = n$. (This is the nonlinear equivalent of the Kalman rank condition for linear controllability.)

Historically the first high-order sufficient condition for STLC, and often the easiest to test first is the Hermes condition [10]: If the system (1) is accessible, $m = 1$, and $S^{((2k))}(f, g)(0) \subseteq S^{((2k-1))}(f, g)(0)$ then the system (1) is STLC. This is special case of Sussmann's condition: If the system (1) is accessible and there exists a weight $\theta \in [0, 1]$ such that

$$(2) \quad L^{(k,l)}(f, g)(0) \subseteq \sum L^{(k',l')}(f, g)(0) \qquad \text{for all } k \text{ odd, } l_1, \ldots, l_m \text{ even}$$

where the sum extends over all (k', l') such that $\theta k' + |l'| < \theta k + |l|$ then the system (1) is STLC (compare theorem 7.3 in [21]).

These sufficient conditions were slighlty generalized by Bianchini-Stefani [1] to the effect that one *does not have to count outside factors* f; more precisely, in the right hand side of the set-inequlity in (2) we may allow the span of all brackets of the form $(ad^\nu f, f^\pi)(0)$ where $\nu \geq 0$ and $X^\pi \in L^{(k',l')}(X,Y)$.

Generalizing the Clebsch-Legendre condition we have the following necessary conditions for STLC: If $m = 1$ and (1) is STLC then $(ad^{2k}g, f)(0) \in S^{((2k-1))}(f,g)(0)$ for all $k \in \mathbf{Z}^+$ (Stefani [18]). If $m = 1$ and (1) is $STLC_\epsilon$ then $(ad^2[f,g], f)(0) \in S^{((1))}(f,g)(0) +$ linear span $\{(ad^\nu f, (ad^3 g, f))(0) : \nu = 0, \ldots, \infty$ (Kawski [12]).

These constitute to the author's knowledge all the presently known easily verifiable conditions for STLC in terms of iterated Lie-products. For other conditions for some kinds of controllability compare e.g. [6, 8]. Also, Bianchini-Stefani recently were able to generalize some of above conditions to the case of controllability about a non-stationary reference trajectory [2].

4 Obtaining algorithms.

The first problem that we need to address is that of finding a suitable set of iterated Lie products that one wants to evaluate for any given system. We recall [4] that each homogeneous subspace $L^{(k,l)}(X,Y)$ is free of rank

$$(3) \quad c(\alpha) = \frac{1}{|\alpha|} \sum_{d|\alpha} \mu(d) \frac{(|\alpha|/d)!}{(\alpha/d)!}$$

where α and $|\alpha|$ stand for (k,l) and $k + \sum l_j$, respectively, and μ denotes the Möbius function and the sum is taken over all positive integers d that divide k and each l_j. Similarily, $L^{((r))}(X,Y)$ is free of rank $\tilde{c}(r) = \frac{1}{r}\sum_{d|r} \mu(d)(m+1)^{r/d}$.

Thus we have to confront an exponentially growing number of Lie-brackets that we possible need to evaluate; for the precise formulation of the complexity of deciding accessibility and controllability in the sense of theoretical computer science compare e.g. Sontag [17] and Kawski [15]. While at present we do not know how to circumvent this problem in a systematic way, we at least know how to solve the folllowing one.

There have been a number of theoretical papers on in particular the Campbell-Baker-Hausdorff formula that use the so-called Dynkin bases, i.e. (not true) bases that consist of iterated Lie products of the form $[Z_n, [Z_{n-1}, [Z_{n-2}, \ldots \ldots, [Z_2, Z_1]\ldots]]]$ where each Z_j is one of the indeterminants X, Y_i. These already significantly reduce the number of all possible products since it is known that all brackets of this form span the Lie algebra; however, they do not form a basis, and indeed their number is still much too large.

Instead we chose to use so-called Hall-bases for a free Lie algebra (as an alternative we could have used the Chen-Fox-Lyndon basis [7]). We briefly remark that neither one is well adapted to the problem of determining controllability (see also the last section), thus leaving open the search for a more suitable basis. We briefly recall the definition [4]:

Definition. A Hall basis of the free Lie algebra $L(Z) = L(Z_1, \ldots, Z_s)$ is any totally ordered subset $\mathcal{B}$ of all iterated Lie products satisfying:

(i) If $u, v \in \mathcal{B}$, and $|u| < |v|$ then $u \prec v$;

(ii) $Z_i \in \mathcal{B}$ for $i = 1, \ldots s$;

(iii) If $|w| = 2$ then $w \in \mathcal{B}$ if and only if $w = [Z_i, Z_j]$ with $Z_i \prec Z_j$;

(iv) If $|w| \geq 3$ then $w \in \mathcal{B}$ if and only if there are $a, b, c \in \mathcal{B}$ with $[b, c] \in \mathcal{B}$, $b \prec c$ and $b \preceq a \prec [b, c]$. (Here, $|u|$ denotes the length of the iterated Lie bracket u.)i (To be even more precise the Hall basis is the homomorphic image of a similarily constructed set $|calB'$ in the free magma $M(Z)$ over $Z_1, \ldots, Z_s$.)

Thus it is easily possible to recursively generate a Hall basis for any free nilpotent Lie algebra of any rank. Moreover, we found a simple algorithm (using anticommutativity and the Jacobi identity repeatedly), that allows us to represent any given iterated Lie bracket as a linear combination of elements of a fixed Hall basis. (The details of this algorithm, as well as those of the other parts of our software will appear elsewhere as a technical note). This reduces the problem of checking any given set of formal Lie brackets for linear relations to the simple computation of the rank of the integer itransformation matrix that that expresses the given brackets as linear combinations of the fixed Hall basis. This is a particularly valuable routine for theoretical purposes in view of the larger problem of finding new bases better adapted to the problems of controllability.

Next, we briefly comment on the actual calculation of iterated Lie products of vector fields and how to use these to determine controllability etc. Lacking any further theoretical results we only can expect partial answers to controllability; in particular, the answers any such software is able to obtain can only depend on a finite number of iterated Lie products. We thus may assume that all computations will involve only brackets of total length less than a fixed bound M. Next we observe, that if we consider truncations of the Taylor series expansions of the the components of the vector fields f, g_j (which are assumed to be real analytic) of order M, then this will not affect the answers we obtain for controllability or accessibility (unless we change our mind and consider longer brackets, in which case we need higher order truncations as well). This leads us to consider as a first step only vector fields with polynomial entries. In order to keep the size of the software manageable, we also require the polynomials to have rational coefficients (theoretical results about the exact consequences of this restriction are still under investigation). In order to further facilitate the program, the first version only works with polynomials with integer coefficients – this is no severe restriction as by a change of time scale one always can change the vector fields with rational coeeficients into ones with integer coefficients without affecting any of accessibility, controllability, or nilpotent approximations.

The actual calculation then is done in a straighforward way leaving improvements that e.g. incorporate the theoretical work by Grossman [9] to future updated implementations. Specifically, we use lists of lists of lists of pairs of integers in the programming language PASCAL to store vector fields (vector field = sum of polynomials, polynomial = sum of monomials, monomial is a product of a rational coefficient and integer powers of the coordinate functions). We symbolically compute Jacobian matrices of partial derivatives and multiply these with vector fields which eventually are to be evaluated at the point $x = 0$.

To determine accessibility of a given system, one needs to input a bound $M < \infty$, and the program will then compute the dimension of the span of $L^{((M))}(f, g)(0)$ (as a linear subspace of the tangent space to $\mathbf{R}^n$ at $x = 0$). If the answer is n then the system is accessible, if the answer is less than n, it is in general inconclusive as consideration of higher order brackets may give a positive answer. Only in the case that $L(f, g)$ is nilpotent of rank less or equal to $M - 1$ we may also obtain a negative answer.

To determine STLC, we first check accessibility, and note that the length of the brackets that one needs to consider in order to decide controllability is at least as long as that of those needed for determining accessibility. Given a positive answer for accessibility, the program verifies if any of the conditions stated in the previous section applies. This basically involves checking if any of the (at $x = 0$) nonvanishing brackets is of type (odd, even, even,..., even), and if yes, then verifying if corresponding tangent vector is linearly dependent on tangent vectors corresponding to lower θ-order formal brackets. Typically, this inolves in each step the computation of an subinterval of $[0, 1]$ of applicable values of θ, and the answer to STLC is positive if these subintervals have a nonempty intersection. Checking the necessary conditions in general is even harder, as each of the subspaces $S^{((2k-1))}(X, Y)$ itself is infinite dimensional. Only in a few cases we may obtain negative answers for STLC via Stefani's or Kawski's condition: (i) If the Lie algebra $L(f, g)$ is nilpotent; (ii) if $k = 1$, in which essentially the Cayley-Hamilton-theorem allows us to stop after checking a finite number of brackets of the form $(ad^\nu f, g_j)$, and (iii) in the case of $k = 2$ a similar conditions is known [12]. Theoretical work on this question is in progress.

Given a set of *lowest θ-order* brackets that positively determine STLC, Hermes [11] has given an algorithm that gives a nilpotent approximating system that still is STLC, provided the given brackets in some sense conform to above sufficient conditions. (This algorithm fails in a modification of the system presented in [13], which was shown to be STLC, yet fails all known general tests for STLC.) This algorithm involves linear coordinate changes in the x-state space (computed from the given set of brackets evaluated at zero), expansion of the components of the vector fields f and g_j into monomials that are homogeneous w.r.t. a family of dilations computed from the length of the brackets in the given set, and finally a truncation of these reordered polynomials after their new leading terms. An implementation of this algorithm into our package is currently under way.

5 Examples.

We begin with an example provided by Bloch [3]. The system models a vertical wheel rolling without slipping on a plane surface with control torques about the vertical and one horizontal axis. Specifically, we use (y, z) as coordinates in the plane to describe the point of contact of the wheel and the supporting plane, and let Θ and Φ denote the rotation angle due to rolling (measured from a fixed reference) and the heading angle of the wheel (measured from the x-axis), respectively. If T_1 and T_2 are the two torques about the rolling axis of the wheel and about the vertical axis through the point of contact, respectively, then the equations of motion are

$$
(4) \quad
\begin{cases}
\ddot{\Theta} &= T_1 \\
\ddot{\Phi} &= T_2 \\
\dot{y} &= \dot{\Theta} \cos \Phi \\
\dot{z} &= \dot{\Theta} \sin \Phi
\end{cases}
$$

We rewrite these equations as a first order system, i.e. we let $x = (x_1, x_2, x_3, x_4, x_5, x_6) = (\dot{\Theta}, \dot{\Phi}, \Theta, \Phi, y, z)$ and use $u = (u_1, u_2)$ as usual for the controls (torques) T_1, T_2, to obtain:

$$
(5) \quad
\begin{cases}
\dot{x}_1 &= u_1 \\
\dot{x}_2 &= u_2 \\
\dot{x}_3 &= x_1 \\
\dot{x}_4 &= x_2 \\
\dot{x}_5 &= x_1 \cos x_4 \\
\dot{x}_6 &= x_1 \sin x_4
\end{cases}
$$

In order to be able to use our our current, polynomial-only version, we take high-order Taylor-expansions of the trigonometric functions. (It can be shown that this will still lead to exact answers for controllability, accessibilty, and nilpotent approximations provided that the order of the Taylor polynomial is sufficiently high, e.g. higher than the length of th longest Lie bracket computed.) In this case, we may take three-term (i.e. fifth order) expansions, and upon writing the system in the standard form

$$(6) \quad \dot{x} = f(x) + u_1 g_1(x) + u_2 g_2(x)$$

we have to work with the vector fields

$$f(x) = x_1\frac{\partial}{\partial x_3} + x_2\frac{\partial}{\partial x_4} + x_1(1 - \frac{x_4^2}{2} + \frac{x_4^4}{24})\frac{\partial}{\partial x_5} + x_1(x_4 - \frac{x_4^3}{6} + \frac{x_4^5}{120})\frac{\partial}{\partial x_6}$$
$$(7) \quad g_1(x) = \frac{\partial}{\partial x_1}$$
$$g_2(x) = \frac{\partial}{\partial x_2}$$

To run our program we may write the input file e.g. in the following way:

$$X = x1\,D3 + x2\,D4 + (x1 - 1/2x1\,x4\hat{\ }2 + 1/24x1\,x4\hat{\ }4)D5$$
$$+(x1\,x4 - 1/6x1\,x4\hat{\ }3 + 1/120x1\,x4\hat{\ }5)D6.$$
$$(8)$$
$$Y = D1.$$
$$Z = D2.$$

Running the program with bound $n = 7$ for the maximal length of the Lie brackets to be computed (as a first guess), the output is:

Accessibility: Yes.

Small-Time Local Controllability: Yes.

Lowest order Lie brackets deciding STLC:

$$Y, Z, [X,Y], [X,Z], [[X,Z],[X,Y]], (\text{ad}^2[X,Z],[X,Y]).$$

Nilpotent STLC-preserving approximating system:[2]

$$(9) \quad \begin{cases} \dot{y}_1 = u_1 \\ \dot{y}_2 = u_2 \\ \dot{y}_3 = y_1 \\ \dot{y}_4 = y_2 \\ \dot{y}_5 = y_1 y_4 \\ \dot{y}_6 = y_1 y_4^2 \end{cases}$$

after the linear coordinate change

$$y = (y_1, \ldots, y_6) = (x_1, x_2, \tfrac{1}{2}(x_3 + x_5), x_4, x_6, -x_5).$$

and truncation

As an example for theoretical applications, we describe how to use this software to obtain some insight into the nature of *good* versus *bad* formal Lie brackets. We recall that the main presently open question for single-input systems is to identify which of the brackets of type (odd,even) are bad (i.e. possible obstruction to controllability, unless neutralized) and which are good, in the sense that they give full controllability. It is known that all brackets of types (odd,2k) and of types $(1, 2k)$ are necessarily obstructions to STLC (unless neutralized). Thus the first interesting case are the brackets of type in $L^{(3,4)}(X, Y)$, which is free of rank 5.

One way to proceed is to generate a control system (on a fairly high dimensional statespace) such that the Lie algebra generated by the fields g and f is isomorphic to the free nilpotent Lie algebra on two generators of length 7. While this is not at all hard to write down (of course this in turn is preferrably done by a computer), all we really need is that the system contains the following one as a subsystem:

---

[2]This routine is still being developed, and shall soon be part of the package.

$$\dot{x}_1 = u \qquad\qquad \dot{y}_1 = x_6$$
$$\dot{x}_2 = x_1 \qquad\qquad \dot{y}_2 = \tfrac{1}{2}x_3^2$$
$$\dot{x}_3 = \tfrac{1}{2}x_1^2 \qquad\qquad \dot{y}_3 = \tfrac{1}{4}x_1^2x_2^2$$
$$(10) \quad \dot{x}_4 = \tfrac{1}{6}x_1^3 \qquad\qquad \dot{y}_4 = x_7$$
$$\dot{x}_5 = \tfrac{1}{24}x_1^4 \qquad\qquad \dot{y}_5 = x_2x_4$$
$$\dot{x}_6 = x_5$$
$$\dot{x}_7 = \tfrac{1}{6}x_1^3x_2$$

We then ask the program to generate a Hall basis for $L^{((7))}(Y,X)$ and evaluate the basis elements corresponding to the homogeneous subspace $L^{(4,3)}(Y,X)$ at $Y = g$ and $X = f$, and again evaluate these vector fields at $x = 0$. The program then will return:

$$[g,[g,[g,[f,[f,[f,g]]]]]](0) \;=\; -\tfrac{\partial}{\partial y_1} + 3\tfrac{\partial}{\partial y_2} - 3\tfrac{\partial}{\partial y_3} + 5\tfrac{\partial}{\partial y_4} + 4\tfrac{\partial}{\partial y_5}$$

$$[[g,[f,g]],[[g,[f,[f,g]]]]](0) \;=\; -\tfrac{\partial}{\partial y_2}$$

$$(11) \quad [[f,g],[[f,g],[g,[f,g]]]](0) \;=\; -\tfrac{\partial}{\partial y_3}$$

$$[[f,g],[g,[g,[f,[f,g]]]]](0) \;=\; \tfrac{\partial}{\partial y_1} - 2\tfrac{\partial}{\partial y_4}$$

$$[[f,[f,g]],[g,[g,[f,g]]]](0) \;=\; -\tfrac{\partial}{\partial y_5}$$

From elementary arguments it is clear that there are at least 3 linearly independent supporting hyperplanes for the intersection of the reachable set of system (10) with the y-subspace, that pass through the origin (these correspond to the inequalities $y_1 \geq 0$, $y_2 \geq 0$, and $y_3 \geq 0$. In other words this means that there are at least three independent bad formal Lie brackets of type $(3,4)$. Moreover, by explicitly constructing families of control variations (compare [12]) that the subsystem consisting of the following reduced vector $(x_1, x_2, x_4, x_7, y_4, y_5)$ is STLC, and hence there are also two good brackets of type $(3,4)$. Obviously, above Hall-basis does not split in the desired fashion into basis for the good and bad subspaces, respectively. (In this case this actually could be achieved by choosing a different strict ordering of the brackets in the recursive construction of the Hall basis.) While in this subspace the Chen-Fox-Lyndon basis ([7]) actually splits in the desired fashion, neither it, nor any Hall basis will split in the case of $L^{(5,4)}$ (which is conjectured to be the direct sum of a 9 dimensional bad and a 5 dimensional good subspace.)

Next, we may ask the program to evaluate a number of our formal Lie brackets on this particular system (10). Upon trying various choices, we came to like the following one:

$$X^{\pi_1} = (ad^2X,(ad^4Y,X))$$

$$X^{\pi_2} = (ad^2(ad^2X,Y),X)(0)$$

$$(12) \quad X^{\pi_3} = (ad^2[X,Y],(ad^2Y,X))(0)$$

$$X^{\pi_4} = [X,[[X,Y],(ad^3Y,X)]](0)$$

$$X^{\pi_5} = [(ad^2X,Y),(ad^3Y,X)]](0)$$

which when evaluated as usual on (f,g) and at $x=0$ (i.e. $X^{\pi_i} \longrightarrow f^{\pi_i}(0)$) gives $f^{\pi_i}(0) = \frac{\partial}{\partial y_i}$; $i=1,\ldots,5$; i.e. in some sense this set of formal brackets is diagonal on our system, and obviously splits into the bases $(W^{\pi_1},W^{\pi_2},W^{\pi_3})$ and (W^{π_4},W^{π_5}) for the bad and good subspaces of formal brackets of type $(3,4)$.

Observe that this computation also immediately lets one read of the iterated integral coefficients corresponding to each of these brackets as they would oocur in a modification (namely by using this partial basis instead of the Hall basis) in Sussmann's directed-product expansion of the Chen-Fliess series [20]. For example the coefficient of W^{π_3} in this product expansion is

$$(13) \quad \frac{1}{2!\,2!} \int_0^t \left(\int_0^{t_1} \int_0^{\sigma_1} u(\sigma_2)\,d\sigma_2\,d\sigma_1 \right)^2 \left(\int_0^{t_1} u(\tau_1)\,d\tau_1 \right)^2 dt_1$$

While clearly the simple calculation as done above for the case of $L^{(3,4)}(X,Y)$ only provides one with a nice basis for one particular homogeneous subspace, it nonetheless is a very helpful tool for obtaining some insight into an *example*, which requires major effort if worked out by pencil and paper only; and it remains to be hoped that upon trying several more *examples* (i.e. other homogeneous subspaces of the form $L^{(\text{odd,even})}(X,Y)$ some combinatorist may discover some patterns which in turn lead to conjectures, theorems, algorithms and proofs.

Software availability: The described software is available free of charge from the author upon mailing him an empty $3\frac{1}{2}''$ or $5\frac{1}{4}''$ floppy disk for a PC-compatible computer, or via *e-mail* or *ftp* which in view of the length of the code may be less preferrable.

References

[1] R. Bianchini and G. Stefani, *Sufficient conditions of local controllability*, Proc. 25th IEEE Conf. Decision and Cntrl., Athens, Greece, (1986).

[2] R. Bianchini and G. Stefani, *Graded approximations and controllability along a trajectory* SIAM J. Control & Opt. **28** no.4 (1990) pp. 903–924.

[3] A. M. Bloch and N. H. McClamroch, *Control of mechnical systems with classical non-holonomic constraints*, Proc. 28th IEEE Conf. Decision and Cntrl., Tampa, Florida, (1989) pp. 201–205.

[4] N .Bourbaki, *Lie Groups and Lie Algebras*, (1975) Paris (Hermann).

[5] A. Bressan, *Local asymptotic approximation of non-linear control systems*, Int.J.Control **41** no.5 (1985) pp.1331–1336.

[6] P. E. Crouch and C. I. Byrnes, *Local accessability, local reachability, and representations of compact groups*, Math. Systems Theory **19** (1986) pp. 43–65.

[7] K. T.Chen, R. H. Fox, and R. C. Lyndon; *Free differential calculus, IV. The quotient groups of the lower central series*; Annals of Mathematics (**68**) no.1 (1958) pp. 81–95.

[8] B. Goncalves, *LOcal controllability in codimenson one*, Systems and Control Letters **6** (1985) pp. 213–217.

[9] R. Grossman and R. G. Larson, *Labeled trees and the algebra of differential operators*, internal report: Center for Pure and Applied Mathematics, UC at Berkeley, PAM-368 (1987).

[10] H. G. Hermes, *Controlled Stability*, Annali di Matematica pura ed applicata IV, **CXIV** (1977) pp. 103-117

[11] H. G. Hermes, *Nilpotent approximations of control systems and distributions*, SIAM J. Control & Opt. **24** no.4 (1986) pp. 731–736.

[12] M. Kawski, *A necessary condition for local controllability*; Differential Geometry: The Interface between Pure and Applied Mathematics; M. Luksic, C. F. Martin, and W. Shadwick, eds., Contemporary Mathemaics vol.68 (1987) pp. 143–155.

[13] M. Kawski, *Control variations with an increasing number of switchings*, Bulletin AMS **18** no.2 (1988) pp. 149–152.

[14] M. Kawski, *High-order small-time local controllability*, in: Nonlinear Controllability and Optimal Control, H. J. Sussmann, ed., (1990) pp. 441–477 (Dekker).

[15] M. Kawski, *The complexity of deciding controllability*, Systems and Control Letters **15** (1990) pp. 9–14.

[16] M. Kawski, *Homogeneous stabilizing feedback laws*, to appear in Control Theory and Advanced Technology (CTAT), **6** no.4 (1990).

[17] E. D. Sontag, *Controllability is harder to decide than accessibility*, SIAM J. Control & Opt. **26** (1988) pp. 1106–1118.

[18] G. Stefani, *On the local controllability of a scalar-input control system*, Proc. 24th IEEE Conf. Decision and Cntrl., Ft.Lauderdale, Florida, (1985).

[19] H. J. Sussmann, *An extension of a theorem of Nagano on transitive Lie algebras* Proc. AMS **45** no.3 (1974) pp. 349–356.

[20] H. J. Sussmann, *A product expansion for the Chen series*, Theory and Applications of Nonlinear Control Systems; C. I. Byrnes and A. Lindquist, eds., (1986) pp. 323–335. (Elsevier Publishers)

[21] H. J. Sussmann, *A general theorem on local controllability*, SIAM J. Control & Opt. **25** no.1 (1987) pp. 158–194.

ALGEBRAIC COMPUTING
AND LINEAR SYSTEM CONTROL

Patrick CHENIN

Laboratoire de Modélisation et de Calcul

LMC-IMAG, BP 53X

F-38041 Grenoble Cedex - France

Abstract

In this paper, some problems and results concerning the application of algebraic computing and linear system control is presented. Special insight is made for structural properties of linear systems, parameter dependancies, systems with delays and programming environment specifications for design.

Résumé

Cet article propose l'énoncé de quelques problèmes et résultats concernant l'utilisation du calcul formel et algébrique dans le domaine des systèmes linéaires de la commande. Sont considérés en particulier les propriétés structurelles des systèmes linéaires, le problème de la dépendance de paramètres , les systèmes à retard et les environnements de programmation pour la conception des systèmes linéaires de la commande.

1- Structural properties of linear sytems

Consider linear systems under state-space representation :

$$\begin{cases} x' & = Ax + Bu \\ y & = Cx \end{cases}$$

and under matrix transfert representation :

$$T = C(\lambda I - A)^{-1}B$$

where dimension are given by :

$$A \in M_{n,n}(I\!\!R), \ B \in M_{n,m}(I\!\!R), \ C \in M_{p,n}(I\!\!R).$$

Most of the results concerning the structure of linear system involve normal forms of the matrix T or normal forms of polynomial matrices obtained from factorizations of T. We recall briefly some of these results. In fact, structure results from Hermite and Smith forms of matrices defined on principal ideal domains. We refer to [31,6]. To perform algorithms, it is necessary to know Euclid's algoritms or GCD algorithms (Greater Common Divisor).

Consider $\mathcal{A}$ some principal ideal domain.

- *Unit matrix :*

  We say that $U \in M_{n,n}(\mathcal{A})$ is unit of $M_{n,n}(\mathcal{A})$ if U admit an inverse in $M_{n,n}(\mathcal{A})$.

- *Hermite form :*

  $H \in M_{p,m}(\mathcal{A})$ is under *right Hermite form* if

$$
H = \begin{bmatrix}
0 & \cdots & 0 & h_{1,j_1} & h_{1,j_1+1} & \cdots & \cdots & \cdots & \cdots & h_{1,m} \\
0 & \cdots & \cdots & \cdots & \cdots & 0 & h_{2,j_2} & \cdots & \cdots & h_{2,m} \\
\vdots & \cdots & \cdots & \cdots & \cdots & \cdots & \cdots & \cdots & \cdots & \vdots \\
0 & \cdots & \cdots & \cdots & \cdots & \cdots & 0 & h_{r,j_r} & \cdots & h_{r,m} \\
0 & \cdots & \cdots & \cdots & \cdots & \cdots & \cdots & \cdots & \cdots & 0 \\
\vdots & \cdots & \cdots & \cdots & \cdots & \cdots & \cdots & \cdots & \cdots & \vdots \\
0 & \cdots & \cdots & \cdots & \cdots & \cdots & \cdots & \cdots & \cdots & 0
\end{bmatrix}
$$

  where :

  h_{i,j_i} are normalized (associate elements),

  $h_{i,k}$ are residue modulo h_{i,j_i} for $1 \leq k < j_i$.

  Left Hermite form is defined in the same way by transposition.

- *Smith form :*

  $S \in M_{p,m}(\mathcal{A})$ is under *Smith form* if

$$
S = \begin{bmatrix}
s_1 & 0 & \cdots & \cdots & \cdots & \cdots & 0 \\
0 & s_2 & \cdots & \cdots & \cdots & \cdots & \vdots \\
\vdots & \cdots & \ddots & \cdots & \cdots & \cdots & \vdots \\
\vdots & \cdots & \cdots & s_r & \cdots & \cdots & \vdots \\
\vdots & \cdots & \cdots & \cdots & 0 & \cdots & \vdots \\
\vdots & \cdots & \cdots & \cdots & \cdots & \ddots & \vdots \\
0 & \cdots & \cdots & \cdots & \cdots & \cdots & 0
\end{bmatrix}
$$

  where :

  s_i divides s_{i+1},

  r is the rank of S.

Every matrix W of $M_{p,m}(\mathcal{A})$ admits a unique right Hermite form : it exists some unit U of $M_{p,p}(\mathcal{A})$ and a right Hermite form H such that :

$$W = UH.$$

Every matrix W of $M_{p,m}(\mathcal{A})$ admits a unique Smith form : it exists some Smith form S and U unit of $M_{p,p}(\mathcal{A})$, V unit of $M_{m,m}(\mathcal{A})$ such that :

$$W = USV.$$

Hermite and Smith forms give invariants numbers of the system; the unit matrices U or U and V are significant for change of basis or constructive aspects of control commands.

Some notations :

- $I\!R[\lambda]$: ring of polynomials with real coefficients,.
- $I\!R(\lambda)$: field of fractions of $I\!R[\lambda]$,
- $I\!R_P(\lambda)$: ring of proper rational fractions, that is :

$$f = \frac{n}{d} \in I\!R_P[\lambda] \Longleftrightarrow \deg(n) \leq \deg(d),$$

- $M_{q,r}(I\!R[\lambda])$: ring of matrices with coefficients in $I\!R[\lambda]$,
- $M_{q,r}(I\!R(\lambda))$: ring of matrices with coefficients in $I\!R(\lambda)$,
- $M_{q,r}(I\!R_P(\lambda))$: ring of matrices with coefficients in $I\!R_P(\lambda)$, where q (resp. r) shows
the number of rows (resp. of columns) of the matrices.
The degree on $I\!R_P(\lambda)$ is defined by :

$$\deg_P(f) = \deg(d) - \deg(n).$$

The transfert matrix T is in $M_{p,m}(I\!R_P(\lambda))$.

Factorization of T. [42,6]
A factorization of the matrix transfert of T is a pair of matrix (N, D) or (N_1, D_1) such
that :
$$T = ND^{-1} = D_1^{-1}N_1$$

with

$$N \in M_{p,m}(I\!R[\lambda]), \ D \in M_{m,m}(I\!R[\lambda]), \ N_1 \in M_{p,m}(I\!R[\lambda]), \ D_1 \in M_{p,p}(I\!R[\lambda]).$$

At any such factorization is associated some realization, that is a state space representation of the system which admits T as matrix transfert. Such realization will be minimal
(observable and controllable) if and only if the pairs (N, D) or (N_1, D_1) are prime. To obtain such factorization, it is sufficient to calculate the *Hermite form* of some concatenation
of N and D or of N_1 and D_1 on $I\!R[\lambda]$. Special choice of N or N_1 (row or column proper)
give controllability or observability canonical forms. The units of $M_{m,m}(I\!R[\lambda])$ are said
unimodular matrices.

Finite zeros and poles. [42]
Suppose the factorization of T is prime; hence, finite poles of the system are the zeros of
the determinant of D or D_1 and finite zeros of the system are the zeros of the polynomials
of the Smith form of N or N_1 as polynomial matrices.

Infinite zeros.
Consider T as element of $M_{m,p}(I\!R_P(\lambda))$. Morse ([32], [36])) used this type of ring. The
units of $M_{m,m}(I\!R_P(\lambda))$ or $M_{p,p}(I\!R_P(\lambda))$ are the "bicausal" rational matrices. The *Smith
form* of T on the ring of rational proper fractions is diagonal matrix with diagonal tems
$s^{-\mu}$ and the integers μ are the infinite zeros orders of the system. The structure at infinity
can be obtain in some cases by the Hermite form of T on the same ring ([14]).

Equations.
In some cases, it is necessary to solve linear polynomial matrix equations : find X such
that $RX = S$, where R, X and S are polynomials (or rational) matrices. Conditions for
existence and algorithms involve Smith or Hermite forms ([36,19]).

To compute Smith and Hermite normal forms, we have to perform row and column elementary operations. These operations involve division algorithm or GCD algorithms with use of Bezout identity. In control problems, we have to consider :

1. complexity : the dimensions of the matrix are generally small, hence the main difficulty is the complexity of intermediate matrices which can have large terms ([25,23, 13]); (parallel algorithms are studied and experiments have been done ([37,38,39]);

2. division and GCD algorithm : for polynomials, these algorithms are well studied but, rings of fraction give rise to a nice problem.

Let Ω a domain of complex plane. We consider $I\!\!R_{P,\Omega}(\lambda)$ the ring of proper fractions where all the zeros of the denominator are in Ω. This kind of ring of fractions is studied for stability system problems and these rings are eucidean rings ([36,32]). The degree depends upon the number of zeros of the numerator in Ω and if the euclidean stathme is well defined it seems no possible to compute it. A. BENAMARA and J.P.GUERIN ([1,2]) had shown that it is possible to compute efficiently GCD and to explicit Bezout identity of two elements of $I\!\!R_{P,\Omega}(\lambda)$ (See also [40]). To complete the solution of the problem, it is necessary to know the exact number of zeros of a polynomial in the region Ω. If Ω is the closed unit circle (or any domain homografically equivalent) it is possible solve this problem, even when the coefficients of the polynomial depend of a parameter ([18,9]). Others methods will be presented in the next paragraph.

It is to observe that it is possible to compute exact *Jordan forms* of matrices ([35]).

2- Algebraic equations

Several problems for linear systems needs solutions of algebraic or semi-algebraic equations, for exemple :

1. study of properties of structured systems;

2. structure of systems depending of parameters.

    They can be studied by elimination methods ([26,41]) or by standard bases ([33,34]). See also ([17]).

First, we want to recall the Dixon's method for implicitization of surfaces of $I\!\!R^3$: this is a particular case of elimination. More generally consider the parametric surface ([5]) :

$$x_i = P_i(u); \; i = 1, ..., n; \; u \in \Omega \subset I\!\!R^{n-1}$$

where the P_i are polynomials. Let

$$(S) = \{x \in I\!\!R^n, u \in \Omega\}.$$

The problem is to construct a polynomial fonction f such that : $f(x) = 0$ for $x \in (S)$. This can be done by introducing :

$$\delta(v, x, u) = \begin{vmatrix} Q_1(u_1, u_2, ..., u_{n-1}) & Q_2(u_1, u_2, ..., u_{n-1}) & \cdots & Q_n(u_1, u_2, ..., u_{n-1}) \\ Q_1(v_1, u_2, ..., u_{n-1}) & Q_2(v_1, u_2, ..., u_{n-1}) & \cdots & Q_n(v_1, u_2, ..., u_{n-1}) \\ \vdots & \vdots & \vdots & \vdots \\ Q_1(v_1, v_2, ..., v_{n-1}) & Q_2(v_1, v_2, ..., v_{n-1}) & \cdots & Q_n(v_1, v_2, ..., v_{n-1}) \end{vmatrix}$$

where $Q_i(u) = x_i - P_i(u)$. If $x \in (S)$, there exist some $u \in R^{n-1}$ such that $Q_i(u) = 0$ and $d(v, x, u) = 0$ for all v . Hence, by linear algebra arguments and some hypothesis, it is possible to construct a matrix $T(x)$ such that the determinant of $T(x)$ "contains" the fonction f. The interest of the method developped (for $n = 3$) in [4] is that :

1. it is shown some generalization (relaxation of the hypothesis, P_i are rational, etc),

2. the method can be implemented with symbolic computation or with numeric computation in which case it is not necessary to compute explicitly the implicit equation and it is possible to find the u-value of the parameter which correspond to a x-point of the surface (S) (*inversion problem*).

The second insight I want to do concern the problem : find the set (S) of points x of $I\!R^n$ such that :
- $x \in [0, 1]^n$,
- $f_i(x) = 0$ for $i = 1, 2, ..., p_1$,
- $f_i(x) \geq 0$ for $i = p_1 + 1, ..., p$,

where the f_i are polynomials. The set (S) is said semi-algebraic. The f_i could be rational.

If $n = 1$ and $p_1 = p = 1$, we have to compute the real roots of a polynomial, this is an extensively studied problem ([10,11,12]). In ([29]), are proposed methods which seems to Collins-Akritas-Loos method but which use the Bezier-Bernstein basis. The last permits to use geometrical properties of the representation :

if f is a polynomial of degre n, write it :

$$f(x) = \sum_{0 \leq i \leq n} P_i B_i^n(x) \quad \text{where} \quad B_k^n(x) = C_n^k x^k (1 - x)^{n-k}.$$

Then, for all x in $[0, 1]$, $f(x)$ is in the convex hull of the P_i and $(x, f(x))$ is in the convex hull of the $(i/n, P_i)$.

```
ProcedureResolution ( f , list, subdi, interval)
    if subdi > nmax
        then
            list := list + subdi, interval
        else
            if 0 ∈ Envconv( {Pᵢ} )
                then subdivision of [0, 1]
                    put : f₁(x) = f(2x) ; f₂(x) = f(2(x - 0.5));
                    resolution ( f₁, list, subdi+1, 2*interval);
                    resolution ( f₂, list, subdi+1, 2*interval + 1);
FinResolution.
```

For the evaluation of f_1 and f_2, we use the subdivision algorithm which permits to compute the $\{P_i\}$ of f_1 and f_2 from those of f. It is possible to use Descartes rules too for testing.

In case $n \geq 2$, the same algorithm can be used to obtain the solution or approximations of the solution ([30]). It is necessary to adapt the test of existence or not of solutions on the hypercube: this is done recursively by testing existence of solutions on the hyperfaces of the hypercube ([30]). Observe that the problem of separating or computing the complex roots of a polynomial which are contained in a domain whose frontier is defined by polynomials is a particular case of the problem considered.

3- Delay system analysis

Let a system with delay. In this case ([27,28]), state space representation can be given by:

$$\begin{pmatrix} \dot{x}(t) \\ y(t) \end{pmatrix} = \begin{pmatrix} A(\nabla) & B(\nabla) \\ C(\nabla) & D(\nabla) \end{pmatrix} \begin{pmatrix} x(t) \\ u(t) \end{pmatrix}$$

where ∇ is the operator "delay" ($\nabla f(t) = f(t - \tau)$). The matrix A,B,C and D are polynomial in ∇.

The problem is to find a law for decoupling. Let d_i ($i \in [1, m]$) the integer associated to the nul row D_i by :

$$d_i = \left\{ \begin{array}{l} \min \{k - 1, 1 \leq k \leq n - 1, C_i A^k B \neq 0\} \\ n - 1 \text{ if for all } k, C_i A^k B = 0 \end{array} \right.$$

and let D^* the matrix such that his row i is D_i^* if $D_i \neq 0$ and $C_i A^{d_i} B$ if $D_i = 0$. Let A^* the matrix such that his row $A_i^* i$ is C_i if $D_i \neq 0$ and $C_i A^{d_i+1} B$ if $D_i = 0$.

The solution of our problem of decoupling ([27,28]) needs the inverse of the polynomial matrix D^* : the decoupling control is obtained, under some conditions, by :

$$u = Fx + Gu^* \text{ with} : G = D^{*-1}\Lambda \text{ and } F = -D^{*-1}A^*.$$

It is easy to program it in symbolic system but, for industrial applications, it is necessary to compute the solutions using floating real numbers (we use the Bareiss algorithm to compute the inverse) : comparing solutions, we can observe that the results are a nice approximation of the exact solutions given by symbolic system ([3]).

4- Programming environment for design of linear control systems

Various CAD software exist for control systems but few of them use algebraic systems. If they do, it is especially for linearization of the system , for exemple ([21]). Some software has been writen for studying systems [7,40]. The objective is to show structural properties of the systems. In ([22]), the interest is given on generation of simulation code.

Another interest for the use of algebraic systems is for learning. This is the case in ([24]) where it is shown how a student can learn linear control system theory with the help of such software.

It seems that few of these systems can run out of his designers. In ([16]), we try to propose some solution. An important goal is to permit the interactive development of the design of the control system, using all facilities that algebraic sytems offer.

5- Conclusion

We have tried to give some ideas of the use of symbolic algebraic system for the understanding and the design of multivariable linear control system. A significant number of theoretical problems have been solved. However, the traduction of them in algebraic computing software is to be developed in future for solution of industrial problems. On the other hand, practical methods for parametric and structured systems and for delay-systems have to be studied.

References

[1] A. BENAMARA Séminaire au Laboratoire d'Automatique de GRENOBLE (mars 1990)

[2] A. BENAMARA and J.P.GUERIN This volume

[3] F. BENISTANT and P. CHENIN "Découplage pour des systèmes à retard: calcul formel et calcul numérique" Rapport Technique LMC-IMAG Grenoble (Avril 1991)

[4] L. BIARD Méthode algorithmique d'implicitisation et d'inversion. Application au lancer de rayons. *Thèse, Université Joseph Fourier, Grenoble*, Novembre 1990

[5] L. BIARD and P. CHENIN Méthode algorithmique d'implicitisation des surfaces rationnelles, *Rapport de Recherche, LMC - IMAG, Grenoble*, Mars 1991

[6] P. CHENIN "Modules et matrices polynomiales pour l'étude des systèmes linaires" in Outils et Modèles mathématiques pour l'automatique, Vol.2, Editions du C.N.R.S 1982.

[7] P. CHENIN "Autored: un logiciel pour l'automatique linéaire" Outils et Modèles mathématiques pour l'automatique, Vol.3, Editions du C.N.R.S 1983

[8] P. CHENIN "Un exemple d'utilisation du calcul formel en automatique: étude structurelle..." Calcul formel et automatique Editions du CNRS (1987).

[9] R. COLEMAN Séminaire Calcul Formel IMAG-Grenoble (Avril 1990)

[10] G.E. COLLINS and A.G. AKRITAS "Polynomial real root isolation using Descartes rules of sign" SYMSAC 1976, pp. 272-275

[11] G.E. COLLINS and R. LOOS "Real zeros of polynomials" Computing 1982, Suppl. 4 pp. 83-94

[12] G.E. COLLINS and J.R. JOHNSON "Quantifier elimination and the sign variation method of real roots" Proc. ACM-SIGSAM 1989 - ISSAC 89 ACM Press pp.264-271

[13] CONTI P. "Hermite canonical form and Smith canonical form of a matrix over a principal ideal domain" SYMSAC 1990

[14] J.M. DION L. DUGARD IEE-CDC 1984

[15] J.M. DION C.COMMAULT SIAM JOURNAL on CONTROL Jan. 1988.

[16] P.O. GARREAU and P.CHENIN Rapport de Recherche LMC-IMAG Grenoble (Avril 1991)

[17] M. GIUSTI This volume

[18] B.GLEYSE "Calcul formel et nombre de racines d'un polynome dans le disque unite: applications en automatique et biochimie" Theses Grenoble-IMAG (1986)

[19] M.L.J. HAUTUS "Linear matrix equations with applications to regulator problems" Outils et Modèles mathématiques pour l'automatique, Vol.3, Editions du C.N.R.S 1983.

[20] W. HIRSCHBERG and D. SCRAMM "Application of NEWEUL in Robot Dynamics" J. Symbolic (1989) 7, 199-204

[21] W. HIRSCHBERG and D. SCRAMM "Application of NEWEUL in Robot Dynamics" J. Symbolic (1989) 7, 199-204

[22] ULF HOLMBERG, M. LILJA and B. MARTENSSON "Integrated different symbolic and numerical toos for linear algebra and linear systems analysis" Linear Algebra in Signals, Systems and Control SIAM Philadelphia 1988

[23] ILIOPOULOS "Gaussian elimination over an euclidean ring" Pric. EUROCAL, BuchbergerB.(Ed) Springer LNCS 204, p29-30, 1985.

[24] J.R. JAMES "Lessons learned in coordinating symbolic and numerical computing in K.B.S. for control design" Coupling Symbolic and Numerical Computing in Expert Systems,II Ed. Kowalik and Kitzmiller North Holand 1988

[25] KANNAN R. and BACHEM A. "Polynomial algorithms for computing the Smith and Hermite Normal Forms of an integer matrix" SIAM J. Computing 8(4),p. 499-507 Nov.1979

[26] Y. LECOURTIER and E. WALTER "The testing of structural properties through symbolic computation in identifiability of parametric modes" Ed. Walter Pergamon Oxford 1987

[27] M. LIU "Decoupling and coefficient assignment for (ABCD) time-delay system" International Journal of Control (1988)

[28] M. LIU "Découplage des systèmes avec retards. Simulation et commande optimale d'une installation de production de granulats" Thèse de Doctorat LAN-ENSM Nantes (1988)

[29] D. MARCHEPOIL and P. CHENIN "Algorithmes de recherche de zros d'une fonction de Bzier" Rapport de Recherche 834 I-M LMC-IMAG Grenoble Nov. 1990

[30] D. MARCHEPOIL and P. CHENIN "Facettisation d'ensembles semi-algbriques" Journes Gomtrie Algorithmique (18-20 juin 1990) Rapport INRIA Sept.1990

[31] M. MARCUS "Introduction to modern algebra" Pure and Applied Mathematics Ed. Dekker 1978

[32] A.S. MORSE "Systems invariants under feedback and cascade control" Proc. Symposium on control, Udine, Springer Verlag (1975)

[33] F. OLLIVIER "Inversibility of rational mappings and structural identifiability in automatics" Proc. ACM-SIGSAM 1989 - ISSAC 89 ACM Press

[34] F. OLLIVIER This volume

[35] P.OZELLO "Calcul des formes de Jordan et Frobenius exactes" Thèse de Doctorat IMAG-Grenoble (1987)

[36] L.PERNEBO "An algebraic theory for the design of controllers for linear multivariable systems" IEEE Transactions on Automatic Control ,Vol AC-26 1981.

[37] J.L.ROCH,P.SENECHAUD,F.SIEBERT,G.VILLARD "Computer algebra on a MIMD machine" Int. Symp. on Agebraic Computation, Roma, Italy 1988

[38] F. SIEBERT-ROCH "Parallel algorithms for Hermite Normal form of an integer matrix" Proc. ACM-SIGSAM 1989,- ISSAC 89 ACM Press

[39] F. SIEBERT-ROCH "Calcul formel et parallelisme" Thèse de Doctorat IMAG-Grenoble 1990

[40] T. UMENO, S. YAMASHITA, O. SAITO and K. ABE "Symbolic computation application for design of linear multivariable control systems" J. Symbolic Computation (1989) 8, 581-588

[41] E. WALTER "Identifiabilité structurelle des systèmes linéaires" Calcul formel et automatique Editions du CNRS (1987).

[42] W.A. WOLOVICH "Linear multivariable systems" Springer Verlag 1974.

Asymptotic Behaviour of Nash Strategy in Linear-Quadratic Games A Computer Algebra Approach

H. Abou-Kandil † and Gerhard Jank ‡
† Ecole Normale Superiéure Cachan - LURPA
61, Av. du Président Wilson - 94235 Cachan Cedex, France
‡ Lehrstuhl II für Mathematik, RWTH Aachen, Templergraben 55
D-5100 Aachen, Federal Republic of Germany

Extended Abstract. It is well known that Riccati type equations occur frequently in control and signal processing problems. For the standard simple case with time invariant linear systems and a quadratic performance index, the asymptotic behavior as $t \to \infty$ of the solutions for the Riccati equation is well established. However for non-standard cases, e.g. nonsquare Riccati matrices, time varying systems or whenever the systems parameters are subject to uncertainties, the asymptotic behavior of the solutions for Riccati type equations cannot be predetermined.
The purpose of this paper is to present a computer algebra based method giving all asymptotic solutions for a Riccati differential matrix equation of the type

$$\dot{W} = \hat{A}W + WA + \hat{Q} + W\hat{S}W \tag{1}$$

It should be noted that all matrices need not to be square, it is only assumed that they are of appropriate dimensions such that matrix multiplication can be achieved. It is then shown that this general type of Riccati equations can be adapted to treat control problems when a Nash strategy is sought for linear systems with quadratic cost functionals.

The first step in the solution procedure is to transform equation (1) to a differential matrix equation of the type

$$\dot{Y} = MY \tag{2}$$

with

$$Y = \begin{pmatrix} U \\ V \end{pmatrix} \quad \text{, using the transformation } W = VU^{-1}$$

For the sake of clarity and to prepare the treatment of Nash games, a matrix M of dimension $3n \times 3n$ is considered, where n is the system's dimension. Assuming that matrix M has only simple eigenvalues, the fundamental system could be written in function of the eigenvectors of M and the general solution of (2) is obtained as

$$Y(t) = (y_1(t), y_2(t), \ldots, y_n(t)) \tag{3}$$

By combining the rows of $Y(t)$ with the eigenvectors of M, a closed form expression for any element w_{ij} of the solution matrix $W = VU^{-1}$ can be obtained. For the cases where $n = 2$ and $n = 3$, it is shown that the limits of w_{ij} as $t \to \infty$ form a finite set of numbers irrespective to the initial conditions. A computer algebra program is used to determine these limits. The same property can be extended to cases where $n > 3$.

The proposed method is then applied to two-players linear-quadratic differential Nash games with an open-loop information pattern. In order to determine the strategy desired, it is well known that coupled Riccati type matrix equations, with coupling in the quadratic terms, have to be solved. Moreover the asymptotic behavior when $t \to \infty$ for such equations cannot be predetermined. The first problem is to transform the coupled equations to a single non-standard Riccati equation with rectangular "weighting matrices". By this means, the computer algebra method developed above could be used to solve this problem. Thus, all asymptotic solutions as $t \to \infty$ are obtained. Finally, an example is given to illustrate the proposed method.

The developed computer algebra program can be easily adapted to tackle other interesting applications in control theory, e.g. robust control problems and Stackelberg optimization schemes.

1. Asymptotic solution for a Riccati equation

Consider the differential matrix equation

$$(1.1) \qquad \qquad \dot{Y} = MY$$

with $Y = \begin{pmatrix} U \\ V \end{pmatrix} \in \mathbf{R}^{3n \times n}$ and $M = \begin{pmatrix} -A & \hat{S} \\ \hat{Q} & \hat{A} \end{pmatrix} \in \mathbf{R}^{3n \times 3n}$.

It is well known that using the transformation $W = VU^{-1}, U \in \mathbf{R}^{n \times n}, W, V \in \mathbf{R}^{2n \times n}, W$ is a solution of the Riccati equation

$$(1.2) \qquad \qquad \dot{W} = \hat{A}W + WA + \hat{Q} + W\hat{S}W$$

From now we assume that matrix M in (1.1) has only simple eigenvalues $\lambda_1, \ldots, \lambda_{3n}$. Then a fundamental system of the equation (1.1) can be written in the form

$$\mathbf{x}_1 e^{\lambda_1 t}, \mathbf{x}_2 e^{\lambda_2 t}, \ldots, \mathbf{x}_{3n} e^{\lambda_{3n} t},$$

where $\mathbf{x}_j \in \mathbf{R}^{3n}, j = 1, \ldots, 3n$, denotes the eigenvectors of M. Herewith we get the general solution of (1.1) by

$$(1.3) \qquad \qquad Y(t) = (\mathbf{y}_1(t), \ldots, \mathbf{y}_n(t))$$

where

$$\mathbf{y}_1(t) = c_{11}\mathbf{x}_1 e^{\lambda_1 t} + \cdots + c_{3n,1}\mathbf{x}_{3n} e^{\lambda_{3n} t}$$
$$\vdots \qquad \vdots \qquad \qquad \vdots$$
$$\mathbf{y}_n(t) = c_{1n}\mathbf{x}_1 e^{\lambda_1 t} + \cdots + c_{3n,n}\mathbf{x}_{3n} e^{\lambda_{3n} t},$$

with $C = (c_{ij}) \in \mathbf{R}^{3n \times n}$.

If we denote the rows of C by $\mathbf{c}_1, \ldots, \mathbf{c}_{3n}$ and the rows of $Y(t)$ in (1.3) by $\mathbf{z}_j, j = 1, \ldots, 3n$, we get

$$(1.4) \qquad \qquad \mathbf{z}_j = \sum_{k=1}^{3n} e^{\lambda_k t} x_{kj} \mathbf{c}_k, \quad j = 1, \ldots, 3n,$$

where x_{kj} denotes the jth component of the eigenvector $\mathbf{x}_k$ of M.
For the present we have, together with (1.1)

$$
(1.5) \qquad U = \begin{pmatrix} \mathbf{z}_1 \\ \cdot \\ \cdot \\ \cdot \\ \mathbf{z}_n \end{pmatrix}, V = \begin{pmatrix} \mathbf{z}_{n+1} \\ \cdot \\ \cdot \\ \cdot \\ \mathbf{z}_{3n} \end{pmatrix},
$$

and if we denote with U_{jk} the complementary algebraic element to the element u_{jk} of the matrix U and put

$$
(1.6) \qquad \mathbf{U}_j = (-1)^j \begin{pmatrix} -U_{j1} \\ U_{j2} \\ \cdot \\ \cdot \\ \cdot \\ (-1)^n U_{jn} \end{pmatrix},
$$

we get for the elements w_{ij} of the solution matrix $W = VU^{-1}$ in (1.2)

$$
(1.7) \qquad w_{ij} = \frac{(\mathbf{U}_j, \mathbf{z}_{n+i})}{(\mathbf{U}_j, \mathbf{z}_j)}, i = 1, \ldots, 2n, \quad j = 1, \ldots, n,
$$

where we make use of the standard scalar product notation.
Together with (1.4),(1.5),(1.6) and (1.7) this yields

$$
(1.8) \qquad w_{ij} = \frac{\sum_{k=1}^{3n} e^{\lambda_k t} x_{k,n+i} (\mathbf{U}_j, \mathbf{c}_k)}{\sum_{k=1}^{3n} e^{\lambda_k t} x_{k,j} (\mathbf{U}_j, \mathbf{c}_k)}
$$

As pointed out in the abstract, the purpose of this paper is to give a method for the computation of the limits of w_{ij} as $t \to \infty$.
We will do this for the cases $n = 2$ and $n = 3$, but with some more effort it could be done for $n > 3$ too. In this particular cases one finds out that these desired limits form a finite set of numbers not depending on the matrix C, i.e. on the initial values of the solution.
These possible limits can be computed easily using a CA-system like MAPLE.

2. The cases where $n \geq 2$

Considering the case $n = 2$ in (1.6) we have, together with (1.4)

$$\mathbf{U}_1 = \sum_{k=1}^{6} e^{\lambda_k t} x_{k2} \begin{pmatrix} c_{k2} \\ -c_{k1} \end{pmatrix} \quad \mathbf{U}_2 = \sum_{k=1}^{6} e^{\lambda_k t} x_{k1} \begin{pmatrix} -c_{k2} \\ c_{k1} \end{pmatrix}.$$

Since $\left(\begin{pmatrix} c_{k2} \\ -c_{k1} \end{pmatrix}, \mathbf{c}_k \right) = 0, \quad k = 1, \ldots, 6$, we obtain from (1.8)

$$(2.1) \qquad w_{ij}(t) = \frac{\sum_{k \neq l} e^{(\lambda_k + \lambda_l)t} x_{k,2+i} x_{l,3-j} d_{lk}}{\sum_{k \neq l} e^{(\lambda_k + \lambda_l)t} x_{k,j} x_{l,3-j} d_{lk}}, \quad i = 1, \ldots 4, \quad j = 1, 2,$$

with $d_{lk} = \left(\begin{pmatrix} c_{l2} \\ -c_{l1} \end{pmatrix}, \mathbf{c}_k \right)$.

For the case $n = 3$ in (1.6) we get after some short computations, together with (1.4)

$$\mathbf{U}_1 = \sum_{i \neq s} e^{(\lambda_i + \lambda_s)t} x_{i2} x_{s3} \mathbf{f}_{is} \quad \mathbf{U}_2 = \sum_{i \neq s} e^{(\lambda_i + \lambda_s)t} x_{i1} x_{s3} \mathbf{f}_{is} \quad \mathbf{U}_3 = \sum_{i \neq s} e^{(\lambda_i + \lambda_s)t} x_{i2} x_{s1} \mathbf{f}_{is},$$

where the summation runs over $i, s = 1, \ldots 9$ and

$$\mathbf{f}_{is} = \begin{pmatrix} c_{i2} c_{s3} - c_{i3} c_{s2} \\ -[c_{i1} c_{s3} - c_{i3} c_{s1}] \\ c_{i1} c_{s2} - c_{i2} c_{s1} \end{pmatrix}.$$

Notice now that $\mathbf{f}_{is} = -\mathbf{f}_{si}$ and that $(\mathbf{f}_{is}, \mathbf{c}_k) = 0$ for $k = i$ or $k = s$. This yields, together with (1.8)
(2.2)

$$w_{lm}(t) = \frac{\sum_{i > s} \sum_{k \neq i,s} e^{(\lambda_i + \lambda_s + \lambda_k)t} x_{k,3+l} D_{mis} d_{isk}}{\sum_{i > s} \sum_{k \neq i,s} e^{(\lambda_i + \lambda_s + \lambda_k)t} x_{k,m} D_{mis} d_{isk}} \quad l = 1, \ldots, 6, \quad m = 1, 2, 3,$$

where all summations have to run from $1, \ldots, 9$, $D_{1is} = x_{i2} x_{s3} - x_{s2} x_{i3}$, $D_{2is} = x_{i1} x_{s3} - x_{s1} x_{i3}$, $D_{3is} = x_{i2} x_{s1} - x_{s2} x_{i1}$, and $d_{isk} = (\mathbf{f}_{is}, \mathbf{c}_k)$.
We conclude with the following

Proposition. *Assuming that matrix M, defined by (1.1), has only simple eigenvalues and that $n=2$ or $n=3$.*

If the sums

$$\lambda_k + \lambda_l \ in \ (2.1) \ or \ \lambda_i + \lambda_s + \lambda_k \ in \ (2.2)$$

have different real parts, respectively, then there is a finite set of limit matrices $\Lambda = \{L_1, \ldots, L_p\}, L_j \in (\mathbf{R} \cup \{\infty\})^{2n \times n}, j = 1, \ldots, p, \quad p \leq \begin{pmatrix} 3n \\ n \end{pmatrix}$
such that for all solutions W of (1.2) there is a matrix $L \in \Lambda$ with

$$\lim_{t \to \infty} W(t) = L.$$

Remark. *The set Λ of limit matrices can be computed by taking n linear independent fundamental solutions in the matrix Y in (1.3). There are $\begin{pmatrix} 3n \\ n \end{pmatrix}$ such possibilities.*

Proof. From the representation of the matrix elements, in (2.1) or (2.2), respectively, of a solution W of the differential equation (1.2), together with the hypothesis of the theorem, there is always exactly one exponential term which is leading with respect to the others. Since only the coefficients d_{lk} and d_{isk} depend on the arbitrary matrix C, and they cancel after taking the limit as $t \to \infty$, we get a limit which can be derived by taking only the two or three relevant fundamental solutions in Y. This gives the desired result.

One possible way for computing these matrices in Λ with a CA-system is to compute a fundamental system for the differential equation (1.1). If the matrix M contains parameters and the eigenvalues can be computed then the proposed method will also work in that case. Taking n linearly independent solutions from the set of fundamental solutions gives a specific solution Y of (1.1). Partitioning this matrix in two blocks U, V and computing $W = VU^{-1}$ one gets one of the limiting matrices, as $t \to \infty$. There are $\binom{3n}{n}$ possibilities to choose the matrix Y in the described way. If the conditions on the eigenvalues are fulfilled it is clear from (2.1), (2.2) that there are no other limiting matrices.

3. Application to Nash games

Consider a two-player linear-quadratic differential game defined by:

$$(3.1) \qquad \dot{x} = Ax + B_1 u_1 + B_2 u_2; \quad x(0) = x_0$$

$$x \in \mathbf{R}^n, \quad u_i \in \mathbf{R}^{r_i}, \quad i = 1, 2$$

and the cost functionals

$$(3.2) \qquad J_1 = \frac{1}{2} x^T(t_f) K_{1f} x(t_f) + \frac{1}{2} \int_0^{t_f} (x^T Q_1 x + u_1^T R_{11} u_1 + u_1^T R_{12} u_2) dt$$

$$J_2 = \frac{1}{2} x^T(t_f) K_{2f} x(t_f) + \frac{1}{2} \int_0^{t_f} (x^T Q_2 x + u_2^T R_{21} u_1 + u_2^T R_{22} u_2) dt.$$

All weighting matrices are symmetric and $R_{ii} > 0, i = 1, 2$.
If an open-loop Nash strategy is sought, it is well known that the solution is given by [1]:

$$(3.3) \quad u_1(t) = -R_{11}^{-1} B_1^T K_1(t) \Phi(t, 0) x_0 \quad u_2(t) = -R_{22}^{-1} B_2^T K_2(t) \Phi(t, 0) x_0$$

where $K_1(t)$ and $K_2(t)$ satisfy the coupled Riccati differential system

$$(3.4) \quad \dot{K}_1 = -A^T K_1 - K_1 A - Q_1 + K_1 S_1 K_1 + K_1 S_2 K_2, \quad K_1(t_f) = K_{1f}$$

$$(3.5) \quad \dot{K}_2 = -A^T K_2 - K_2 A - Q_2 + K_2 S_2 K_2 + K_2 S_1 K_1, \quad K_2(t_f) = K_{2f}$$

where

$$S_1 = B_1 R_{11}^{-1} B_1^T \quad \text{and} \quad S_2 = B_2 R_{22}^{-1} B_2^T.$$

$\Phi(t, 0)$ is the transition matrix satisfying

$$(3.6) \qquad \dot{\Phi}(t, 0) = (A - S_1 K_1 - S_2 K_2) \Phi(t, 0), \quad \Phi(t, t) = I.$$

Many algorithms has been developed to solve this kind of coupled Riccati matrix equation [2 - 6]. In reference [5] explicit expressions for $K_1(t)$ and $K_2(t)$ are obtained under the assumption that $Q_2 = \alpha Q_1$ where α is a scalar. As $t_f \to \infty$ it is interesting to know the asymptotic behavior of the solution $K_1(t), K_2(t)$. This case is widely studied in economics [7]. Generally, constant matrices are desired, however, depending on the game under consideration, this could not be guaranteed. The purpose of this paper is to give a systematic method using symbolic computation to obtain all possible limits of solutions as $t_f \to \infty$ of the coupled Riccati equations. Let τ be the remaining time before t_f, i.e. $\tau = t_f - t$. Writing the necessary conditions for an open loop Nash strategy in function of τ, we obtain [5]

$$(3.7) \qquad \frac{d}{d\tau} \begin{pmatrix} \chi(\tau) \\ \psi_1(\tau) \\ \psi_2(\tau) \end{pmatrix} = \begin{pmatrix} -A & S_1 & S_2 \\ Q_1 & A^T & 0 \\ Q_2 & 0 & A^T \end{pmatrix} \begin{pmatrix} \chi(\tau) \\ \psi_1(\tau) \\ \psi_2(\tau) \end{pmatrix} = MY(\tau).$$

Rewriting equation (3.7) in partitioned form

$$(3.8) \qquad \frac{d}{d\tau} \begin{pmatrix} U \\ V \end{pmatrix} = \begin{pmatrix} -A & -\hat{S} \\ \hat{Q} & \hat{A} \end{pmatrix} \begin{pmatrix} U \\ V \end{pmatrix}$$

with $\hat{S} = (S_1 \ S_2)$; $\hat{Q} = \begin{pmatrix} Q_1 \\ Q_2 \end{pmatrix}$; $\hat{A} = \begin{pmatrix} A^T & 0 \\ 0 & A^T \end{pmatrix}$; $U = \chi$; $V = \begin{pmatrix} \psi_1 \\ \psi_2 \end{pmatrix}$

and

$$Y = \begin{pmatrix} \chi \\ \psi_1 \\ \psi_2 \end{pmatrix} = \begin{pmatrix} U \\ V \end{pmatrix}.$$

It is easy to see that (3.8) is identical to (1.1). In fact, the problem is reformulated to avoid the coupling in the Riccati equations.

3. Example

If we start with the matrix

$$M = \begin{pmatrix} 0 & -1 & 1 & 0 & 0 & 0 \\ -2 & 1 & 0 & 0 & 0 & 2 \\ 0.1 & 0 & 0 & 2 & 0 & 0 \\ 0 & 0 & 1 & -1 & 0 & 0 \\ 0 & 0 & 0 & 0 & 0 & 2 \\ 0 & 0.2 & 0 & 0 & 1 & -1 \end{pmatrix}$$

with the eigenvalues $2.1437, -0.9431, -2.1324, 1.0408, -2.0156, 0.9066$.
The 15 possible limit matrices for the solution W of the associated Riccati
equation are:

$$\begin{pmatrix} .01866933697 & -.02281890147 \\ -.02717721047 & -.02319792598 \\ -.1805187340 & -.002486109923 \\ .0007538515521 & .09080026057 \end{pmatrix} \begin{pmatrix} -.1128255227 & -.08610995427 \\ .1371207777 & .05588191542 \\ .8636421525 & .5000889868 \\ -1.097854698 & -.4379815633 \end{pmatrix}$$

$$\begin{pmatrix} 1.288991328 & .5886119668 \\ .6290545470 & .2926592788 \\ 1.468065282 & .7910096838 \\ .7422395850 & .4476918747 \end{pmatrix} \begin{pmatrix} -2.320441380 & -1.148678728 \\ 2.278489392 & 1.086564593 \\ -1.279975403 & -.5316761502 \\ 1.316692803 & .7241874658 \end{pmatrix}$$

$$\begin{pmatrix} -3.348076925 & -1.643299854 \\ -1.709140137 & -.8327596433 \\ 11.13202737 & 5.442464015 \\ 5.418918595 & 2.698669129 \end{pmatrix} \begin{pmatrix} .2735898954 & -.2939208752 \\ -.3456910673 & .3155340139 \\ -2.204765743 & 2.150252621 \\ 2.130555014 & -2.174192812 \end{pmatrix}$$

$$\begin{pmatrix} 4.592299018 & -4.886765698 \\ 2.335500177 & -2.535849346 \\ 5.754994343 & -6.314763629 \\ 2.670377130 & -2.748280842 \end{pmatrix} \begin{pmatrix} -1.874995993 & 1.991049308 \\ 1.839412890 & -2.008272271 \\ -1.070601859 & .9440962031 \\ 1.066093748 & -1.042163477 \end{pmatrix}$$

$$\begin{pmatrix} -.2610019208 & .2746048546 \\ -.1668956955 & .1253893723 \\ .7591998399 & -1.001854547 \\ .4508337895 & -.3878490891 \end{pmatrix} \begin{pmatrix} 2.564815344 & -1.526122408 \\ 1.076774174 & -.4494556784 \\ 2.018163924 & -.1208029895 \\ 2.416949429 & -2.328213289 \end{pmatrix}$$

$$\begin{pmatrix} -2.023879926 & .9416388282 \\ 1.990827303 & -.9410256667 \\ -.9920112003 & 1.498043556 \\ .9923330570 & -1.562066737 \end{pmatrix} \begin{pmatrix} -.5034627347 & .1239714115 \\ -.08580415971 & .1757690483 \\ 2.103488110 & -.1666896100 \\ -.3109933898 & -.8611488736 \end{pmatrix}$$

$$\begin{pmatrix} -1.433692854 & 5.10158065 \\ 1.873263963 & -1.769672508 \\ -.6048496354 & 4.226957258 \\ 1.175563826 & -.2705620983 \end{pmatrix} \begin{pmatrix} 1.004213266 & 1.060644119 \\ .4854580948 & .5306767088 \\ 2.061561914 & -.1927370653 \\ 1.029450277 & -.02837246964 \end{pmatrix}$$

$$\begin{pmatrix} -2.298065195 & -.9909598759 \\ 2.365317779 & 1.698575686 \\ -1.550239769 & -2.436638313 \\ 1.227369148 & .09458843452 \end{pmatrix}$$

References

[1] Starr, A.W. and Y.C. Ho, Nonzero-sum differential games, Journal of Optimization Theory and Applications,vol. 3, pp. 184 - 206, 1969.

[2] Cruz, J.B., Jr. and C.I. Chen, Series Nash solution of two person nonzero-sum linear-quadratic games, Journal of Optimization Theory and Applications, vol. 7 , pp. 240 -257, 1971.

[3] Simaam, M. and J. B. Cruz, Jr., On the solution of open-loop Nash Riccati equations in linear-quadratic differential games, Int. Journal of Control, vol. 18, pp. 57 - 63, 1973.

[4] Özgüner, U. and W.R. Perkins., A series solution to the Nash strategy for large scale interconnected systems, Automatica, vol. 13, pp. 313 - 315, 1977.

[5] Abou-Kandil, H. and P. Bertrand, Analytic solution for a class of linear-quadratic open-loop Nash games, Int. Journal of Control, vol. 43, pp. 997 - 1002, 1986.

[6] Jodar, L. and H. Abou-Kandil, A resolution method for Riccati differential systems coupled in their quadratic terms, SIAM Journal on Mathematical Analysis, vol. 19, pp. 1425 - 1430, 1988.

[7] Neck, R. and E. Dockner, Open-loop equilibria for an LQ-differential game of stabilization policies, Proceedings of the 10th IFAC World Congress, Munich, FRG, vol.5, pp. 208 - 214, 1987.

OPTIMAL POLE PLACEMENT BY WEIGHT SELECTION

Driss Mehdi and Eric Ostertag
ENSPS-LAII, 7 Rue de l'Université, F-67000 Strasbourg

Abstract

In this paper a recursive method of optimal pole placement of by weight selection is presented. First, a degree of stability by cost criterion modification is introduced. The method allows the allocation of both real and imaginary parts in the case of a complex conjugate pair.

Key words : Optimal Pole Placement, Weight Selection, Performance Modification

0. Introduction

It is well known that optimal regulators minimizing a quadratic cost criterion (LQ technique) have the advantages that the poles of the closed loop system are garanteed to be stable, but their exact location and thus the final performance of the closed loop system (transient or harmonic response) are not known. On the other hand, when a regulator is designed by pure pole placement techniques, the choice of these poles in order to achieve a given closed loop performance is uneasy for large order systems. Several methods have been reported, which utilize the LQ technique to achieve at the same time a desired pole placement. Solheim (1972, 1974) has given methods to design an optimal regulator by successive eigenvalue shifting for both continuous and discrete time systems. Amin (1984) has improved Solheim's methods to include the recursion of the control weighting in the discrete time case. This has led him to overcome the drawbacks of the algorithm given by Solheim. Recently, Saif (1989) has improved Amin's result (1985) in the continuous time case by using a result obtained by Medanic (1988), which consists in introducing a relative degree of stability by perturbing the state weighting matrix in the cost criterion, unlike the method given by Anderson *et al.* (1971). The algorithm of Saif is powerful in the sense that it allows to achieve, in the case of a complex conjugate pair, assignment of both the real and the imaginary part.

In the present work, a recursive procedure for determining the weighting matrices appearing in the quadratic cost criterion associated with an optimal regulator and giving rise to a desired set of closed loop eigenvalues is presented. This work is a discrete time version of the work of Saif (1989) and Medanic (1988) and is an improvement of Amin's work in the sense that it allows assignment of both the real and the imaginary part in the case of a complex conjugate pair.

Let us consider the dynamical system governed by the state space equation

$$x_{k+1} = Ax_k + Bu_k, \quad \text{discrete time system} \tag{1}$$

$$\dot{x}(t) = Ax(t) + Bu(t), \quad \text{continuous time system}$$

where x_k and u_k, or $x(t)$ and $u(t)$, denote respectively the state vector of dimension n and the control vector of dimension n_u. The matrices A and B are of appropriate dimensions. The objective here is to design a control law

$$u_k = -Kx_k \tag{2}$$

resp. $u(t) = -Kx(t)$

which gives a set Λ_s of desired eigenvalues to the closed loop system

$$x_{k+1} = (A - BK) \, x_k = F_s \, x_k \tag{3}$$

resp. $\dot{x}(t) = (A - BK) \, x(t) = F_s x(t))$

and where (2) is a solution to the minimization of the cost criterion

$$J = \sum_{k=0}^{\infty} \; x_k^T Q \, x_k + u_k^T R \, u_k \tag{4}$$

resp. $J = \int_0^{\infty} (\|x(t)\|_Q^2 + \|u(t)\|_R^2 \,)\,dt$

It is well known that the optimal control which minimizes (4) is given by :

$$u_k = -(R + B^T P_s B)^{-1} B^T P_s A = - K \, x_k \tag{5}$$

resp. $u(t) = - R^{-1}B^T P_s x(t) = - Kx(t)$

where $P_s \geq 0$ is the stabilizing solution of the algebraic Riccati equation (ARE) :

$$P = A^T P (I + SP)^{-1} A + Q \tag{6}$$

resp. $PA + A^TP - PSP + Q = 0$

where S stands for the matrix $B R^{-1} B^T$.

The relationship between P_s and the eigenstructure of F_s on one side and the eigenstructure and the spectrum of the Hamiltonian matrix

$$H = \begin{bmatrix} A + SA^{-T}Q & - SA^{-T} \\ -A^{-T}Q & A^{-T} \end{bmatrix} \tag{7}$$

resp. $H = \begin{bmatrix} A & -S \\ - Q & -A^{-T} \end{bmatrix}$

on the other side, is well known (Pappas 1980, Kucera, 1972).

The Jordan decomposition of H can be written as follows

$$T^{-1}H T = \begin{bmatrix} \Lambda_s & 0 \\ 0 & \Lambda_u \end{bmatrix} \text{ with } T = \begin{bmatrix} T_{11} & T_{12} \\ T_{21} & T_{22} \end{bmatrix} \tag{8}$$

where Λ_s and Λ_u contain respectively the stable and the unstable eigenvalues of H. If we assume that the pair [A, B] is stabilizable and that there is a full rank factorization D^TD of Q such that the pair [A, D] is detectable, then the nonnegative solution of (6) is given by

$$P_s = T_{21} T_{11}^{-1} \tag{9}$$

In addition the nonpositive definite solution is given by

$$P_u = T_{22} T_{12}^{-1} \tag{10}$$

provided that T_{12}^{-1} exists.

The closed loop system (3) obtained by the use of the optimal control law (5) that minimizes the cost criterion (4) with the weighting matrices Q and R will be referred to hereafter as the "nominal system (A, B, Q, R)". We will denote $\Lambda_s(A,B,Q,R)$ its stable spectrum.

1. Shifting all eigenvalues

Given the nominal system (A,B,Q,R), the problem of confining all eigenvalues into a disk of radius r^2 centered at the origin of the z plane or, in other terms, of assigning a relative degree of stability r^2 to this system, can be accomplished by modifying the state weighting matrix Q in accordance with the following theorem

Theorem 1 : Let $\Lambda_s(A,B,Q,R)$ be the stable spectrum of the nominal system and assume that a relative degree of stability r^2 (α for continuous systems) is introduced and that Q is perturbed by

$$\Delta Q = (r^2 - 1) (P_u - \tilde{P}_s) \tag{11}$$

resp. $\qquad \Delta Q = 2\alpha (P_u - \tilde{P}_s)$

where P_u is the unstable solution of (6) and $\tilde{P}_s$ the stable solution of the ARE :

$$\tilde{P} = \frac{1}{r^2} A^T \tilde{P} (S \tilde{P} + I)^{-1} A + \frac{1}{r^2} (Q + (r^2 - 1) P_u) \tag{12}$$

resp $\qquad \tilde{P} (A + \alpha I) + (A + \alpha I)^T \tilde{P} - \tilde{P} S \tilde{P} + Q - 2\alpha P_u = 0$

It follows then, that
$$\Lambda_s(A,B,Q + \Delta Q,R) = r^2 \Lambda_s(A,B,Q,R) \tag{13}$$
resp. $\qquad \Lambda_s(A,B,Q + \Delta Q,R) = \Lambda_s(A,B,Q,R) - 2\alpha$

2. Partial pole placement

Consider the system (A,B,Q,R) and its associated Hamiltonian matrix (7). Consider also the modal matrix M of A, i.e. $M^{-1}AM = \Lambda = diag (\lambda_1, \dots ,\lambda_n)$ where the λ_i (i=1,...,n) are the eigenvalues of A. Let

$$H^* = \Gamma^{-1}H\Gamma = \begin{bmatrix} \Lambda + S^* \Lambda^{-T} Q^* & -S^* \Lambda^{-T} \\ -\Lambda^{-T}Q^* & \Lambda^{-T} \end{bmatrix}$$

resp. $\qquad H^* = \Gamma^{-1}H\Gamma = \begin{bmatrix} \Lambda & -S^* \\ -Q^* & -\Lambda^{-T} \end{bmatrix}$

with $\qquad S^* = M^{-1}S\,M^{-T} \;\; ; \;\; Q^* = M^T Q M \;\; \text{and} \;\; \Gamma = \begin{bmatrix} M & 0 \\ 0 & M^{-T} \end{bmatrix}$

It is obvious that H and H^* have the same eigenspectrum and a simple matrix manipulation shows that

$$\det(zI - H^*) = \det(zI - \hat{H}) \times \det(zI - \Lambda_2) \times \det(zI - \Lambda_2^{-T})$$

resp. $\qquad \det(zI - H^*) = \det(zI - \hat{H}) \times \det(zI - \Lambda_2) \times \det(zI + \Lambda_2)$

where $\hat{H}$ is the Hamiltonian matrix corresponding to the agregated system $(\hat{A}, \hat{B}, \hat{Q}, R)$ with

$$\Lambda = \begin{bmatrix} \Lambda_1 & 0 \\ 0 & \Lambda_2 \end{bmatrix} \qquad \text{and } \Lambda_1 \in R^{m \times m} \qquad\qquad (24)$$

$$Q = C^T \hat{Q} C \qquad\qquad (25.1)$$

$$C = [I_m \;\vdots\; 0]\, M^{-1} \qquad\qquad (25.2)$$

$$\hat{A} = C A C^+ \qquad\qquad (25.3)$$

$$\hat{B} = C B \qquad\qquad (25.4)$$

$$\hat{S} = \hat{B} R^{-1} \hat{B}^T \qquad\qquad (25.5)$$

$$C^+ = C^T (C C^T)^{-1} \qquad\qquad (25.6)$$

From the previous result, it becomes clear that pole placement can be achieved partially by placing the poles of successive agregated systems with lower dimension, not exceeding two. Equation 25.1 gives the relation between the state weighting matrix $\hat{Q}$ of the agregated system and that of the overall system. In the continuous time case, Saif has constructed on this base an algorithm which achieves a successive optimal pole placement. we have developed for the discrete time case the following algorithm :

Algorithm : Discrete time case

Step 1: Initialization

i = 1,

$A_1 = A$, $B_1 = B$, $R_1 = I$ (or any real positive definite matrix)

step 2

a) In order to shift a real pole λ_i to a new real location λ_d :

I) take $\hat{A} = \lambda_i$. The weighting matrix, here a scalar, is given by

$$\hat{Q}_i = \frac{\lambda_i}{\hat{S}_i} ((\lambda_d + \frac{1}{\lambda_d}) - (\lambda_i + \frac{1}{\lambda_i})) \tag{26}$$

II) Calculate the stabilizing solution $\hat{P}_{si}$ of (6) corresponding to the agregated system ($\hat{A}_i$, $\hat{B}_i$, $\hat{Q}_i$, R_i). The desired state weighting matrix is then $\hat{Q}_{di} = \hat{Q}_i$

III) Calculate $\hat{K}_i$ as follows :

$$\hat{K}_i = (R_i + \hat{B}_i^T \hat{P}_{si} \hat{B}_i)^{-1} \hat{B}_i^T \hat{A}_i \hat{P}_{si} \tag{27}$$

IV) Calculate the weighting matrix Q_{di} and the optimal gain K_i for the higher dimensional system, as follows

$$Q_{di} = C^T \hat{Q}_{di} C ; \qquad K_i = \hat{K}_i C, \qquad P_{si} = C^T \hat{P}_{si} C$$

b) In order to shift a complex pair $\alpha \pm j\beta$ to a new complex pair $\sigma \pm j\gamma = \rho e^{\pm j\varphi}$:

I) take

$$\hat{A}_i = L^{-1} C A_i C^+ L = \begin{bmatrix} \alpha & -\beta \\ \beta & -\alpha \end{bmatrix} \tag{28}$$

and

$$\hat{B}_i = L^{-1} C B_i \tag{29}$$

with

$$L = \frac{1}{2} \begin{bmatrix} 1 & j \\ 1 & -j \end{bmatrix}$$

II) Find the appropriate $\hat{Q}_i = qI$ to achieve the desired angle φ using the solution of the following equation

$$\left(4\frac{F_2}{C_1^2}\cos^2\varphi - 1\right)\rho^4 + 2\cos\varphi\left(\frac{F_1}{C_1} - 2\frac{F_2}{C_1^2}C_o\right)\rho^3$$

$$+ \left(8\frac{F_2}{C_1^2}\cos^2\varphi + \frac{F_2}{C_1^2}C_o - \frac{F_1 C_o}{C_1} + F_o - 4\cos^2\varphi\right)\rho^2 + 2\cos\varphi\left(\frac{F_1 C_1 - 2F_2 C_o}{C_1^2}\right)\rho$$

$$+ \left(4\frac{F_2}{C_1^2}\cos^2\varphi - 1\right) = 0 \qquad (30)$$

with $C_o = \dfrac{2\alpha(\alpha^2 + \beta^2 + 1)}{\alpha^2 + \beta^2}$ $\qquad$ $C_1 = \dfrac{\alpha(e_1 + e_2)}{\alpha^2 + \beta^2}$

$F_o = \dfrac{(e_1 + e_2)(1 + \alpha^2 + \beta^2)}{\alpha^2 + \beta^2}$ $\qquad$ $F_2 = \dfrac{(e_1 e_2 - e_3^2)}{\alpha^2 + \beta^2}$

$F_1 = \dfrac{(e_1 + e_2)(1 + \alpha^2 + \beta^2)}{\alpha^2 + \beta^2}$

where the e_i's are the entries of $\hat{S}_i$. Compute and store the solution ρ' of (30). Compute q using the following equation :

$$q = \frac{1}{C_1}\left[2\rho'\cos\left(1 + \frac{1}{\rho'^2}\right) - C_o\right]$$

III) Calculate the value $r = \sqrt{\rho/\rho'}$ and calculate the non stabilizing solution $\hat{P}_{ui}$ of (6) corresponding to the system $(\hat{A}_i, \hat{B}_i, \hat{Q}_i, R_i)$. Then with

$$\overline{Q}_i = \frac{1}{r^2}\left(\hat{Q}_i + (r^2 - 1)\hat{P}_{ui}\right)$$

calculate the stabilizing solution $\check{P}_{si}$ of (12). The appropriate weighting matrix is then given by

$$\hat{Q}_{di} = \hat{Q}_i + (r^2 - 1)(P_{ui} - \check{P}_{si})$$

IV) Calculate $\hat{K}_i$ as follows :

$$\hat{K}_i = (R_i + \hat{B}_i \, \bar{\hat{P}}_s \, \hat{B}_i^T)^{-1} \, \hat{B}_i^T \, \hat{A}_i \, \bar{\hat{P}}_{si}$$

V) Calculate the weighting matrix Q_{di} and the optimal gain K_i for the higher dimensional system, as follows

$$Q_{di} = C^T \, L^{-T} \, \hat{Q}_{di} \, L^{-1} \, C, \qquad\qquad K_i = \hat{K}_i \, L^{-1} \, C, \qquad P_{si} = C^T L^{-T} \, \bar{\hat{P}}_{si} \, L^{-1} \, C$$

c) In order to shift a complex pair $\alpha \pm j\beta$ to two new real locations υ and η :

I) Use (28) and (29) to compute $\hat{A}_i$ and $\hat{B}_i$.

II) Use

$$F_2 \, q^2 + F_1 q + F_o = 0$$

with

$$F_2 = \frac{e_1 \, e_2 - e_3^2}{\alpha^2 + \beta^2} \; ; \; F_1 = \frac{(1 + \alpha^2 + \beta^2)(e_1 + e_2)}{\alpha^2 + \beta^2} - \frac{\alpha(e_1 + e_2)}{\alpha^2 + \beta^2} \, \alpha^2$$

$$F_o = \frac{(1 + 4\alpha^2 + \beta^2)}{\alpha^2 + \beta^2} - 2 - \frac{2\alpha(1 + \alpha^2 + \beta^2)}{\alpha^2 + \beta^2} \, (\eta + \frac{1}{\eta}) \, (\eta + \frac{1}{\eta})^2$$

to find the appropriate $\hat{Q}_i = qI$ to achieve the allocation of one desired real pole, for instance η. Calculate and store the value υ'.

III) Calculate the stabilizing solution $\hat{P}_{si}$ of (6) corresponding to the agregated system $(\hat{A}_i, \hat{B}_i, \hat{Q}_i, R_i)$. The desired state weighting matrix is then $\hat{Q}_{di} = \hat{Q}_i$

IV) Use (27) to calculate $\hat{K}_i$

V) Calculate the weighting matrix Q_{di} and the optimal gain K_i for the higher dimensional system, as follows

$$Q_{di} = C^T L^{-T} \, \hat{Q}_{di} \, L^{-1} \, C, \qquad\qquad K_i = \hat{K}_i L^{-1} C, \qquad P_{si} = C^T L^{-T} \, \hat{P}_{si} \, L^{-1} C$$

Step 3

Let $\quad A_{i+1} = A_i - B \, K_i$

$$R_{i+1} = R_i + B^T P_{si} B$$

Step 4

If all the eigenvalues have been shifted the recursive process terminates and the desired weighting matrix Q_d for the overall system as well as the optimal gain K and the associated matrix P_s are then found in the following way :

$$Q_d = \sum Q_i \qquad\qquad K = \sum K_i \qquad P_s = \sum P_{si}$$

Furthermore, the closed loop system matrix is given by A_{i+1}. If, on the contrary, not all the eigenvalues have been placed, let $i = i + 1$ and go to step 2.

3. Illustrative example :

The previously described method will be illustrated on the following example of a discrete time system of fifth order, defined by the following state space representation :

$$A = \begin{bmatrix} 0 & 1 & 0 & 1 & 0 \\ 0 & 0 & 1 & 0 & 4 \\ -2 & -1 & 4 & 2 & -2 \\ 1 & 0 & 0 & 0 & 1 \\ 10 & -1 & 0 & 4 & -2 \end{bmatrix}$$

$$B = \begin{bmatrix} 0 & 0 \\ 0 & -1 \\ 1 & 1 \\ 0 & 0 \\ 1 & 0 \end{bmatrix}$$

The corresponding set of open loop poles is :
$$\Lambda(A) = \{-2.54 \pm 3.04j, \ 3.64 \pm 0.70j, \ -0.20\}$$
If the set of the desired poles of the closed loop system is chosen to be
$$\Lambda(F_s) = \{-0.64 \pm 0.37j = 0.75 \ e^{\pm 5\frac{\pi}{6}}, \ -0.01, \ 0.1, \ 0.2\}$$
the algorithm described hereabove, for the control weighting matrix R taken as being equal to the identity matrix, gives the following results :
State weighting matrix :

$$Q = \begin{bmatrix} 1587.56 & 477.49 & -524.48 & 644.25 & 39.11 \\ 477.49 & 121.06 & -141.15 & 109.92 & 58.57 \\ -524.48 & -141.15 & 171.87 & -183.15 & -11.13 \\ 644.25 & 109.92 & -183.15 & 404.91 & 50.01 \\ 39.11 & 58.57 & -11.13 & 50.01 & -11.62 \end{bmatrix}$$

Optimal state feedback matrix :

$$K = \begin{bmatrix} 7.05 & -1.00 & 1.26 & 3.48 & -0.87 \\ -5.45 & -1.61 & 1.01 & -2.40 & -3.03 \end{bmatrix}$$

Solution of the Riccati equation :

$$P_s = \begin{bmatrix} 2705.37 & 805.83 & -771.02 & 1108.62 & 287.75 \\ 805.83 & 294.23 & -215.99 & 308.67 & 160.74 \\ -771.02 & -215.99 & 254.34 & -285.05 & -9.77 \\ 1108.62 & 308.57 & -285.05 & 662.78 & 148.48 \\ 287.75 & 160.74 & -9.77 & 148.48 & 254.25 \end{bmatrix}$$

Closed loop system matrix

$$F_s = \begin{bmatrix} 0 & 1 & 0 & 1 & 0 \\ -5.45 & -1.60 & 2.01 & -2.40 & 0.97 \\ -3.59 & 1.60 & 1.72 & 0.92 & 1.90 \\ 1 & 0 & 0 & 0 & 1 \\ 2.95 & 0.00 & -1.26 & 0.52 & -1.13 \end{bmatrix}$$

A simple verification shows that the eigenvalues of F_s are indeed equal to the desired $\Lambda(F_s)$. The difference between the solutions of the Riccati equation, obtained respectively by the above algorithm and by solving it directly with the Q and R matrices given here, is equal to the following "error" matrix :

$$Error = \begin{bmatrix} -8.62 & -2.67 & 2.35 & -4.16 & -0.88 \\ -2.79 & -0.79 & 0.80 & -1.01 & -0.18 \\ 2.89 & 0.88 & -0.78 & 1.39 & 0.35 \\ -3.77 & -1.08 & 1.08 & -1.53 & -0.24 \\ 0.16 & 0.09 & -0.04 & 0.18 & 0.1 \end{bmatrix} \times 10^{-11}$$

4. Conclusion

In this paper a degree of stability is achieved by perturbing the state weighting matrix of a quadratic cost criterion. This result is used afterwards to set up an algorithm to achieve an optimal pole placement by weight selection. The controller and the state weighting matrix obtained result in a control law that minimizes a quadratic cost criterion and at the same time allows to achieve a placement of a desired set of closed loop eigenvalues. The algorithm proposed is recursive and allows to assign both real and imaginary part in the case of a complex conjugate pair.

5. Reference

ANDERSON, B.D.O. and Moore, J.B., 1971, Linear optimal control. (Englewood Cliffs, N.J. : Prentice Hall).

AMIN, M. H., 1984, Optimal discrete systems with prescribed eigenvalues. Int. J. Control, Vol. 40, N° 4, 783-794. 1985, Optimal pole shifting for continuous multivariable linear systems. Int . J. Control, Vol. 41, N° 4, 701-707.

Kucera, V., 1972, A contribution to matrix quadratic equations. I.E.E.E. Transactions on Automatic Control, Vol. 17, pp 557-560.

MEDANIC, J., Tharp, H.S., Perkins, W.R., 1988, Pole placement by performance criterion modification. IEEE, AC-33, N° 5, 469-472.

PAPPAS, T., Laub, A.J. , Sandell, N.R., 1980, On the numerical solution of the discrete time algebraic Riccati equation. IEEE, Vol AC-25, N°4, 631-641.

SAIF, M., 1989, Optimal linear regulator pole placement by weight selection. Int.J. Control, 1989, Vol. 50, N° 1, 399-414.

SOLHEIM, O.A.,1972, Design of optimal control system with prescribed eugenvalues. Int. J. Control, Vol. 15, N°1, 143-160. 1974, Design of optimal discrete optimal control system with prescribed eigenvalues. Int J. Control, Vol. 19, N° 2, 417-427.

ALGEBRAIC COMPUTATIONS FOR MULTI-DELAY SYSTEMS ANALYSIS [1]

Maciej Szymkat

Institute of Automatics, Academy of Mining and Metallurgy

al.Mickiewicza 30, 30-059 Kraków, Poland,

Abstract. In this paper some results concerning the application of symbolic calculations in computer aided design of multi-delay control systems are presented. All considerations relate to the D2M package originally developed at the Institute of Automatics, AMM, Cracow, Poland. This package performs the transformation of graphically defined model structures and symbolically described input-output relations into MATLAB[2] source code files which contain a combined model representation in an algorithmic form. The main attention is paid to generation of characteristic quasipolynomial. The advantages of the proposed approach are illustrated with the example of a distillation column design.

1. Introduction

D2M (*Diagram-to-Model*) is a computer tool aimed at the integration of control system design process. Roughly speaking, the **D2M** package is a MATLAB programing environment preprocessor oriented on abstract model specifications. Since MATLAB itself is numerically oriented the adding of the front-end tool of the **D2M** kind creates new possibilities of manipulations on model data. It will be shown in what follows that providing the front-end facility with certain features enables the user to include symbolic operations like formal parameter substitution into design process.

The **D2M** package consists of three major modules:

- **SE** (*Structure Editor*) — a graphical editor for the specification of model structure in the form of hierarchical block diagrams,

- **DE** (*Description Editor*) — a symbolically oriented tool for specification of block input-output relation descriptions,

- **MG** (*Model Generator*) — a MATLAB source code generator.

The above modules may be executed separately. The **DE** and **MG** modules can also be run within the **SE** graphical environment. In such a case the data corresponding to the currently selected objects are transferred with no user action. All revise/edit functions are performed simultaneously with movements of various cursors, or they may be

---

[1]This work was supported by the Polish Ministry of Education under the contract R.P.I.02 ASO 2.1.
[2]MATLAB is the trademark of The MathWorks, Inc.

initiated through the user-friendly interface. An on-line help facility and key macro-definitions mechanism are available. The user carries out the numerical analysis of the model (time and frequency domains) using the powerful numerical and graphical procedures from MATLAB programming environment. Extensions for other standard tools are being developed. The current version 1.21 of the package is running on PC XT/AT/386 and compatibles.

The symbolic approach to the mathematical descriptions makes it possible to apply the **D2M** package to a broad area of multivariable and possibly nonlinearly parametrized linear systems. The model generator **MG** uses an original (non-matrix) graph reduction and algebraization method producing a model dependent algorithm in each run. Such an approach minimizes the numerical effort and round-off errors. In this paper the main stress will be put on the specific use of the **D2M** capabilities allowing the analytical representation of complex multi-delay systems. The efficiency of algebraization algorithms will be illustrated with a practical design example.

2. Computer representation of delay systems

To begin with we will present an example of a simple delay-system. Following [1] let us consider a chemical reactor with unreacted feed recycle. The linearized model of the dynamics of such a process consists of three SISO blocks with the following transfer functions

$$R(s) = \frac{K}{\tau s + 1} \tag{1}$$

$$S(s) = e^{-Ts} \tag{2}$$

$$G(s) = g \tag{3}$$

corresponding to the reactor, separator and output gain block respectively. The exponential function in (2) represents the pure time-delay introduced by the separation process.

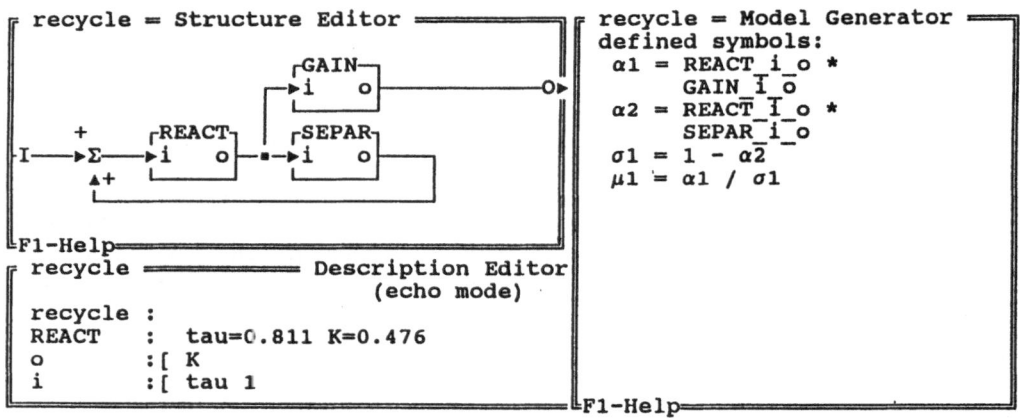

Figure 1. D2M screen dump.

The particular elements of the formal description of this model in the **D2M** package may be identified in a computer screen dump shown in Figure 1. As its seen there **D2M** uses the extended ASCII code semigraphic characters for schematics presentation. The algebraic representation displayed in the right-hand side window (Model Generator) is used in order to generate the final program output, i.e., MATLAB source code files.

Larger systems may contain more delay blocks with possibly different and independent delays, and their overall structure may be much more complex. In this case (called multi-delay) the operator representation involves exponential functions with exponents corresponding to various sums of individual delays.

First approaches towards computer representation of the delay systems employed the Padé approximations, for example in the CC package. This could produce no analytical results, and was also poor from the numerical point of view. The newer tools allow accurate pure time-delay representation in the frequency domain representation. The Multivariable Frequency Domain Toolbox for MATLAB, compare [2], uses the so called frequency response matrices, containing the values corresponding to a given frequency sequence. This, for instance, makes the stability analysis fully accessible. However, the problem of explicit expression of the characteristic quasipolynomial remains hard to deal with. The quasipolynomial is defined as a linear combination of exponential functions with polynomial coefficients.

In what follows we will particularly concentrate on the symbolic generation of characteristic quasipolynomial. If the representation of the system were introduced in time-domain matrix formulation, this problem could have been solved using the method presented in [3]. The results presented in later sections concern the model specifications based on a frequency-domain block diagram formulation.

3. The distillation column model

The methanol distillation column with conventional two-point control is a classical example of the multi-delay system. A detailed description of its model can be found in [4,pp.224-229]. The matrix PI-controller with partial loop decoupling for this system was proposed in [5]. Our considerations will take this system as the subject of the following case study.

The structure of the system is presented in Figure 2. The depicted block diagram was created using **D2M** print facility.

The transfer functions representing column dynamics read as follows

$$G_i = \frac{g_i \, e^{-sd_i}}{T_i s + 1}, \quad i = 1, 2, \tag{4}$$

$$G_{ij} = \frac{g_{ij} \, e^{-sd_{ij}}}{T_{ij} s + 1}, \quad i = 1 \text{ and } j = 2 \quad \text{or} \quad i = 2 \text{ and } j = 1. \tag{5}$$

The remaining blocks represent the controller dynamics for main loops

$$I_i = \frac{1}{s}, \tag{6}$$

$$K_{i,1} = a_i, \quad K_{i,2} = b_i, \tag{7}$$

for $i = 1, 2$, and for partial loop decoupling

$$K_{ij,1} = \frac{g_{ij}a_j}{g_i}, \tag{8}$$

$$K_{ij,2} = \frac{g_{ij}(b_j + (T_i - T_{ij})a_j)}{g_i}, \tag{9}$$

$$H_{ij} = e^{-s(d_{ij} - d_i)} \tag{10}$$

for $i = 1, j = 2$ and $i = 2, j = 1$.

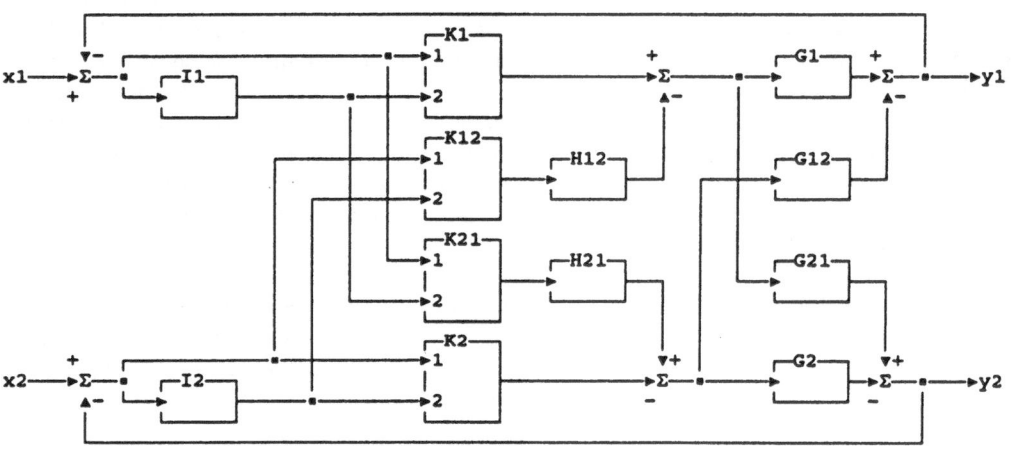

Figure 2. Block diagram of the distillation column control system.

4. Block diagram algebraization

The structure of the model may be described by the following connection matrix

$$
\begin{bmatrix}
1 & 0 & 0 & 0 & 0 & 1 & 0 & 0 & 0 & 0 & 0 & 0 \\
-I_1 & 1 & 0 & 0 & 0 & 0 & 0 & 0 & 0 & 0 & 0 & 0 \\
-K_{1,1} & -K_{1,2} & 1 & 0 & 0 & 0 & 0 & 0 & 0 & 0 & 0 & 0 \\
-K_{21,1} & -K_{21,2} & 0 & 1 & 0 & 0 & 0 & 0 & 0 & 0 & 0 & 0 \\
0 & 0 & -1 & 0 & 1 & 0 & 0 & 0 & 0 & H_{12} & 0 & 0 \\
0 & 0 & 0 & 0 & -G_1 & 1 & 0 & 0 & 0 & 0 & -G_{12} & 0 \\
0 & 0 & 0 & 0 & 0 & 0 & 1 & 0 & 0 & 0 & 0 & 1 \\
0 & 0 & 0 & 0 & 0 & 0 & -I_2 & 1 & 0 & 0 & 0 & 0 \\
0 & 0 & 0 & 0 & 0 & 0 & -K_{2,1} & -K_{2,2} & 1 & 0 & 0 & 0 \\
0 & 0 & 0 & 0 & 0 & 0 & -K_{12,1} & -K_{12,2} & 0 & 1 & 0 & 0 \\
0 & 0 & 0 & H_{21} & 0 & 0 & 0 & 0 & -1 & 0 & 1 & 0 \\
0 & 0 & 0 & 0 & -G_{21} & 0 & 0 & 0 & 0 & 0 & -G_2 & 1 \\
\end{bmatrix} . \tag{11}
$$

The calculation of the characteristic quasipolynomial for the system under consideration requires calculation of the determinant of the above matrix. This may be achieved using standard Laplace expansion yielding its explicit form

$$
\begin{aligned}
1+ \\
I_1 \quad & *(K_{1,2}*(G_1 \ *(1 \quad +I_2 \ *K_{2,2} \ *G_2 \quad +K_{2,1} \ *G_2)- \\
& \qquad\quad G_{12} \ *(I_2 \ *K_{2,2} \ *G_{21} \ +K_{2,1} \ *G_{21}))- \\
& \quad K_{21,2} *(H_{12} \ *(G_1 \ *(I_2 \ *K_{12,2}*H_{21} \ *G_2 \quad -K_{12,1}*H_{21} \ *G_2)+ \\
& \qquad\qquad\quad G_{12} \ *(I_2 \ *K_{12,2}*H_{21} \ *G_{21} \ +K_{12,1}*H_{21} \ *G_{21}))- \\
& \qquad\quad G_{12} \ *H_{21})+ \\
K_{1,1} \ *G_1 \quad & *(1 \quad +I_2 \ *K_{2,2} \ *G_2 \quad +K_{2,1} \ *G_2)- \\
G_{12} \quad & *(I_2 \ *K_{2,2} \ *G_{21} \ +K_{2,1} \ *G_{21}))- \\
K_{21,1} \ *H_{12} \ *G_1 \quad & *(I_2 \ *K_{12,2}*H_{21} \ *G_2 \quad -K_{12,1}*H_{21} \ *G_2)+ \\
G_{12} \quad & *(I_2 \ *K_{12,2}*H_{21} \ *G_{21} \ +K_{12,1}*H_{21} \ *G_{21}))- \\
G_{12} \quad & *H_{21})- \\
H_{12} \quad & *(I_2 \ *K_{12,2}*G_{21} \ -K_{12,1}*G_{21})+ \\
I_2 \quad & *K_{2,2} \ *G_2+ \\
K_{2,1} \quad & *G_2.
\end{aligned}
\tag{12}
$$

There exist two basic approaches to automatic generation of the symbolic determinant formulas of the above type. First, looking quite attractive at first glance, is to use an elimination-type procedure leading to triangularization of the connection matrix. This corresponds to *node-by-node* signal flow graph reduction process. It is no doubt that this would be the right solution from the point of view of pure numerical efficiency. On the other hand such an approach suffers from a serious disadvantage, that could be initially overlooked. Since the operands of all algebraic operations involved here are in principle rational or meromorphic functions, not just numbers, the resulting expression would be a finite continued fraction. It turns out that the conversion to the zero-pole-form often produces redundant common numerator and denominator factors. This is exactly the case for many numerically oriented software tools, see [6]. There is a possibility to get rid of the redundant factors, using Euclid algorithm or iterative roots calculations, but this significantly decreases the overall efficiency. For larger systems the accuracy of such calculations is often harmed by the round-off error propagation. Even having succeeded here one should be able to distinguish the common factors inherent to the problem from the ones introduced during the calculations.

The alternative method is the direct determinant expansion corresponding to *path-by-path* signal flow graph reduction process. In practice, due to the sparseness of a typical connection matrix this method shows accepatble efiiciency and above all yields reliable results. The "parenthesized" Mason rule based algorithm implemented in the **D2M** package belongs to this class of methods. More details on it can be found in [7]. The D2M-generated algorithm for matrix (11) determinant consists of the following sequence of algebraic operations:

$$\alpha_1 \ = H_{12} * [\text{-}1]$$
$$\alpha_2 \ = G_1 * [\text{-}1]$$
$$\alpha_3 \ = G_2 * [\text{-}1]$$
$$\alpha_4 \ = K_{1,2}{}^* G_{21}$$
$$\alpha_5 \ = I_1 * \alpha_4$$
$$\alpha_6 \ = K_{1,1} * G_{21}$$
$$\beta_1 \ = \alpha_5 + \alpha_6$$
$$\alpha_7 \ = G_{12} * \beta_1$$
$$\alpha_8 \ = I_1 * K_{1,2}$$
$$\beta_2 \ = \alpha_8 + K_{1,1}$$
$$\alpha_9 \ = \alpha_2 * \beta_2$$
$$\sigma_1 \ = 1 - \alpha_9$$
$$\alpha_{10} = \alpha_3 * \sigma_1$$
$$\beta_3 \ = \alpha_7 + \alpha_{10}$$
$$\alpha_{11} = [\text{-}1] * \beta_3$$
$$\alpha_{12} = K_{2,1} * \alpha_{11}$$
$$\alpha_{13} = H_{21} * \alpha_3$$
$$\alpha_{14} = K_{21,2} * \alpha_{13}$$
$$\alpha_{15} = I_1 * \alpha_{14}$$
$$\alpha_{16} = K_{21,1} * \alpha_{13}$$
$$\beta_4 \ = \alpha_{15} + \alpha_{16}$$

$$\alpha_{17} = \alpha_2 * \beta_4$$
$$\alpha_{18} = K_{21,2} * H_{21}$$
$$\alpha_{19} = I_1 * \alpha_{18}$$
$$\alpha_{20} = K_{21,1} * H_{21}$$
$$\beta_5 \ = \alpha_{19} + \alpha_{20}$$
$$\alpha_{21} = G_{12} * \beta_5$$
$$\sigma_2 \ = 1 - \alpha_{21}$$
$$\alpha_{22} = G_{21} * \sigma_2$$
$$\beta_6 \ = \alpha_{17} + \alpha_{22}$$
$$\alpha_{23} = \alpha_1 * \beta_6$$
$$\alpha_{24} = K_{12,1} * \alpha_{23}$$
$$\beta_7 \ = \alpha_{12} + \alpha_{24}$$
$$\alpha_{25} = K_{2,2} * \alpha_{11}$$
$$\alpha_{26} = K_{12,2} * \alpha_{23}$$
$$\beta_8 \ = \alpha_{25} + \alpha_{26}$$
$$\alpha_{27} = I_2 * \beta_8$$
$$\beta_9 \ = \beta_7 + \alpha_{27}$$
$$\alpha_{28} = [\text{-}1] * \beta_9$$
$$\beta_{10} = \alpha_{21} + \alpha_9$$
$$\beta_{11} = \alpha_{28} + \beta_{10}$$
$$\sigma_3 \ = 1 - \beta_{11}$$

In **D2M** the symbols used to define the algebraic relations between the subsequently introduced expressions are given names starting with a Greek letter:

- α — for multiplication,
- β — for addition,
- σ — for calculation of the unit complement,
- μ — for division.

The omitted parts concern solely the simultaneously generated algorithm for Mason rule numerator. The total number of multiplicative operations in the above sequence is 23 (the sign changes are not counted), additive operations with one unitary operand – 3, and other additive operations – 11. This means a considerable improvement as compared to respectively 56, 3 and 21 for formula (12). The main reason for the better performance is the usage of subexpressions repetitions, clearly visible in (12). The other is the multiplication of the additively constructed factors, however it does not appear in the specific example under consideration.

The generation of the most simplified form of the calculation sequence is itself an interesting problem. The **D2M** package applies a two-stage simplification procedure to the initially generated expression tree. In the first stage the commutatively identical factors are being collected, and respective repetitions cancelled. In the second stage a similar procedure is used for whole branches of the expression tree. Unfortunately, the results of this process cannot be determined uniquely. They depend on a specific sequence of partial simplifications. The possible extensions may refine the final result, however, the combinatorial explosion would be hard to avoid.

It is important to note here that in the D2M-produced algorithms the multiplicative operations are accomplished only for the final step or for the evaluation of the terms of additive operations. Otherwise, only the pointers to the respective factors are being collected.

Parallelly to the graphical definition of the model structure in the **D2M** package the symbolic descriptions of the block input-output relations are specified. This task is handled by the **DE** module. It allows the use of general transfer functions whose coefficients are algebraic expressions containing parameters (global and local), standard functions and other parenthesized subexpressions. The program performs their lexical and syntactic check. It carries out some symbolic simplifications and numerical evaluations as well. In fact the mathematical formulas may be specified almost as they were presented in the model description section.

The numerical part of model analysis is carried out within the MATLAB programming environment. The major advantage of this tool is the possibility of interactive dialogue with the user allowing the use of powerful high-level language. The user input may be interpreted conversationally causing implicit dynamic memory allocation or execution of advanced built-in functions, or it may be prepared in the source code files (m-files) invoked during the MATLAB session. The m-files can be usual command scripts for immediate execution or procedures with formal parameters. Both kinds can access globally declared user variables. The MATLAB framework has enough flexibility to adopt various preprocessing mechanisms. This is the solution implemented (among others) in the **D2M** package. The **MG** module automatically produces the MATLAB source code files corresponding to the model specification and the generated formal algebraization.

5. Generation of the characteristic quasipolynomial

The MATLAB programming environment is strongly oriented on matrix operations. This feature, of course, enables the wide use of MATLAB for linear systems analysis – both in the state space and operator domain representations. But when it comes to delay systems this is hardly sufficient, especially from the point of view of formal calculations. Certain difficulties arise with exponential functions. Leaving these functions in the initial specification would restrict the evaluation possibilities to specific frequency values. Anyway, with the **D2M** automated preprocessing these difficulties seem to be fully resolved. The symbolic approach to the model description helps to obey the above mentioned limitations and to extend the range of model analysis. The quasipolynomial representation may be obtained with the method presented below. It will only involve manipulations on polynomials (represented as coefficient vectors).

The crucial step in the proposed method is the formal substitution of extra parameters h_i instead of exponential functions. In the model under consideration this may be achieved by defining, for example,

$$h_1 = e^{-sd_1} \tag{13}$$

$$h_2 = e^{-sd_2} \tag{14}$$

$$h_3 = e^{-sd_{21}} \tag{15}$$

$$h_4 = e^{-sd_{12}} \tag{16}$$

$$h_5 = e^{-s(d_{21}-d_2)} \tag{17}$$

$$h_6 = e^{-s(d_{12}-d_1)}. \tag{18}$$

In this way all the polynomials involved in the transfer calculations become affine combinations of the newly introduced parameters. Let us assume that the needed characteristic quasipolynomial has the following form

$$D(h_1, h_2, h_3, \ldots, h_n) = p_0 + \sum p_{i_1 i_2 \cdots i_k} h_{i_1} h_{i_2} h_{i_3} \cdots h_{i_k}, \tag{19}$$

where the summation runs over all k satisfying $1 \leq k \leq n$ and all k-element subsets of $\{1, 2, 3, \ldots, n\}$, and $p_{i_1 i_2 \cdots i_k}$ stand for unknown coefficients being polynomials themselves. Let us introduce evaluation symbols

$$D_0 = D(0, 0, 0, \ldots, 0), \tag{20}$$

$$D_i = D(\delta_{i1}, \delta_{i2}, \delta_{i3}, \ldots, \delta_{in}), \tag{21}$$

where δ_{ij} is Kronecker symbol ($\delta_{ii} = 1$, $\delta_{ij} = 0$ for $i \neq j$),

$$D_{i_1 i_2} = D(\delta_{i_1 1} + \delta_{i_2 1}, \delta_{i_1 2} + \delta_{i_2 2}, \delta_{i_1 3} + \delta_{i_2 3}, \ldots, \delta_{i_1 n} + \delta_{i_2 n}), \tag{22}$$

and so on. Each symbol of this kind represents a polynomial also.

In the **D2M** package all algebraic operations are performed with complete preservation of symbolic polynomial factors integrity. This means that no simplifications are made to the factors which result from any numerical operations and no numerators or denominators are normalized. To achieve this effect the user of the **D2M** package is supplied with specially designed polynomial manipulation procedures (tf_mul, tf_add, tf_cpl, tf_div, tf_fct and tf_nrm) replacing the original MATLAB supported functions. This is why each evaluation symbol (20-22) may be effectively computed in a single run of the D2M-produced algorithm.

Finally, the needed polynomial coefficients of the quasipolynomial (19) may be obtained from the following relations

$$p_0 = D_0, \tag{23}$$

$$p_i = D_i - p_0, \tag{24}$$

$$p_{i_1 i_2} = D_{i_1 i_2} - p_{i_1} - p_{i_2} - p_0, \tag{25}$$

$$p_{i_1 i_2 i_3} = D_{i_1 i_2 i_3} - p_{i_1 i_2} - p_{i_1 i_3} - p_{i_2 i_3} - p_{i_1} - p_{i_2} - p_{i_3} - p_0. \tag{26}$$

In practice, only few $p_{i_1 i_2 \cdots i_k}$ values are different from zero. There exist methods for selecting the nonzero coefficients. The simplest one seems to be the analysis of the discrete system obtained by fixing the frequency value. For the model under consideration the final formula may be written in the following form

$$
\begin{aligned}
D(h_1, h_2, h_3, h_4, h_5, h_6) = {} & D_0 + (D_1 - D_0)h_1 + (D_2 - D_0)h_2 + \\
& + (D_{1,2} - D_1 - D_2 + D_0)h_1 h_2 + (D_{3,4} - D_0)h_3 h_4 + \\
& + (D_{3,6} - D_0)h_3 h_6 + (D_{4,5} - D_0)h_4 h_5 + \\
& + (D_{1,2,5,6} - D_{1,2})h_1 h_2 h_5 h_6 + \\
& + (D_{3,4,5,6} - D_{3,4} - D_{3,6} - D_{4,5} + 2D_0)h_3 h_4 h_5 h_6. \quad (27)
\end{aligned}
$$

Only nine evaluations of the algorithm presented in the previous section are thus needed in order to determine all the polynomial coefficients of the characteristic quasipolynomial.

6. Example of stability analysis

The explicit form of the charactersitic quasipolynomial may be used while solving various design problems concerning delay systems, including stability and optimal control. The most important advantage is the possibility of using existing analytical results. Let us take as an example the generalized Mikhailov-Leonhard criterion, compare [8, p.181]. The authors of the monograph [8] suggest to plot the values of the characteristic quasipolynomial on the complex plane transformed onto unitary disc using the Poincaré transformation

$$z \rightarrow \frac{Lz}{\sqrt{1+|z|^2}}. \tag{28}$$

We propose another transformation

$$z \rightarrow \frac{e^{i \arg(z)}}{1 + \frac{K}{\ln(1+|z|)}}. \tag{29}$$

where K is a scaling constant set equal to 3 in the investigated case. The transformation (29) makes it possible to emphasize the high-frequency part of the plot.

The particular values assumed for further calculations are

$$\begin{array}{lll}
g_1 = 13, & T_1 = 17, & d_1 = 1, \\
g_{12} = 18, & T_{12} = 21, & d_{12} = 4, \\
g_{21} = 7, & T_{21} = 11, & d_{21} = 7, \\
g_2 = 20, & T_2 = 14, & d_2 = 3, \\
a_1 = 0.02 & b_1 = 0.15, \\
a_2 = 0.04 & b_2 = 0.21756.
\end{array} \tag{30}$$

All the plant data used here are the same as the ones taken in [4]. The previously described method resulted in obtaining the characteristic quasipolynomial in the form

$$\begin{aligned}
D(s) = \ & 54978s^6 + 14777s^5 + 1461s^4 + 63s^3 + s^2 + \\
& + (6306.3s^5 + 2606.33s^4 + 358.924s^3 + 20.189s^2 + 0.3965s)e^{-s} + \\
& + (17090.3s^5 + 6514.4s^4 + 833.248s^3 + 43.552s^2 + 0.8s)e^{-3s} + \\
& + (1960.36s^4 + 1030.531s^3 + 186.898s^2 + 13.43597s + 0.3172)e^{-4s} + \\
& - (3983.156s^5 + 1383.97s^4 + 175.156s^3 + 9.59176s^2 + 0.19215s)e^{-8s} + \\
& - (2790.268s^5 + 2433.99s^4 + 373.689s^3 + 20.7183s^2 + 0.38769s)e^{-10s} + \\
& - (1383.68s^4 + 894.828s^3 + 173.352s^2 + 12.7151s + 0.30744)e^{-11s} + \\
& + (202.155s^4 + 192.247s^3 + 40.1902s^2 + 3.00648s + 0.074495)e^{-18s}. \tag{31}
\end{aligned}$$

The frequency plot obtained for the above function is shown in Figure 2.

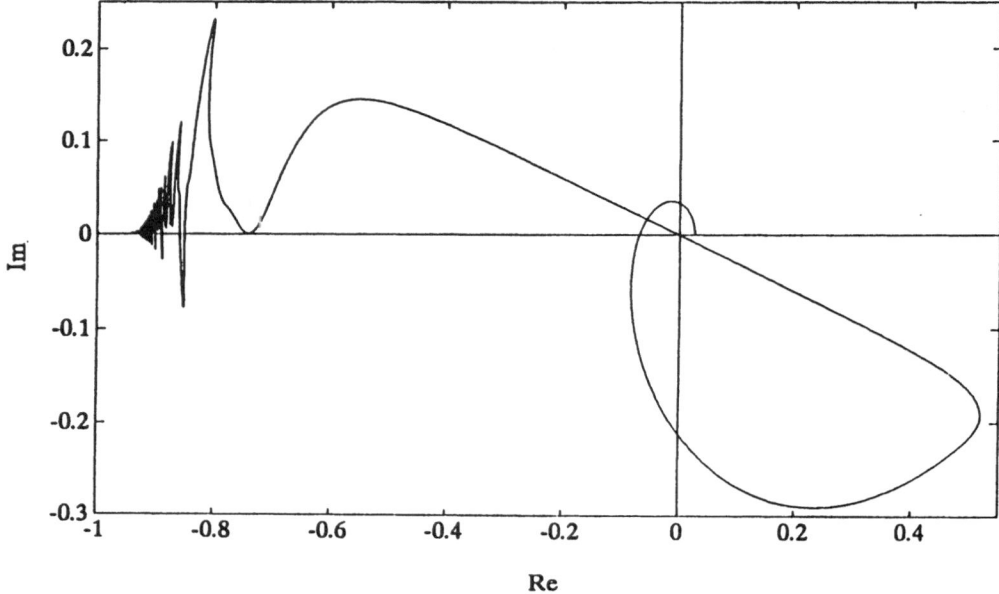

Figure 3. The Mikhailov plot of the characteristic quasipolynomial.

It is easily seen that the parameter values correspond to the stability boundary, since the curve crosses the origin.

7. Conclusions

The approach presented here illustrates the interesting possibility that arises with the advent of symbolically oriented computer tools for control design. The analytical methods which were reckoned as too complex or too cumbersome because of the effort needed for formula manipulation are now fully accessible. On the other hand the requirements for the model specification are also minimized. The example of **D2M** package user interface proves that even a microcomputer tool can manage the task of interactive graphically supported dialogue.

These trends create new problems both for software developers and control engineers. The latter should introduce the new design methodologies including newly available techniques such as formal expression substitution and evaluation. The research reported here is an attempt at showing such possibilities in the area of multi-delay systems. The above mentioned transition would be eased if the certain existing standard tools with powerful numerical and graphical capabilities (as MATLAB) were "enriched" with symbolically oriented front-end preprocessors. The experience with the **D2M** package implementation may be interesting for further developments in this field.

References

[1] M.M.Denn, R.Lavie : *Dynamics of the plant with time-delay.* The Chemical Engineering Journal, vol.24, pp. 55-59, 1982.

[2] J.M.Boyle, M.P.Ford, J.M.Maciejowski: *Multivariable toolbox for use with MATLAB,* IEEE Control Systems Magazine, vol. 9, no.1, pp. 59-65, 1989.

[3] M.Busłowicz: *The algorithm for characteristic quasipolynomial calculation for the linear stationary system with incommensurable time-delays based on the state equation.* Archiwum Automatyki i Telemechaniki, vol. XXVI, no. 1, pp. 125-131, 1981 (in Polish).

[4] W.H.Ray : *Advanced process control,* McGraw-Hill, New York 1981.

[5] J.Pułaczewski : *Application of the method of moments to the control of two-dimensional delay plants.* Pomiary – Automatyka – Kontrola, special section: Chemoautomatyka, no. 11, pp. 41–44, 1988 (in Polish).

[6] M.Rimer, D.K.Frederick, C.Y.Huang : *Solutions of the Grumman F-14 benchmark control problem.* IEEE Control Systems Magazine, vol.10, no. 4, pp. 36-40, 1987.

[7] M.Szymkat: *Signal flow graph algebraization in the D2M package.* To be presented at 5th IFAC/IMACS Symposium on Computer Aided Design in Control Systems – CADCS 91, Swansea, UK July 1991.

[8] H.Górecki, S.Fuksa, P.Grabowski, A.Korytowski : *Analysis and synthesis of time delay systems.* PWN, Warsaw - J.Wiley, Chichester, 1989.

Lecture Notes in Control and Information Sciences

Edited by M. Thoma and A. Wyner

Lecture Notes in Control and Information Sciences

Edited by M. Thoma and A. Wyner

Lecture Notes in Control and Information Sciences

Edited by M. Thoma and A. Wyner